# INTRODUCTION
# TO THE
# MICROBIAL WORLD

# Roger Y. Stanier
Institut Pasteur   Paris 15e, France

# Edward A. Adelberg
Yale University School of Medicine   New Haven, Connecticut   06510

# John L. Ingraham
University of California   Davis, California   95616

# Mark L. Wheelis
University of California   Davis, California   95616

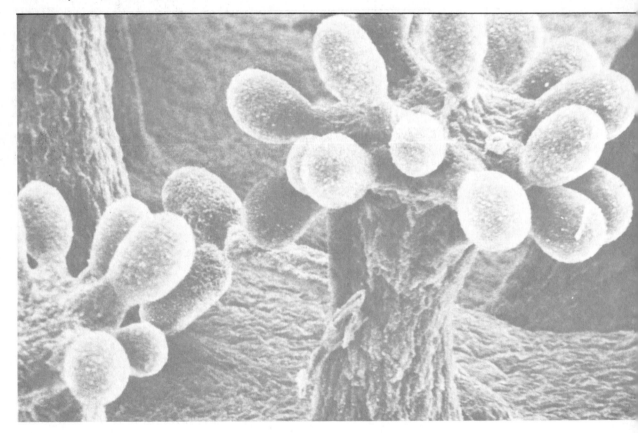

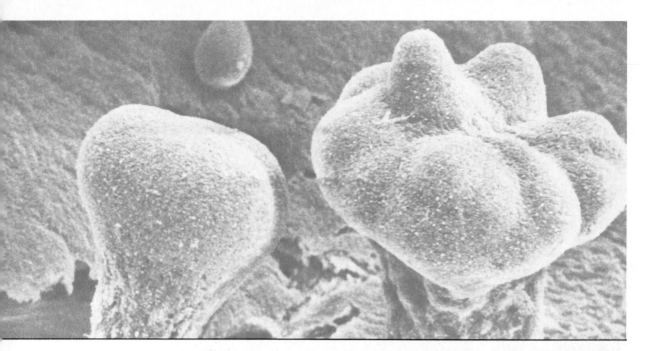

# INTRODUCTION TO THE MICROBIAL WORLD

PRENTICE-HALL, INC.,
ENGLEWOOD CLIFFS, NEW JERSEY 07632

*Library of Congress Cataloging in Publication Data*

Main entry under title:

Introduction to the microbial world.

    Includes index.
    1. Microbiology.  I. Stanier, Roger Y.
[DNLM: 1. Microbiology.  QW4.3 I615]
QR41.I54     576     78–59675
ISBN 0–13–488049–8

INTRODUCTION TO THE MICROBIAL WORLD
*Roger Y. Stanier, Edward A. Adelberg, John L. Ingraham, Mark L. Wheelis*

| | |
|---|---|
| *Editorial and Production Supervision by:* | Nancy Milnamow |
| *Interior and Cover Design by:* | Lorraine Mullaney |
| *Cover Illustration by:* | Alice Brickner |
| *Manufacturing Buyer:* | Phil Galea |

Printed in the United States of America

10  9  8  7  6  5  4  3  2  1

PRENTICE-HALL INTERNATIONAL, INC., *London*
PRENTICE-HALL OF AUSTRALIA PTY. LIMITED, *Sydney*
PRENTICE-HALL OF CANADA, LTD., *Toronto*
PRENTICE-HALL OF INDIA PRIVATE LIMITED, *New Delhi*
PRENTICE-HALL OF JAPAN, INC., *Tokyo*
PRENTICE-HALL OF SOUTHEAST ASIA PTE. LTD., *Singapore*
WHITEHALL BOOKS LIMITED, *Wellington, New Zealand*

# CONTENTS

V

# THE NATURE OF THE MICROBIAL WORLD     41

# MICROBIAL METABOLISM     62

# MICROBIAL GROWTH     102

# PREFACE

In 1955–1957, when the authors planned the first edition of *The Microbial World*, it was explicitly designed as "an attempt to present a modern synthesis of microbiological knowledge in a form intelligible to the beginner." It contained a special section—"The Biological Background"—for students not previously exposed to biology. The treatments of physiology and metabolism were low-keyed, and the embryo science of bacterial genetics was covered in a little more than 20 pages.

In later editions, each of which has involved a nearly complete rewriting of the book, the authors largely lost sight of the stated goal. "The Biological Background" was already jettisoned in the second edition (1963), and some knowledge of both organic chemistry and genetic principles was presupposed in the considerably enlarged sections that dealt with metabolism and genetics. The fourth and current edition of *The Microbial World* (1976) cannot pretend to provide an introduction for the beginner, in a course which spans either a semester or a quarter. Nevertheless, the authors have frequently regretted this progressive transmutation of their book, believing that its style and viewpoint, if not its scope and content, are very well-suited to a beginner. Extensive discussion of the problem has led to the presently greatly abridged and simplified *Introduction to the Microbial World* which is not intended to replace *The Microbial World* but to complement it. Once again, over 20 years after the first edition, we hope to present anew "a modern synthesis of microbiological knowledge in a form intelligible to the beginner."

R. Y. STANIER

xi

# THE
# HISTORY
# OF
# MICROBIOLOGY

Microbiology is the study of organisms that are too small to be clearly perceived by the unaided human eye, called *microorganisms*. If an object has a diameter of less than 0.1 mm, the eye cannot perceive it at all, and very little detail can be perceived in an object with a diameter of 1 mm. Roughly speaking, therefore, organisms with a diameter of 1 mm or less are microorganisms and fall into the broad domain of microbiology. Microorganisms have a wide taxonomic distribution; they include some metazoan animals, protozoa, many algae and fungi, bacteria, and viruses. The existence of this microbial world was unknown until the invention of microscopes, optical instruments that serve to magnify objects so small that they cannot be clearly seen by the unaided human eye.

## the discovery of the microbial world

**Figure 1.1**
Antony van Leeuwenhoek (1632–1723). In this portrait, he is holding one of his microscopes. Courtesy of the Rijksmuseum, Amsterdam.

The discoverer of the microbial world was a Dutch merchant, Antony van Leeuwenhoek (Figure 1.1). His scientific activities were fitted into a life well filled with business affairs and civic duties. He had little formal education and never attended a university. This was probably no disadvantage scientifically, since the scientific training then available would have provided little basis for his life's work; more serious handicaps, insofar as the communication of his discoveries went, were his lack of connections in the learned world and his ignorance of any language except Dutch. Nevertheless, through a fortunate chance, his work became widely known in his own lifetime, and its importance was immediately recognized. About the time that Leeuwenhoek began his observations, the Royal Society had been established in England for the communication and publication of scientific

2

work. The Society invited Leeuwenhoek to communicate his observations to its members and a few years later (1680) elected him as a Fellow. For almost 50 years, until his death in 1723, Leeuwenhoek transmitted his discoveries to the Royal Society in the form of a long series of letters written in Dutch, which were translated and published, thus becoming quickly disseminated.

Leeuwenhoek's microscopes (Figure 1.2) bore little resemblance to the instruments with which we are familiar. The almost spherical lens (a) was mounted between two small metal plates. The specimen was placed on the point of a blunt pin (b) attached to the back plate and was brought into focus by manipulating two screws (c) and (d), which varied the position of the pin relative to the lens. During this operation the observer held the instrument with its other face very close to his eye and squinted through the lens. No change of magnification was possible, the magnifying power of each microscope being an intrinsic property of its lens. Despite the simplicity of their construction, Leeuwenhoek's microscopes were able to give clear images at magnifications that ranged, depending on the focal length of the lens, from about 50 to nearly 300 diameters. The highest magnification that he could obtain was consequently somewhat less than one-third of the highest magnification that is obtainable with a modern light microscope. Leeuwenhoek constructed hundreds of such instruments; a few survive today.

Leeuwenhoek's place in scientific history depends not so much on his skill as a microscope maker, essential though this was, as on the extraordinary range and skill of his microscopic observations. Indeed, it would be easy to fill a page with a mere list of his major discoveries about the structure of higher plants and animals. His greatest claim to fame rests, however, on his discovery of the microbial world: the world of "animalcules," or little animals, as he and his contemporaries called them. A new

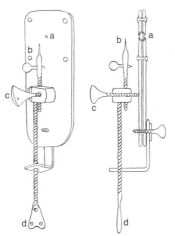

**Figure 1.2**
A drawing to show the construction of one of Leeuwenhoek's microscopes: (a) lens, (b) mounting pin, (c) and (d) focusing screws. After C. E. Dobell, *Antony van Leeuwenhoek and His Little Animals*. New York: Russell and Russell, Inc., 1932.

dimension was thus added to biology. All the main kinds of unicellular micro-organisms that we know today—protozoa, algae, yeasts, and bacteria—were first described by Leeuwenhoek, often with such accuracy that it is possible to identify individual species from his accounts of them. In addition to the diversity of this microbial world, Leeuwenhoek emphasized its incredible abundance. For example, in one letter describing for the first time the characteristic bacteria of the human mouth, he wrote:

I have had several gentlewomen in my house, who were keen on seeing the little eels in vinegar; but some of them were so disgusted at the spectacle, that they vowed they'd never use vinegar again. But what if one should tell such people in future that there are more animals living in the scum on the teeth in a man's mouth, than there are men in a whole kingdom?

Although Leeuwenhoek's contemporaries marveled at his scientific discoveries, the microscopic exploration of the microbial world that he had so brilliantly begun was not appreciably extended for over a century after his death. The principal reasons for this long delay seem to have been technical ones. Simple microscopes of high magnification are both difficult and tiring to use, and the manufacture of the very small lenses is an operation that requires great skill. Consequently, most of Leeuwenhoek's contemporaries and immediate successors used microscopes that suffered from serious optical defects. The major optical improvements that were eventually to lead to compound microscopes of the quality that we use today began about 1820 and extended through the succeeding half century. These improvements were closely followed by resumed exploration of the microbial world and resulted, by the end of the nineteenth century, in a detailed knowledge of its constituent groups. In the meantime, however, the science of microbiology had been developing in other ways, which led to the discovery of the roles the microorganisms play in the transformations of mattter and in the causation of disease.

## the controversy over spontaneous generation

After Leeuwenhoek had revealed the vast numbers of microscopic creatures present in nature, scientists began to wonder about their origin. From the beginning there were two schools of thought. Some believed that the animalcules were formed spontaneously from nonliving materials, whereas others (Leeuwenhoek included) believed that they were formed from the "seeds" or "germs" of these animalcules. The belief in the spontaneous formation of living beings from nonliving matter is known as the doctrine of *spontaneous generation* and has had a long existence. In ancient times it was considered self-evident that many plants and animals can be generated spontaneously under special conditions. The doctrine of spontaneous generation was accepted without question until the Renaissance.

As knowledge of living organisms accumulated, it gradually became evident that the spontaneous generation of plants and animals simply does not occur. A decisive step in the abandonment of the doctrine as applied to animals took place as the result of experiments performed about 1665 by an Italian physician Francesco Redi. He showed that the maggots that develop in putrefying meat are the larval stages of flies and will never appear if the meat is protected by placing it in a vessel closed with fine gauze so that flies are unable to deposit their eggs on it. For technical reasons, it is far more difficult to show that microorganisms are not generated spontaneously, and as time went on the proponents of the doctrine came to center their claims more and more on the mysterious appearance of these simplest forms of life in organic infusions.

One of the first to provide strong evidence that microorganisms do not arise spontaneously in organic infusions was the Italian naturalist Lazzaro Spallanzani, in the middle of the eighteenth century. He showed that heating can prevent the appearance of animalcules in infusions. Spallanzani concluded that animalcules can be carried into infusions by air and that this is the explanation for their supposed spontaneous generation in well-heated infusions. Earlier workers had closed their flasks with corks, but Spallanzani was not satisfied that any mechanical plug could completely exclude air, and he resorted to hermetic sealing. He observed that after sealed infusions had remained barren for a long time, a tiny crack in the glass would be followed by the development of animalcules. His final conclusion was that to render an infusion *permanently* barren, it must be sealed hermetically and boiled. Animalcules could never appear unless new air entered the flask.

Spallanzani's beautiful experiments showed that even very perishable plant or animal infusions do not undergo putrefaction or fermentation when they have been rendered free of animalcules. In the beginning of the nineteenth century, François Appert found that one can preserve foods by enclosing them in airtight containers and heating the containers. He was able in this way to preserve highly perishable foodstuffs indefinitely, and "appertization," as this original canning process was called, came into extensive use for the preservation of foods long before the scientific issue had been finally settled.

In the late eighteenth century, the work of J. Priestley, H. Cavendish, and A. Lavoisier laid the foundations of the chemistry of gases. One of the gases first discovered was oxygen, which soon was recognized to be essential for the life of animals. In the light of this knowledge, it seemed possible that the hermetic sealing recommended by Spallanzani and practiced by Appert was effective in preventing the appearance of microbes and the decomposition of organic matter, not because it excluded air-carrying germs but because it excluded oxygen.

### the experiments of Pasteur

By 1860 some scientists had begun to realize that there is a *causal relationship* between the development of microorganisms in organic infusions and the chemical changes that take place in these infusions: *microorganisms are the agents that bring about the chemical changes*. The great pioneer in these studies was Louis Pasteur (Figure 1.3). However, the acceptance of this concept was conditional on the demonstration that spontaneous generation does not occur. Stung by the continued claims of adherents to the doctrine of spontaneous generation, Pasteur finally initiated a series of experiments that effectively ended the controversy.

Pasteur first demonstrated that air does contain microscopically observable organisms. He aspirated large quantities of air through a tube that contained a plug of guncotton, which served as a filter. The plug was then dissolved and the sediment examined microscopically. The sediment contained considerable numbers of small round or oval bodies, indistinguishable from microorganisms. Pasteur next confirmed the fact that heated air can be supplied to a boiled infusion without giving rise to microbial development. He went on to show that the addition of a piece of germ-laden guncotton to a sterile infusion invariably provoked microbial growth. These experiments showed Pasteur how germs can enter infusions and led him to what was perhaps his most elegant experiment on the subject. This was the demonstration that infusions will remain sterile indefinitely in open flasks, provided that the neck of the flask is drawn out and bent down in such a way that the germs from the air cannot ascend it. Pasteur's swan-necked flask is illustrated in Figure 1.4. If the neck of such a flask was broken off, the infusion

**Figure 1.3**
Louis Pasteur (1822–1895). Courtesy of the Institut Pasteur, Paris.

**Figure 1.4**
The swan-necked flask used by Pasteur during his studies on spontaneous generation. The construction of the neck permitted free access of air to the flask contents but prevented entry of microorganisms present in the air.

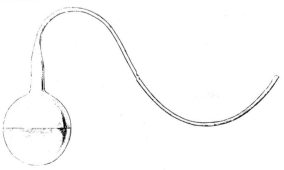

rapidly became populated with microbes. The same thing happened if the sterile liquid in the flask was poured into the exposed portion of the bent neck and then poured back.

### the experiments of Tyndall

The proponents of spontaneous generation maintained a stubborn rear-guard action for some years. The English physicist John Tyndall, an ardent partisan of Pasteur, then undertook a series of experiments designed to refute their claims, in the course of which he established an important fact that had been overlooked by Pasteur, and in part accounted for the conflicting claims of the spontaneous generationists.

Using infusions prepared from meat and fresh vegetables, Tyndall obtained satisfactory sterilization by heating them for 5 minutes in a bath of boiling brine. However, with infusions prepared from dried hay, this sterilization procedure proved completely inadequate. Worse still, when he then attempted to repeat his earlier experiments with other types of infusions, he found that these infusions could no longer be sterilized by immersion in boiling brine, even if the period of heating was for as long as an hour. After many experiments, Tyndall finally realized what had happened. Dried hay contained spores of bacteria that were many times more resistant to heat than any microbes with which he had previously dealt, and as a result of the presence of the hay in his laboratory, the air had become thoroughly infected with these spores. Once he had grasped this point, he proceeded to test the actual limits of heat resistance of the spores and found that even boiling infusions for as long as $5\frac{1}{2}$ hours would not render them sterile with certainty. From these results he concluded that bacteria have phases, one relatively *thermolabile* (destroyed by boiling for 5 minutes) and one *thermoresistant* to an almost incredible extent. These conclusions were almost immediately confirmed by a German botanist, Ferdinand Cohn, who demonstrated that the hay bacteria can produce microscopically distinguishable resting bodies (*endospores*), which are highly resistant to heat. Recognition of the tremendous heat resistance of bacterial spores was essential to the development of adequate procedures for sterilization.

## the discovery of the role of microorganisms in transformations of organic matter

During the long controversy over spontaneous generation, a correlation between the growth of microorganisms in organic infusions and the onset of chemical changes in the infusion itself was established. In 1837 three men, C. Cagniard-Latour, Th. Schwann, and F. Kützing, independently proposed that the yeast which appears during alcoholic fermentation is a microscopic plant, and that the conversion of sugars to ethyl alcohol and carbon dioxide characteristic of the alcoholic fermentation is a physiological function of the yeast cell. This theory was bitterly attacked by such leading chemists of the time as J. J. Berzelius, J. Liebig, and F. Wohler, who held the view that fermentation and putrefaction are purely chemical processes. With the demonstration by the laboratory synthesis of urea in 1828 that organic compounds were not the exclusive products of living activity, these chemists rightly felt that a large body of natural phenomena had now become amenable to analysis in physicochemical terms. The conversion of sugars to alcohol and carbon dioxide appeared to be a relatively simple chemical process. Accordingly, the chemists did not look with favor on the attempt to interpret this process as the result of the action of a living organism.

Pasteur, himself a chemist by training, eventually convinced the scientific world that *all fermentative processes are the results of microbial activity.* This work had a practical origin. The distillers of Lille, France, where the manufacture of alcohol from beet sugar was an important local industry, had encountered difficulties and called on Pasteur in 1857 for assistance. Pasteur found that their troubles were caused by the fact that the alcoholic fermentation had been in part replaced by another kind of fermentative process, which resulted in the conversion of the sugar to lactic acid. When he examined microscopically the contents of fermentation vats in which lactic acid was being formed, he found that the cells of yeast characteristic of the alcoholic fermentation had been replaced by much smaller rods and spheres. If a trace of this material was placed in a sugar solution, a vigorous lactic fermentation ensued, and the small spherical and rod-shaped organisms multiplied. Successive transfers of minute amounts of material to fresh flasks of the same medium always resulted in the production of a lactic fermentation and in growth of the microorganisms. Using similar methods, Pasteur studied a considerable number of fermentative processes during the following 20 years. He was able to show that fermentation is invariably accompanied by the development of microorganisms. Furthermore, he showed that each particular chemical type of fermentation, as defined by its principal organic end products (for example, the lactic, the alcoholic, and the butyric fermentations), is accompanied by the development of a *specific type of microorganism.*

### the discovery of anaerobic life
During his studies on the butyric fermentation, Pasteur discovered another fundamental biological phenomenon: the *existence of forms of life that can live only in the absence of free oxygen.* While examining microscopically fluids that were undergoing a butyric fermentation, Pasteur observed that the bacteria at the margin of a flattened drop, in close contact with the air, became immotile, whereas those in the center of the drop remained motile. This observation suggested that air had an inhibitory effect on the microorganisms in question, an inference that Pasteur quickly confirmed

by showing that passage of a current of air through the fermenting fluid could retard, and sometimes completely arrest, the butyric fermentation. He thus concluded that some microorganisms can live only in the absence of oxygen, a gas previously considered essential for the maintenance of all life. He introduced the terms *aerobic* and *anaerobic* to designate, respectively, life in the presence and in the absence of oxygen.

Pasteur was thus the first to realize that the breakdown of organic compounds in the absence of oxygen can also be used by some organisms as a means of obtaining energy; as he put it, "Fermentation is life without air." Some strictly anaerobic microorganisms, such as the butyric acid bacteria, are dependent on fermentative mechanisms to obtain energy. Many other microorganisms, including certain yeasts, are *facultative anaerobes,* which have two alternative energy-yielding mechanisms at their disposal. In the presence of oxygen they employ aerobic respiration, but they can employ fermentation if no free oxygen is present in their environment. This was beautifully demonstrated by Pasteur, who showed that sugar is converted to alcohol and carbon dioxide by yeast in the absence of air but that in the presence of air little or no alcohol is formed; carbon dioxide is the principal end product of this aerobic reaction.

The further development of knowledge about the nature of fermentation resulted from an accidental observation made in 1897 by H. Buchner. In attempting to preserve an extract of yeast, prepared by grinding yeast cells with sand, Buchner added a large quantity of sugar and was surprised to observe an evolution of carbon dioxide accompanied by the formation of alcohol. A soluble enzymatic preparation, able to carry out alcoholic fermentation, was thus discovered. Buchner's discovery inaugurated the development of modern biochemistry; the detailed analysis of the mechanism of cell-free alcoholic fermentation was eventually to show that this complex metabolic process can be interpreted as resulting from a succession of chemically intelligible reactions, each catalyzed by a specific enzyme. Today, the belief that even the most complex physiological process can be similarly understood in physicochemical terms is accepted as a matter of course by all biologists. In this sense, the intuition of the nineteenth-century chemists who battled against the biological theory of fermentation has proved to be a correct one.

## the discovery of the role of microorganisms in the causation of disease

It was shown in 1813 that specific fungi can cause diseases of wheat and rye, and in 1845 M. J. Berkeley proved that the great Potato Blight of Ireland, a natural disaster that deeply influenced Irish history, was caused by a fungus. The first recognition that fungi may be specifically associated with a disease of animals came in 1836 through the work of A. Bassi in Italy on a fungal disease of silkworms. A few years later J. L. Schönlein showed that certain skin diseases of man were caused by fungal infections. Despite these indications, very few medical scientists were willing to entertain the notion that organisms as small and as apparently simple as the bacteria could act as agents of disease.

### surgical antisepsis

The introduction of anesthesia about 1840 made possible a very rapid development of surgical methods. However, with the elaboration of surgical technique, a problem that had always existed was becoming increasingly serious, that is, *surgical sepsis,* or the infections that followed surgical intervention and often resulted in death. Pasteur's studies had shown the presence of microorganisms in the air and had indicated various ways in which their access to organic infusions could be prevented. A young British surgeon, Joseph Lister, deeply impressed by Pasteur's work, reasoned that surgical sepsis might well result from microbial infection of the tissues exposed during operation. He decided to develop methods for preventing the access of microorganisms to surgical wounds. By the scrupulous sterilization of surgical instruments, by the use of disinfectant* dressings, and by the conduct of surgery under a spray of disinfectant, he succeeded in greatly reducing the incidence of surgical sepsis. Lister's procedures of antiseptic surgery, developed about 1864, were initially greeted with considerable scepticism but, as their striking success was recognized, gradually became common practice. This work provided powerful *indirect* evidence for the germ theory of disease.

### the bacterial etiology of anthrax

The discovery that bacteria can act as specific agents of infectious disease in animals was made through the study of anthrax, a serious infection of domestic animals that is transmissible to man. In the terminal stages of a generalized anthrax infection, the rod-shaped bacteria responsible for the disease occur in enormous numbers in the bloodstream. These objects were first observed as early as 1850, and their presence in the blood of infected animals was reported by a series of investigators during the following 15 years.

The conclusive demonstration of the bacterial causation, or *etiology,* of anthrax was provided in 1876 by Robert Koch (Figure 1.5), a German country doctor. His experiments were conducted in his home, using very primitive equipment and small experimental animals. He showed that mice could be infected with material from a diseased domestic animal. He

**Figure 1.5**
Robert Koch (1843–1910). Courtesy of VEB George Thieme, Leipzig.

* A *disinfectant* is a compound that kills infectious agents; an *antiseptic* is a compound that prevents the growth of infectious agents.

transmitted the disease through a series of 20 mice by successive inoculation; at each transfer, the characteristic symptoms were observed. He then proceeded to cultivate the causative bacterium by introducing minute, heavily infected particles of spleen from a diseased animal into drops of sterile serum. Observing hour after hour the growth of the organisms in this culture medium, he saw the rods change into long filaments within which ovoid, refractile bodies eventually appeared. He showed that these bodies were spores, which had not been seen by previous workers (Figure 1.6). When spore-containing material was transferred to a fresh drop of sterile serum, the spores germinated and gave rise once more to typical rods. In this fashion, he transferred cultures of the bacterium eight successive times. The final culture of the series, injected into a healthy animal, again produced the characteristic disease, and from this animal the organisms could again be isolated in culture.

This series of experiments fulfilled the criteria that had been laid down 36 years before by J. Henle as logically necessary to establish the causal relationship between a specific microorganism and a specific disease. In generalized form, these criteria are: (1) the microorganism must be present in every case of the disease; (2) the microorganism must be isolated from the diseased host and grown in pure culture; (3) the specific disease must be reproduced when a pure culture of the microorganism is inoculated into a healthy susceptible host; and (4) the microorganism must be recoverable once again from the experimentally infected host. Since Koch was the first to apply these criteria experimentally, they are now generally known as *Koch's postulates*.

**Figure 1.6**
The first photomicrographs of bacteria, taken by Robert Koch in 1877. (a) Unstained chains of vegetative cells of *Bacillus anthracis*. (b) Unstained chains of *B. anthracis;* the cells contain refractile spores. (c) A stained smear of *B. anthracis* from the spleen of an infected animal. Note the rod-shaped bacilli and the larger tissue cells.

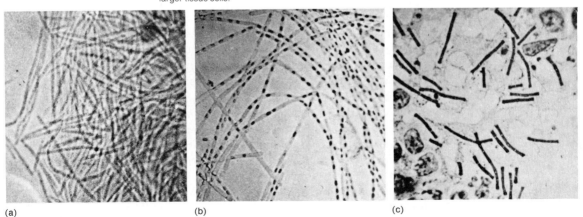

(a)                    (b)                    (c)

Koch carried out another series of experiments that demonstrated the *biological specificity* of disease agents. He showed that another spore-forming bacterium, the hay bacillus, does not cause anthrax upon injection, and he also differentiated bacteria that cause other infections from the anthrax organism. From these studies he concluded that "only one kind of bacillus is able to cause this specific disease process, while other bacteria either do not produce disease following inoculation, or give rise to other kinds of disease."

In the meantime, Pasteur and J. Joubert, unaware of Koch's work, undertook a study of anthrax, which confirmed Koch's conclusions and provided additional demonstrations that the bacillus, and not some other agent, was the specific cause of the disease.

This work on anthrax abruptly ushered in the golden age of medical bacteriology, during which newly established institutes, created in Paris and in Berlin for Pasteur and Koch, respectively, became the world centers of bacteriological science. The German school concentrated primarily on the isolation, cultivation, and characterization of the causative agents for the major infectious diseases of man. The French school turned almost immediately to a more subtle and complex problem: the experimental analysis of how infectious disease takes place in the animal body and how recovery and immunity are brought about. Within 25 years most of the major bacterial agents of human disease had been discovered and described, and methods for the prevention of many of these diseases, either by artificial immunization or by the application of hygienic measures, had been developed. It was by far the greatest medical revolution in all human history.

### the discovery of viruses

One of the early technical contributions from Pasteur's new institute was the development of filters able to retain bacterial cells and thus to yield bacteria-free filtrates. Infectious fluids were often tested for the presence of disease-producing bacteria by passing them through such filters; if the filtrate was no longer able to produce infection, the presence of a bacterial agent in the original fluid was indicated. In 1892 a Russian scientist, D. Iwanowsky, applied this test using an infectious extract from tobacco plants infected with mosaic disease. He found to his surprise that the filtrate remained fully infectious. His specific discovery was soon confirmed, and within a few years other workers found that many major plant and animal diseases are caused by similar, filter-passing, submicroscopic agents. A whole class of infectious entities, much smaller than any previously known organisms, was thus discovered. The true nature of these *viruses*, as they came to be known, remained obscure for many decades, but eventually it was established that they are a distinctive group of biological objects entirely different in structure and mode of development from all cellular organisms (see Chapter 7).

# the development of pure culture methods

Pasteur possessed an intuitive skill in the handling of microorganisms and was able to reach correct conclusions even when working with cultures that contained a mixture of microbial forms. However, there are pitfalls in working with mixed microbial populations, and not all the scientists who began to study microorganisms in the middle of the nineteenth century were as skillful as Pasteur and Koch. It was frequently claimed that microorganisms had a large capacity for variation with respect both to their *morphological form* and to their *physiological function*. This belief became known as the doctrine of *pleomorphism*.

## the origin of the belief
## in pleomorphism

Let us consider what happens when a nutrient solution is inoculated with a mixed microbial population. The principle of natural selection at once begins to operate, and the microbe that can grow most rapidly under the conditions provided soon predominates. As a result of its growth and chemical activities, the composition of the medium changes; after some time, conditions no longer permit growth of the originally predominant form. The environment may now be favorable for the growth of a second kind of microorganism, also originally introduced into the medium but hitherto unable to develop, which gradually replaces the first as the predominant form in the culture. In this fashion one may obtain the *successive development of many different microbial types* in a single culture flask seeded with a mixed population. It is often possible to maintain the predominance of the form that first develops by repeated transfer of the mixed population at short intervals into a fresh medium of the same composition; this was essentially the device used by Pasteur in his studies on fermentation.

If one does not recognize the possibility of such microbial successions, it is easy to conclude that the chemical and morphological changes observable over the course of time in a single culture inoculated with a mixed population reflect *transformations and activities of a single kind of microorganism*. Between 1865 and 1885, claims for the pleomorphism of microorganisms, based on such observations, were frequently made.

## the first pure cultures

Around 1870 it began to be realized that a sound understanding of the form and function of microorganisms could be obtained only if the complications inherent in the study of mixed microbial populations were avoided by the use of pure cultures. *A pure culture is one that contains only a single kind of microorganism.* The leading advocates of the use of pure cultures were two great mycologists (students of fungi), A. de Bary and O. Brefeld.

Much of the pioneering work on pure culture techniques was done by Brefeld, working with fungi. He introduced the practice of isolating single cells, as well as the cultivation of fungi on solid media, for which purpose he added gelatin to his culture liquids. His methods of obtaining pure cultures worked admirably for the fungi but were found to be unsuitable when applied to the smaller bacteria.

Koch realized very early that the development of simple methods for obtaining pure cultures of bacteria was vital for the growth of the new science. A promising approach had already been suggested by the earlier observations of J. Schroeter, who had noted that on such solid substrates as potato, starch paste, bread, and egg albumen, isolated bacterial growths, or *colonies*, arose. The colonies differed from one another, but within each colony the bacteria were of one type. At first Koch experimented with the use of sterile, cut surfaces of potatoes, which he placed in sterile, covered glass vessels and then inoculated with bacteria. However, potatoes have disadvantages: the cut surface is moist, which allows motile bacteria to spread freely over it; the substrate is opaque, and hence it is often difficult to see the colonies; and most important of all, the potato is not a good nutrient medium for many bacteria. Koch perceived that it would be far better if one could solidify well-tried liquid media with some clear substance. In this fashion, a translucent gel could be prepared on which developing bacterial colonies would be clearly visible. With this in mind, he added gelatin as a hardening agent. Once set, the gelatin surface was seeded by picking up a minute quantity of bacterial cells (the *inoculum*) on a platinum needle, previously sterilized by passage through a flame, and drawing it several times lightly across the surface of the jelly. Different bacterial colonies soon appeared, each of which could be purified by repetition of the streaking process. This became known as the *streak method* for isolating bacteria. Shortly thereafter, Koch discovered that instead of streaking the bacteria over the surface of the already solidified gelatin, he could mix them with the melted gelatin. When the gelatin set, the bacteria were immobilized in the jelly and there developed into isolated colonies. This became known as the *pour plate method* for isolating bacteria.

Gelatin, the first solidifying agent used by Koch, has several disadvantages. It is a protein highly susceptible to microbial digestion and liquefaction. Furthermore, it changes from a gel to a liquid at temperatures above 28°C. A new solidifying agent, *agar*, was soon introduced. Agar is a complex polysaccharide, extracted from red algae. A temperature of 100°C is required to melt an agar gel, so it remains solid throughout the entire temperature range over which bacteria are cultivated. However, once melted, it remains a liquid until the temperature falls to about 44°C, a fact that makes possible its use for the preparation of cultures by the pour plate method. It produces a stiff and transparent gel. Finally, it is attacked by relatively few bacteria, so the problem of its liquefaction rarely arises. For these reasons, agar has rapidly replaced gelatin as the hardening agent of choice for bacteriological work. All modern attempts to find an equally satisfactory synthetic substitute for agar have failed.

## microorganisms as geochemical agents

Although the role played by microorganisms as agents of infectious disease was the central microbiological interest in the last decades of the nineteenth century, some scientists carried forward the work initiated by Pasteur through his early investigations on the role of microorganisms in fermentation. This work had clearly shown that microorganisms can serve as specific agents for large-scale chemical transformations and indicated that the microbial world as a whole might well be responsible for a wide variety of other geochemical changes.

The establishment of the cardinal roles that microorganisms play in the biologically important cycles of matter on earth—the cycles of carbon, nitrogen, and sulfur—was largely the work of two men, S. Winogradsky and M. W. Beijerinck. In contrast to plants and animals, microorganisms show an extraordinarily wide range of physiological diversity. Many groups are specialized for carrying out chemical transformations that cannot be performed at all by plants and animals, and thus play vital parts in the turnover of matter on earth.

One example of microbial physiological specialization is provided by the autotrophic bacteria, discovered by Winogradsky. These bacteria can grow in completely inorganic environments, obtaining the energy necessary for their growth by the oxidation of inorganic compounds, and use carbon dioxide as the source of their cellular carbon. Winogradsky found that there are several physiologically distinct groups among the autotrophic bacteria, each characterized by the ability to use a particular inorganic energy source; for example, the sulfur bacteria oxidize inorganic sulfur compounds, the nitrifying bacteria, inorganic nitrogen compounds.

Another discovery, to which both Winogradsky and Beijerinck contributed, was the role that microorganisms play in the fixation of atmospheric nitrogen, which cannot be used as a nitrogen source by most living organisms. They showed that certain bacteria, some symbiotic in higher plants and others free-living, can use gaseous nitrogen for the synthesis of their cell constituents. These microorganisms accordingly help to maintain the supply of combined nitrogen, upon which all other forms of life depend.

# the growth of microbiology in the twentieth century

During the last decades of the nineteenth century, microbiology became a solidly established discipline with a distinctive set of concepts and techniques. During the same period a science of general biology also emerged. It was largely the creation of Charles Darwin, who imposed a new order and coherence in the heretofore anecdotal materials of natural history by interpreting them in terms of the theory of evolution through natural selection. Logically, microbiology should have taken its place, alongside other specialized biological disciplines, in the framework of post-Darwinian general biology. In fact, however, this did not occur. The major interests of microbiology in this period were the characterization of agents of infectious disease, the study of immunity and its functions in the prevention and cure of disease, and the analysis of the chemical activities of microorganisms. All these problems were both conceptually and experimentally remote from the dominant interests of biology in the early twentieth century; the organization of the cell and its role in reproduction and development; and the mechanisms of heredity and evolution in plants and animals.

However, microbiology did contribute significantly to the development of the new discipline of biochemistry. The discovery of cell-free alcoholic fermentation by Buchner (as discussed earlier) provided the key to the chemical analysis of biological processes. In the first two decades of the twentieth century, parallel studies on metabolism of muscle and of yeast gradually revealed their fundamental similarity. A few years later the analysis of animal and microbial nutrition revealed another unexpected common denominator: the "vitamins" required in traces by animals proved chemically identical with the "growth factors" required by some bacteria and yeasts. For technical reasons, the detailed study of the functions of these substances was conducted in large measure with microorganisms. These studies, spanning the period from 1920 to 1935, demonstrated the fundamental similarities of all living systems at the metabolic level: a doctrine proclaimed by biochemists and microbiologists under the slogan "the unity of biochemistry."

The second great advance of biology in the early twentieth century—the creation of the discipline of genetics—had no immediate impact on microbiology. Indeed, it long seemed doubtful whether the mechanisms of inheritance operative in plants and animals likewise functioned in bacteria. The first important contact between genetics and microbiology occurred in 1941, when G. Beadle and E. Tatum succeeded in isolating a series of biochemical mutants from the fungus *Neurospora*. This opened the way to the analysis of the consequences of mutation in biochemical terms, and *Neurospora* joined the fruit fly and the maize plant as a material of choice for genetic research.

In 1943 an analysis by M. Delbrück and S. Luria of mutation in bacteria provided the technical and conceptual basis for genetic work on these microorganisms. Soon afterward several mechanisms of genetic transfer were shown to exist in bacteria, all significantly different from the mechanism of sexual recombination in plants and animals. In 1944 the work of O. T. Avery, C. M. McLeod, and M. McCarty on one such process of bacterial genetic transfer known as *transformation* revealed the chemical nature of the hereditary material to be DNA.

The confluence of microbiology, genetics, and biochemistry between 1940 and 1945 brought to an end the long isolation of microbiology from the main currents of biological thought. It also sets the stage for the second major revolution in biology, to which microbiologists made many contributions of fundamental importance: the advent of molecular biology.

# THE
# METHODS
# OF
# MICROBIOLOGY

As a result of the small size of microorganisms, the amount of information that can be obtained about their properties from the examination of *individuals* is limited; for the most part, the microbiologist studies *populations,* containing millions or billions of individuals. Such populations are obtained by growing microorganisms under more or less well-defined conditions, as *cultures.* A culture that contains only one kind of microorganism is known as a *pure culture.* A culture that contains more than one kind of microorganism is known as a *mixed culture.*

At the heart of microbiology there accordingly lie two kinds of operations: *isolation,* the separation of a particular microorganism from the mixed populations that exist in nature; and *cultivation,* the growth of microbial populations in artificial environments (culture media) under laboratory conditions. These two operations come into play irrespective of the kind of microorganism with which the microbiologist deals; they are basic to the study of viruses, bacteria, fungi, algae, protozoa, and even small invertebrate animals. Furthermore, these two operations have been extended in recent years to the study in isolation of cell or tissue lines derived from higher plants and animals (*tissue culture*). The unity of microbiology as a science, despite the biological diversity of the organisms with which it deals, is derived from this common operational base.

## pure culture technique

Microorganisms are ubiquitous, so the preparation of a pure culture involves not only the isolation of a given microorganism from a mixed natural microbial population, but also the maintenance of the

THE METHODS OF MICROBIOLOGY

isolated individual and its progeny in an artificial environment to which the access of other microorganisms is prevented. Microorganisms do not require much space for development; hence an artificial environment can be created within the confines of a test tube, a flask, or a petri dish, the three kinds of containers most commonly used to cultivate microorganisms. The culture vessel must be rendered initially *sterile* (free of any living microorganism), and after the introduction of the desired type of microorganism, it must be protected from subsequent external contamination. The primary source of external contamination is the atmosphere, which always contains floating microorganisms. The form of a petri dish, with its overlapping lid, is specifically designed to prevent atmospheric contamination. Contamination of tubes and flasks is prevented by closure of their orifices with an appropriate stopper. This is usually a plug of cotton wool, although metal caps or plastic screw caps are now often employed, particularly for test tubes.

The external surface of a culture vessel is, of course, subject to contamination, and the interior of a flask or tube can become contaminated when it is opened to introduce or withdraw material. This danger is minimized by passing the orifice through a flame, immediately after the stopper has been removed and again just before it is replaced.

The *inoculum* (i.e., the microbial material used to seed or inoculate a culture vessel) is commonly introduced on a straight metal wire or loop, which is rapidly sterilized just before use by heating in a flame. Transfers of liquid cultures can also be made with a pipette, sterilized in a paper wrapping or in a glass or metal container, which keeps both inner and outer surfaces free of contamination until the time of use.

### the isolation of pure cultures by plating methods

Pure cultures of microorganisms that form discrete colonies on solid media (e.g., most bacteria, yeasts, and many fungi and unicellular algae) may be most simply obtained by one of the modifications of the plating method. This method involves the separation and immobilization of individual organisms on or in a nutrient medium solidified with agar or some other appropriate jelling agent. Each viable organism gives rise, through growth, to a colony from which transfers can be readily made.

The *streaked plate* is in general the most useful plating method. A sterilized bent wire is dipped into a suitable diluted suspension of organisms and is then used to make a series of parallel, non-overlapping streaks on the surface of an already solidified agar plate. The inoculum is progressively diluted with each successive streak, so that even if the initial streaks yield confluent growth, well-isolated colonies develop along the lines of later streaks (Figure 2.1). Alternatively, isolations can be made with *poured plates:* successive dilutions of the inoculum are placed in sterile petri dishes and mixed with the molten but cooled

**Figure 2.1**
Isolation of a pure culture by the streak method. A petri dish containing nutrient agar was streaked with a suspension of bacterial cells. As a result of subsequent growth, each cell has given rise to a macroscopically visible colony.

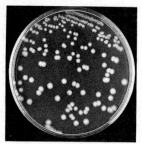

agar medium, which is then allowed to solidify. Colonies subsequently develop embedded in the agar.

The isolation of strictly anaerobic bacteria by plating methods poses special problems. Provided that the organisms in question are not rapidly killed by exposure to oxygen, plates may be prepared in the usual manner and then incubated in closed containers, from which the oxygen is removed either by chemical absorption or combustion. For the more oxygen-sensitive anaerobes, a modification of the pour plate method, known as the *dilution shake culture,* is to be preferred. A tube of melted and cooled agar is inoculated and mixed, and approximately one-tenth of its contents is transferred to a second tube, which is then mixed and used to inoculate a third tube in a similar fashion. After 6 to 10 successive dilutions have been prepared, the tubes are rapidly cooled, and sealed by pouring a layer of sterile petroleum jelly and paraffin on the surface, thus preventing access of air to the agar column (Figure 2.2). To make a transfer, the petroleum jelly-paraffin seal is removed with a sterile needle, and the agar column is gently blown out of the tube into a sterile petri dish by passing a stream of gas through a capillary pipette inserted between the tube wall

**Figure 2.2**

Isolation of a pure culture of anaerobic bacteria by the dilution shake method. A complete series of dilution shakes is shown. Note the confluent growth in the more densely seeded tubes (at right), and the well-isolated colonies in the two final tubes of the series (at left). After the agar had solidified, each tube was sealed with a mixture of sterile vaseline and paraffin to prevent the access of atmospheric oxygen, which inhibits the growth of anaerobic bacteria.

and the agar. The column is sectioned with a sterile knife to permit examination and transfer of colonies.

In isolating from a mixed natural population it is often possible, provided one's technique is good, to prepare a first plate or dilution shake series in which many of the colonies that develop are well separated from one another. Can one then pick material from such a colony, transfer it to an appropriate medium, and call it a pure culture? Although this is often done, a culture so isolated may be far from pure. Microorganisms vary greatly in their nutritional requirements, and consequently no single medium and set of growth conditions will permit the growth of all the microorganisms present in a natural population. Indeed, it is probable that only a very small fraction of the microorganisms initially present will be able to form colonies on any given medium. Hence, for every visible colony on a first plate, there may be a thousand cells of other kinds of microorganisms that were also deposited on the agar surface but have failed to give macroscopically visible growth, although they may still be viable. The probability is high that some of these organisms will be picked up and carried over when a transfer is made. *One should never pick from a first plate for the preparation of a pure culture.* Instead, a second plate should be streaked from a cell suspension prepared from a well-isolated colony. If all the colonies on this second plate appear identical, a well-isolated colony can be used to establish a pure culture.

**Figure 2.3**
Exponential (logarithmic) death of bacteria. The same data are plotted semilogarithmically in (a) and arithmetically in (b); *N* is the number of surviving bacteria.

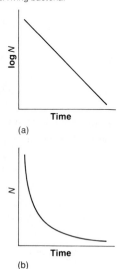

## the theory and practice of sterilization

Sterilization is a treatment that *frees the treated object of all living organisms.* It can be achieved by exposure to lethal physical or chemical agents or, in the special case of solutions, by filtration.

To understand the basis of sterilization by lethal agents, it is necessary to describe briefly the kinetics of death in a microbial population. The only valid criterion of death in the case of a microorganism is *irreversible loss of the ability to reproduce;* this is usually determined by quantitative plating methods, survivors being detected by colony formation. When a pure microbial population is exposed to a lethal agent, the kinetics of death are nearly always *exponential:* the number of survivors decreases geometrically with time. This reflects the fact that all the members of the population are of similar sensitivity; probability alone determines the actual time of death of any given individual. If the logarithm of the number of survivors is plotted as a function of the time of exposure, a straight line is obtained (Figure 2.3); its negative slope defines the *death rate.*

The death rate tells one only what *fraction* of the initial population survives a given period of treatment. To determine the *actual number* of survivors, one must also know the *initial population size,* as illustrated graphically in Figure 2.4. Accordingly, for the establishment of procedures of

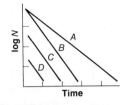

**Figure 2.4**
Relationship of death rate and population size to the time required for the destruction of bacterial cultures; *N* is the number of surviving bacteria. Cultures *B*, *C*, and *D* have identical death rates. Culture *A* has a lower death rate.

sterilization, two factors have to be taken into account: the death rate and the initial population size.

In the practice of sterilization the microbial population to be destroyed is almost always a *mixed* one. Since microorganisms differ widely in their resistance to lethal agents, the significant factors become the initial population size and the death rate of the *most resistant* members of the mixed population. As Tyndall found (see Chapter 1), these are almost always the highly resistant endospores of certain bacteria. Consequently, spore suspensions of known resistance are the objects commonly used to assess the reliability of sterilization methods.

Taking into account the kinetics of microbial death, we can formulate the practical goal of sterilization by a lethal agent in a slightly more refined way: *the probability that the object treated contains even one survivor should be infinitesimally small*. For example, if we wish to sterilize a liter of a culture medium, this goal will be achieved for all practical purposes if the treatment is one that will leave no more than one survivor in $10^6$ liters; under such circumstances, the probability of failure is very small indeed. Procedures of routine sterilization are always designed to provide a very wide margin of safety.

### sterilization by heat

Heat is the most widely used lethal agent for purposes of sterilization. Objects may be sterilized by dry heat, applied in an oven in an atmosphere of air, or by moist heat provided by steam. Of the two methods, sterilization by dry heat requires a much greater duration and intensity. Heat conduction is less rapid in air than in steam. In addition, bacteria can survive in a completely desiccated state, and in this state the intrinsic heat resistance of vegetative bacterial cells is greatly increased, almost to the level characteristic of spores. Consequently, the death rate is much lower for dry cells than for moist ones.

Dry heat is used principally to sterilize glassware or other heat-stable solid materials. The objects are wrapped in paper or otherwise protected from subsequent contamination and exposed to a temperature of 170°C for 90 minutes in an oven.

Steam must be used for the heat sterilization of aqueous solutions. Treatment is usually carried out in a metal vessel known as an *autoclave*, which can be filled with steam at a pressure greater than atmospheric. Sterilization can thus be achieved at temperatures considerably above the boiling point of water; laboratory autoclaves are commonly operated at a steam pressure of 15 lb/in.² above atmospheric pressure, which corresponds to a temperature of 120°C. Even bacterial spores that survive several hours of boiling are rapidly killed at 120°C. Small volumes of liquid (up to about 3 liters) can be sterilized by exposure for 20 minutes; if larger volumes are to be sterilized, the time of treatment must be extended.

### sterilization by other methods

Some substances used in preparing culture media are too heat labile to be sterilized by autoclaving. For such substances, reliable alternative methods of chemical sterilization would be extremely useful. The best available candidate is *ethylene oxide,* a liquid that boils at 10.7°C. However, ethylene oxide is both explosive and toxic for humans, so special precautions must be taken in its handling. For these reasons, ethylene oxide sterilization has not become a routine laboratory procedure. It is, however, used industrially for the sterilization of plastic petri dishes and other plastic objects that melt at temperatures greater than 100°C. A number of other chemicals can be effectively used as disinfectants if one is not concerned about destruction of labile organic compounds or with leaving a toxic residue. These include chlorine gas, commonly employed to treat municipal water supplies and sewage, as well as mercury salts, hypochlorite, and some detergents.

As an aid to the use of disinfectants in sterilizing dry objects such as working surfaces, ultraviolet (UV) light is sometimes used. Since intense UV light can burn exposed skin, and can cause eye damage and skin cancer, its use is usually confined to enclosed spaces.

The principal laboratory method used to sterilize solutions of heat-labile materials is *filtration* through filters capable of retaining microorganisms. Microorganisms are retained in part by the small size of the filter pores and in part by adsorption on the pore walls during their passage through the filter.

## the principles of microbial nutrition

To grow, organisms must draw from the environment all the substances that they require for the synthesis of their cell materials and for the generation of energy. These substances are termed *nutrients.* A culture medium must therefore contain appropriate quantities of all necessary nutrients. However, microorganisms are extraordinarily diverse in their specific physiological properties, and correspondingly in their specific nutrient requirements. Nevertheless, the design of a culture medium can and should be based on scientific principles, which we shall outline as a preliminary to the description of culture media.

The chemical composition of cells, broadly constant throughout the living world, indicates the major material requirements for growth. *Water* accounts for some 80 to 90 percent of the total weight of cells and is always therefore the major essential nutrient, in quantitative terms. The solid matter of cells contains, in addition to *hydrogen* and *oxygen* (derivable metabolically from water), *carbon, nitrogen, phosphorus,* and *sulfur,* in order of decreasing abundance. These six elements are constituents of the macromolecules of the cell, and account for about 95 percent of the cellular dry weight. Many other elements are included in the residual fraction. Nutritional studies show that *potassium, magnesium, calcium, iron, manganese, cobalt, copper, molybdenum,* and *zinc* are required by nearly all organisms, principally as cofactors for certain enzymes.

All the required metallic elements can be supplied as nutrients in the form of the *cations of inorganic salts*. Potassium, magnesium, calcium, and iron are required in relatively large amounts and should always be included as salts in culture media. The quantitative requirements for manganese, cobalt, copper, molybdenum, and zinc (referred to as *trace elements*) are very small, and they are often present in adequate amounts as contaminants of the major inorganic constituents of media. One nonmetallic element, phosphorus, can also be used as a nutrient when provided in inorganic form, as phosphate salts.

The needs for *carbon, nitrogen, sulfur,* and *oxygen* cannot be so simply described because organisms differ with respect to the *specific chemical form* under which these elements must be provided as nutrients.

### the requirements for carbon

Organisms that perform photosynthesis and bacteria that obtain energy from the oxidation of inorganic compounds typically use the most oxidized form of carbon, $CO_2$, as the sole or principal source of cellular carbon. The conversion of $CO_2$ to organic cell constituents is a reductive process, which requires a net input of energy. In these physiological groups, accordingly, a considerable part of the energy derived from light or from the oxidation of reduced inorganic compounds must be expended for the reduction of $CO_2$ to the level of organic matter.

All other organisms obtain carbon principally from organic nutrients, which are commonly at the same general oxidation level as organic cell constituents. In addition to meeting the biosynthetic needs of the cell for carbon, organic substrates must supply the energetic requirements of the cell. Organic substrates thus usually have a *dual nutritional role:* they serve at the same time as a source of carbon and as a source of energy. Many microorganisms can use *a single organic compound* to supply completely both these nutritional needs. Others, however, cannot grow when provided with only one organic compound, and they need a variable number of additional organic compounds (termed *growth factors*), whose roles are described below.

When the organic carbon requirements of *individual* microorganisms are examined, some show a high degree of versatility, whereas others are extremely specialized. Certain bacteria of the *Pseudomonas* group, for example, can use any one of over 90 different organic compounds as sole carbon and energy source. At the other end of the spectrum are methane-oxidizing bacteria, which can use only two organic substrates, methane and methanol, and certain cellulose-decomposing bacteria, which can use only cellulose.

### the requirements for nitrogen and sulfur

Nitrogen and sulfur occur in the organic compounds of the cell principally in reduced form. Most photosynthetic organisms and many nonphotosynthetic bacteria and fungi can meet the needs for nitrogen and sulfur by reducing nitrates and sulfates. Other microorganisms are unable to bring about a reduction of one or both of these anions and must be supplied with the elements in a reduced form. The requirement for a reduced nitrogen source is relatively common and can be met by the provision of nitrogen as ammonium salts. A requirement for reduced sulfur is rarer; it can be met by the provision of sulfide.

The nitrogen and sulfur requirements can often also be met by organic nutrients that contain these two elements in reduced organic combination (amino acids or more complex protein degradation products, such as peptones). Such compounds may also simultaneously provide organic carbon and energy sources.

Some bacteria can also reduce the most abundant natural nitrogen source, atmospheric $N_2$, by the process termed *nitrogen fixation*.

### growth factors

Any organic compound that an organism requires as a precursor or constituent of its organic cell material, but which it cannot synthesize from simpler carbon sources, must be provided as a nutrient. Organic nutrients of this type are known collectively as *growth factors*. They fall into three classes, in terms of chemical structure and metabolic function:

1. *Amino acids*, required as constituents of proteins.
2. *Purines* and *pyrimidines*, required as constituents of nucleic acids.
3. *Vitamins*, a diverse collection of organic compounds that are precursors of coenzymes.

Because growth factors fulfill specific needs in biosynthesis, they are required in only small amounts, relative to the principal cellular carbon source, which must serve as a general precursor of cell carbon. Some 20 different amino acids enter into the composition of proteins, so the need for any specific amino acid that the cell is unable to synthesize is obviously not large. The same argument applies to specific need for a purine or a pyrimidine: five different compounds of these classes enter into the structure of nucleic acids. The quantitative requirements for vitamins are even smaller, since the various coenzymes of which they are precursors have catalytic roles and consequently are present at levels of a few parts per million (which corresponds to a few hundred molecules per cell).

### the roles of oxygen in nutrition

As an elemental constituent of water and of organic compounds, oxygen is a universal component of cells and is always provided in large amounts in the major nutrient, water. However, many organisms also require *molecular oxygen* ($O_2$). These are organisms that are dependent on *aerobic respiration* for the fulfillment of their energetic needs and for which molecular oxygen functions as a terminal electron acceptor (see Chapter 4). Such organisms are termed *obligate aerobes*.

At the other physiological extreme are those microorganisms that obtain energy by *fermentation* (see Chapter 4) and for which this chemical form of the element is not a nutrient. Indeed, for many of these physiological groups, molecular oxygen is a toxic substance, which either kills them or inhibits their growth. Such organisms are *obligate anaerobes*.

Some microorganisms are *facultative anaerobes*, able to grow either in the presence or in the absence of molecular oxygen. In metabolic terms, facultative anaerobes fall into two subgroups. Some, like the lactic acid bacteria, have an exclusively *fermentative* (see Chapter 4) energy-yielding metabolism but are not sensitive to the presence of oxygen. Others (e.g., many yeasts, coliform bacteria) can shift from a *respiratory* (see Chapter 4) to a fermentative mode of metabolism. Such facultative anaerobes use $O_2$ as a terminal oxidizing agent when it is available but can also obtain energy in its absence by fermentative reactions.

Among microorganisms that are obligate aerobes, some grow best at partial pressures of oxygen considerably below that (0.2 atm) present in air. They are termed *microaerophilic*. This probably reflects the possession of enzymes that are inactivated under strongly oxidizing conditions and can thus be maintained in a functional state only at low partial pressures of $O_2$.

### nutritional categories among microorganisms

Perhaps the most useful nutritional classification is that based on two parameters: the nature of the *energy source* and of the *principal carbon source*. With respect to energy source, there is a basic dichotomy between photosynthetic organisms that are able to use light as an energy source, termed *phototrophs*, and organisms that are dependent on a chemical energy source, termed *chemotrophs*. Organisms able to use $CO_2$ as a principal carbon source are termed *autotrophs*; organisms dependent on an organic carbon source are termed *heterotrophs*. By means of these criteria, four major nutritional categories can be distinguished:

1. *Photoautotrophs*, using light as the energy source and $CO_2$ as the principal carbon source. They include most photosynthetic organisms: higher plants, algae, and many photosynthetic bacteria.

2. *Photoheterotrophs*, using light as the energy source and an organic compound as the principal carbon source. This category includes certain of the photosynthetic bacteria.

3. *Chemoautotrophs*, using a chemical energy source and $CO_2$ as the principal carbon source. Energy is obtained by the oxidation of *reduced inorganic compounds*. Only bacteria belong to this nutritional category.

4. *Chemoheterotrophs*, using a chemical energy source and an organic substance as the principal carbon source. The clear-cut distinction between energy source and carbon source, characteristic of the three preceding categories, loses its clarity in the context of chemoheterotrophy, where *both carbon and energy can usually be derived from the metabolism of a single organic compound*. The chemoheterotrophs include all metazoan animals, protozoa, fungi, and the great majority of bacteria. Certain further subdivisions within this very complex nutritional category can be made. One is based on the *physical state in which organic nutrients enter the cell*. The *osmotrophs* (for example, bacteria and fungi) take up all nutrients in dissolved form; the *phagotrophs* (for example, protozoa) can take up solid food particles by the mechanism termed *phagocytosis* (see p. 53).

It must be emphasized that the marked nutritional versatility of many microorganisms makes the application of this system of nutritional categories to some degree arbitrary. For example, many photoautotrophic algae can also grow in the dark, as chemoheterotrophs. Chemoheterotrophy is likewise an alternate nutritional mode for certain photoheterotrophs and chemoautotrophs. More or less by convention, such organisms are assigned to the category characterized by the simplest nutritional requirements: thus, phototrophy takes precedence over chemotrophy, autotrophy over heterotrophy. The qualifications *obligate* and *facultative* are often used to indicate the absence (or presence) of nutritional versatility. Thus, an *obligate photoautotroph* is strictly dependent on light for its energy source and on $CO_2$ for its principal carbon source, but a *facultative photoautotroph* is not.

## the construction of culture media

In constructing a culture medium for any microorganism, the primary goal is to provide a balanced mixture of the required nutrients, at concentrations that will permit good growth. It might seem at first sight reasonable to make the medium as rich as possible, by providing all nutrients in great excess. However, this approach is not a wise one. In the first place, many nutrients become growth inhibitory or toxic as the concentration is raised. This is true of organic substrates, such as salts of fatty acids (e.g., acetate) and even of sugars if the concentration is high enough. Some inorganic constituents may also become inhibitory if provided in excess; many algae are very sensitive to the concentration of inorganic phosphate. Second, even if growth can occur in a concentrated medium, the metabolic activities of the growing microbial population will eventually change the nature of the environment to the point where it becomes highly unfavorable and the population becomes physiologically abnormal or dies. This may be brought about by a drastic change in the hydrogen ion concentration (pH), by the accumulation of toxic organic metabolites, or, in the case of strict aerobes, by the depletion of oxygen. Since the usual goal of the microbiologist is to study the properties and behavior of *healthy* microorganisms, it is wise to limit the total growth of cultures by providing a limiting quantity of one nutrient; in the case of chemoheterotrophs, the principal carbon source is usually selected for this purpose. Examples of the appropriate concentrations of nutrients will be provided in the various media described below.

The rational point of departure for the preparation of media is to compound a *mineral base*, which provides all those nutrients that can be supplied in inorganic form. This base can then be supplemented, as required, with a carbon source, an energy source, a nitrogen source, and any required growth factors; these supplements will, of course, vary with the nutritional properties of the particular organism that one wishes to grow. A medium composed entirely of chemically defined nutrients is termed a *synthetic medium*. One that contains ingredients whose chemical composition is not precisely defined is termed a *complex medium*. Such media often contain extracts of tissues or cells, for example, meat or yeast.

We may illustrate these principles by considering the composition of four media of increasing chemical complexity, each of which is suitable for the cultivation of certain kinds of chemotrophic bacteria (Table 2.1). All four media share a common mineral base. Medium 1 is sup-

**Table 2.1**
Four Media of Increasing Complexity

| COMMON INGREDIENTS | ADDITIONAL INGREDIENTS | | | |
|---|---|---|---|---|
| | MEDIUM 1 | MEDIUM 2 | MEDIUM 3 | MEDIUM 4 |
| Water, 1 liter<br>$K_2HPO_4$, 1 g<br>$MgSO_4 \cdot 7H_2O$, 200 mg<br>$FeSO_4 \cdot 7H_2O$, 10 mg<br>$CaCl_2$, 10 mg<br>Trace elements (Mn, Mo, Cu, Co, Zn) as inorganic salts, 0.02–0.5 mg of each | $NH_4Cl$, 1 g | Glucose,[a] 5 g<br>$NH_4Cl$, 1 g | Glucose, 5 g<br>$NH_4Cl$, 1 g<br>Nicotinic acid, 0.1 mg | Glucose, 5 g<br>Yeast extract, 5 g |

[a] If the media are sterilized by autoclaving, the glucose should be sterilized separately and added aseptically. When sugars are heated in the presence of other ingredients, especially phosphates, they are partially decomposed to substances that are very toxic to some microorganisms.

plemented with $NH_4Cl$ at a concentration of 1 g/liter but has no added source of carbon. However, if it is incubated aerobically, the $CO_2$ of the atmosphere will be available as a carbon source. In the dark the only organisms that can grow in this medium are chemoautotrophic nitrifying bacteria, such as *Nitrosomonas,* which obtain carbon from $CO_2$ and energy from the aerobic oxidation of ammonia; the ammonia also provides them with a nitrogen source.

Medium 2 is additionally supplemented with glucose at a concentration of 5 g/liter. Under aerobic conditions, it will support the growth of many bacteria and fungi, since glucose can commonly be used as a carbon and energy source for aerobic growth. If incubated in the absence of oxygen, it can also support the development of many facultatively or strictly anaerobic bacteria, able to derive carbon and energy from the fermentation of glucose. Note, however, that this medium is not a suitable one for any microorganism that requires growth factors.

Medium 3 is additionally supplemented with one vitamin, nicotinic acid. It can therefore support the growth of all those organisms able to develop in medium 2, together with others, such as the bacterium *Proteus vulgaris,* that require nicotinic acid as a growth factor.

For the three media so far described, the chemical nature of every ingredient is known; thus, they are good examples of synthetic media. Medium 4 is a complex medium in which the $NH_4Cl$ and nicotinic acid of Medium 3 have been replaced by a nutrient of unknown composition, yeast extract. Medium 4 can support the growth of a great many chemoheterotrophic microorganisms, both aerobic and anaerobic, having no growth-factor requirements, relatively simple ones, or highly complex ones. The yeast extract provides a variety of organic nitrogenous constituents that can fulfill the general nitrogen requirements, and it also contains most of the organic growth factors likely to be required by microorganisms. Complex media are, accordingly, useful for the cultivation of a wide range of microorganisms, including ones whose precise growth-factor requirements are not known. Even when the growth-factor requirements of a microorganism have been precisely determined, it is often more convenient to grow the organism in a complex medium, particularly if the growth-factor requirements are numerous.

The media described in Table 2.1 can support the development of microorganisms only if certain other requirements for growth are also met. These include a suitable temperature of incubation, favorable osmotic conditions, and a hydrogen ion concentration within the range tolerated by the organism in question. Suitable chemical adjustments may be required to accommodate the osmotic conditions and hydrogen ion concentration to the needs of some microorganisms for which these media are satisfactory with respect to their content of nutrients.

## the control of pH*

Although a given medium may be suitable for the *initiation* of growth, the subsequent development of a bacterial population may be severely limited by chemical changes that are brought about by the growth and metabolism of the organisms. For example, in glucose-containing media, organic acids that may be produced as a result of fermentation may become inhibitory to growth.

In contrast, the microbial decomposition or utilization of anionic components of a medium tends to make the medium more alkaline. The decomposition of proteins and amino acids may also make a medium alkaline as a result of ammonia production.

To prevent excessive changes in hydrogen ion concentration either *buffers* or *insoluble carbonates* are often added to the medium. A buffer is a mixture of a weak acid and its salt, the pH of which is determined by the ratio of the two. It resists radical pH change since it contains both acidic and basic components. For example, a mixture of monobasic and dibasic phosphates (e.g., $KH_2PO_4$ and $K_2HPO_4$) will neutralize either a strong acid:

$$K_2HPO_4 + HCl = KH_2PO_4 + KCl$$

or a strong base:

$$KH_2PO_4 + KOH = K_2HPO_4 + H_2O$$

The phosphates are used widely in the preparation of media because they are the only inorganic agents that buffer in the physiologically important range around neutrality and that are relatively nontoxic to microorganisms. In addition, they provide a source of phosphorus, which is an essential element for growth. In high concentrations, phosphate becomes inhibitory, so the amount of phosphate buffer that can be used in a medium is limited by the tolerance of the particular organism being cultivated. Generally, about 5 g of potassium phosphates per liter of medium can be tolerated by bacteria and fungi.

When a great deal of acid is produced by a culture, the limited amounts of phosphate buffer that may be used become in-

---

*The pH scale provides a convenient method of expressing the degree of acidity (hydrogen ion concentrations) in aqueous solutions and is almost invariably used by biologists and biochemists. The pH value of a given solution is the *logarithm of the reciprocal of the hydrogen ion concentration* (expressed in moles per liter), or

$$pH = \log \frac{1}{[H^+]}$$

For example, a solution of acid, which is 0.1 $N$ with respect to hydrogen ions, has a pH value of 1.0. This follows from the fact that log (1/0.1) equals the logarithm of 10, which is 1.0. The pH value of pure water is 7.0.

sufficient for the maintenance of a suitable pH. In such cases, carbonates may be added to media to neutralize the acids as they are formed. In the presence of hydrogen ions, carbonate is transformed to bicarbonate, and bicarbonate is converted further to carbonic acid, which decomposes spontaneously to $CO_2$ and water. Of the insoluble carbonates, finely powdered chalk ($CaCO_3$) is the most generally employed. When the pH of the liquid drops below approximately 7.0, the carbonate is decomposed with the evolution of $CO_2$. It thus acts as a neutralizing agent for any acids that may appear in a culture by converting them to their calcium salts.

The addition of $CaCO_3$ to agar media used for the isolation and cultivation of acid-forming bacteria helps to preserve neutral conditions. Furthermore, since the acid-forming colonies dissolve the precipitated chalk and become surrounded by clear zones, they can be easily recognized against the opaque background of the medium (see Figure 12.14, p. 304).

In some instances, neither buffers nor insoluble carbonates can be used to maintain a relatively constant pH in a culture medium. In these cases, therefore, it is necessary to adjust the pH of the culture, either periodically or continuously, by the aseptic addition of strong acids or bases. In some laboratories and in industrial plants, elaborate mechanical devices are used for this purpose. With their aid a continuous titration of the medium is feasible, and the pH is kept nearly constant.

### the control of oxygen concentration

Oxygen is an essential nutrient for all obligate aerobes. They can be grown easily on the surface of agar plates and in shallow layers of liquid medium. In unshaken liquid cultures, growth usually occurs at the surface. Below the surface, however, conditions become anaerobic, and growth is impossible. To obtain large populations in liquid cultures, *it is therefore necessary to aerate the medium.* Various types of shaking machines that constantly agitate, and thus aerate, the medium are available for laboratory use. Another method of aeration is the continuous passage of a stream of very fine bubbles of air through a culture.

### techniques for cultivation
### of obligate anaerobes

Many of the more sensitive, strictly anaerobic microorganisms are rapidly killed by contact with molecular oxygen. The exposure of cultures or inocula to air should accordingly be minimized or avoided completely. Furthermore, many strict anaerobes can initiate growth only in media of low oxidation-reduction potential.* The use of media that

---

*The oxidation-reduction potential is the relative voltage required to remove an electron from a compound. Hence a low oxidation-reduction potential indicates reducing, rather than oxidizing, conditions.

are *pre-reduced* by the inclusion of such reducing agents as cysteine, thiogly-collate, $Na_2S$, or sodium ascorbate is therefore a factor of cardinal importance for the cultivation of many anaerobes. Once prepared, such media must, of course, be protected from exposure to air during both storage and use. During inoculation or sampling, this can be achieved by passage of a stream of $O_2$-free $CO_2$ or $N_2$ into the orifice of the opened culture vessel.

Liquid cultures of strict anaerobes are usually prepared in tubes or flasks completely filled with medium and closed by rubber stoppers or screw caps. Isolation in solid media can be undertaken by several methods. Organisms able to tolerate a transient exposure to air can be isolated on streaked plates incubated anaerobically, or in shake tubes. For the isolation of more oxygen-sensitive anaerobes it is best to use a *roll tube*, in which the molten agar is distributed in a thin layer over the inside of the walls of the tube. The tube is gassed continuously with an oxygen-free mixture during manipulation, and then closed with a rubber stopper.

### the provision of carbon dioxide

A problem frequently encountered in the cultivation of photoautotrophs and chemoautotrophs is the provision of carbon dioxide in sufficient amounts. Although the diffusion of carbon dioxide from the atmosphere into the culture medium will permit growth to occur, the carbon dioxide concentration in the atmosphere is very low (0.03 percent in the open atmosphere, somewhat higher inside a building), and the growth rates of autotrophs are often limited by the availability of carbon dioxide under these conditions. The solution is to gas the cultures with air that has been artificially enriched with carbon dioxide and contains from 1 to 5 percent of this gas. In the case of autotrophs that can be grown under anaerobic conditions in stoppered bottles (e.g., the purple and green sulfur bacteria), the requirement for $CO_2$ can be met by the incorporation of $NaHCO_3$ in the medium.

### the provision of light

For the cultivation of phototrophic microorganisms, *light* is an essential requirement. A relatively uncontrolled and discontinuous illumination may be obtained by the exposure of cultures to daylight. Direct exposure to sunlight should be avoided because the intensity may be too high and the temperature may rise to a point where growth is prevented. The growth of many phototrophic microorganisms is much more rapid under conditions of continuous illumination, so artificial light sources are advantageous. The *emission spectrum* of the lamp employed is important. Fluorescent light sources have the practical advantage of producing relatively little heat, so maintenance of a suitable temperature is not difficult. However, their emission spectra are deficient, compared to sunlight, in the longer wavelengths of the visible spectrum and the near infrared region. Fluorescent light sources are satisfactory for the cultivation of algae and blue-green bacteria, which perform photosynthesis with light of wavelengths shorter than 700 nm, but provide little photosynthetically effective light for purple and green bacteria, which use wavelengths in the range 750 to 1,000 nm. The most suitable artificial light sources for the latter photosynthetic bacteria are incandescent lamps, and if high intensities are used, the dissipation of heat may become a problem.

# selective media

It is clear that no single medium or set of conditions will support the growth of all the different types of organisms that occur in nature. Conversely, any medium that is suitable for the growth of a specific organism is, to some extent, *selective* for it. In a medium inoculated with a variety of organisms, only those that can grow in it will reproduce, and all others will be suppressed. Further, if the growth requirements of an organism are known, it is possible to devise a set of conditions that will specifically favor the development of this particular organism. Alternatively, media may be supplemented with selectively toxic compounds. Some dyes (e.g., crystal violet), antibiotics, detergents, or bile salts may be used for this purpose. These techniques permit the isolation from a mixed natural population of even a minor component of the total population. Microorganisms can be selectively obtained from natural habitats (e.g., soil or water) either by *direct isolation* or by *enrichment*.

### direct isolation

If a mixed microbial population is spread over the surface of a selective medium solidified with agar (or some other gelling agent), every cell in the inoculum capable of development will grow and eventually form a colony. The spatial dispersion of the microbial population on a solid medium considerably reduces the competition for nutrients: under these circumstances, even organisms that grow relatively slowly will be able to produce colonies. Direct plating on a selective medium is the technique of choice when one wishes to isolate a considerable diversity of microorganisms, all able to grow under the conditions of culture employed.

### enrichment

If a mixed microbial population is introduced into a liquid selective medium, there is a direct competition for nutrients among the members of the developing population. Liquid enrichment media therefore tend to select the microorganism of highest growth rate.

The selectivity of an enrichment culture is not solely determined by the chemical composition of the medium used. The outcome of enrichment in a given medium can be significantly modified by variation of such other factors as temperature, pH, ionic strength, illumination, aeration, or source of inoculum.

An enrichment medium that is not initially highly selective may acquire greatly increased selectivity for a particular type of microorganism, as a result of chemical changes produced by this organism during its development. Thus, fermentative bacteria and yeasts are typically more tolerant than other organisms of the organic end products which they themselves produce from carbohydrates; in a carbohydrate-rich medium their development will therefore tend to suppress competing microorganisms.

In the isolation of spore-forming bacteria competition from nonsporulating bacteria can be almost wholly eliminated by a pretreatment of the inoculum. *Pasteurization* of the inoculum, i.e., brief exposure to a high temperature (2 to 5 minutes at 80°C), will destroy most vegetative cells, leaving the much more heat-resistant spores unaffected.

Enrichment culture, developed by Winogradsky and Beijerink, is one of the most powerful techniques available to the microbiologist. An almost infinite number of permutations and combinations of the different environment variables, nutritional and physical, can be developed for the specific isolation of microorganisms from nature. Enrichment techniques provide a means for isolating known microbial types at will from nature, even if they comprise a minor fraction of the microflora of the inoculum, by taking advantage of their specific requirements. Enrichment media can be indefinitely elaborated, as a means of obtaining for study hitherto undescribed organisms capable of growing in the environments devised by the scientist.

## microscopy

In order to understand the indispensable role played by the microscope in the study of microorganisms, it is necessary to appreciate the intrinsic limitations of the eye as a magnifying instrument. The apparent size of an object viewed by the human eye is directly related to the angle that the object subtends at the eye: hence, if its distance from the eye is halved, its apparent size is doubled. However, the eye cannot focus on objects brought closer to it than approximately 25 cm; this is, accordingly, the distance of maximal effective magnification. In order to be seen at all, an object must subtend an angle at the eye of about 1 minute or greater; and for a distance of 25 cm, this corresponds to a particle with a diameter of approximately 0.1 mm.

Most cells (and hence most unicellular microorganisms) are too small to be detected by the unaided human eye. In order to detect them, and to observe their form and structure, the use of a microscope is therefore essential. The function of the magnifying lens system of this instrument, interposed between the specimen and the eye, is greatly to increase the apparent angle subtended at the eye by objects within the microscopic field. In addition to this factor of *magnification,* two other factors, *contrast* and *resolution,* are of great importance. In order to be perceived through the microscope, an object must possess a certain *degree of contrast* with its surrounding medium; and in order to produce a clear magnified image, the microscope must possess a *resolving power* sufficient to permit the perception as separate objects of closely adjacent points in the image.

### the light microscope

As discussed in Chapter 1, Leeuwenhoek discovered the microbial world through the use of simple microscopes containing a single lens of short focal length. The development and improvement of the more complex *compound microscopes* now employed required almost two centuries of research in applied optics.

A modern compound microscope contains three separate lens systems (Figure 2.5): the *condenser,* interposed between the light source and the specimen, collimates the light rays in the plane of the microscopic field; the *objective* produces a magnified image of the microscopic field within the microscope; and the *ocular* further enlarges this image.

Single lenses have two inherent optical defects. They fail to bring the whole microscopic field into simultaneous focus, and they produce colored fringes around objects in the field. These defects can be largely eliminated by placing additional correcting lenses adjacent to a primary magnifying lens. Consequently, both the ocular and objective lenses of a modern compound microscope are multiple ones, designed to minimize these aberrations.

### resolving limit

The physical properties of light set a fixed limit to the effective magnification obtainable with a light microscope. The distance between two points that can just be distinguished from one another is known as the *resolving limit,* and it is a function of the wavelengths of the illuminating light.* Under the best obtainable conditions the maximal resolution of the light microscope approaches 200 nm. In other words, two adjacent points closer together than 200 nm cannot be resolved into separate images. This corresponds to a useful magnification of approximately 1,000×.

---

*Resolving limit is also inversely proportional to the refractive index of the material between the specimen and the objective lens. Hence high-power objectives are designed to be used with *immersion oil* (with a refractive index of about 1.52) instead of air (whose refractive index is 1.00) between the cover slip and the lens. The use of oil also results in a brighter image with less glare.

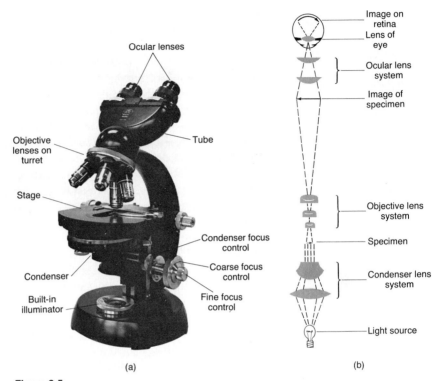

(a)                                                    (b)

**Figure 2.5**
The modern compound microscope (a) with the principle parts identified. Schematic representation (b) of the optical system of a compound microscope. The light path shown is generalized; no attempt is made to show refractive events at individual lens elements. Light produced by the bulb is directed into the condenser lens system which either collimates it (*Köhler* illumination) or focuses it on the specimen (*critical* illumination). After passing through the specimen, the light traverses the objective lens system, which forms within the microscope tube an enlarged image of the specimen. This image is further enlarged by the ocular lens system which, acting in conjunction with the lens of the microscopist's eye, forms the final image on the retina.

## contrast and its enhancement in
## the light microscope

When a small biological object, such as a microbial cell, is observed in the living state, it is normally suspended in an aqueous medium, compressed into a thin layer between a slide and cover slip (a *wet mount*). The perception of such an object depends on the fact that it displays some degree of contrast with the surrounding aqueous medium, as a result of the fact that less light is transmitted through it than through the medium. With the exception of intensely pigmented cell structures (e.g., chloroplasts), biological objects transmit almost as much light as water. However, contrast can be greatly increased by staining procedures: treatment with dyes that bind selectively either to the whole cell or to certain cell components. Most staining treatments kill cells; and as a preliminary to staining, the cells are usually *fixed,* by heat or by a chemical treatment designed to minimize postmortem changes of structure.

Another advantage of staining procedures is to provide *specific information about the internal structure or the chemical properties of cells.* Thus, staining methods specific for deoxyribonucleic acid can reveal the structure and location of the nucleus; and a variety of special staining methods can be used to demonstrate intracellular deposits of reserve materials. The Gram stain (see p. 57) and the acid-fast stain (see p. 307) are used to obtain information on the composition of the wall layers of bacterial cells. So-called *negative stains,* which do not enter the cell, are sometimes useful to reveal surface layers of very low refractive index, such as the capsules and slime layers that often surround microbial cells. These can be made visible by adding India ink to the suspending medium; since the carbon particles of the ink cannot penetrate the capsular layer, it is revealed as a clear zone surrounding the cell.

The relatively low contrast of living cells as viewed with a conventional light microscope can be greatly increased by the use of an instrument with a modified optical system, known as the *phase contrast microscope.* By this optical modification, the degree of contrast of cells or intracellular structures is greatly increased, making staining unnecessary.

### the transmission electron microscope

The development of the electron microscope has revolutionized our knowledge of biological structure. It was based on the discovery that an electromagnetic field acts on a beam of electrons in a way analogous to the action of a glass lens on a beam of light. An electron beam has the properties of an electromagnetic wave of very short wavelength. The resolving limit of the electron microscope is consequently several orders of magnitude lower than that of the light microscope, and it thus permits the use of far higher effective magnifications (to approximately 100,000×). The path of electrons through a transmission electron microscope is analogous to the path of light rays through a light microscope. The image is rendered visible by allowing it to impinge on a phosphorescent screen. Since electrons can travel only in a high vacuum, the entire electron path through the instrument must be evacuated; consequently, specimens must be completely dehydrated prior to examination.

In a transmission electron microscope, contrast results from the differential scattering of electrons by the specimen, the degree of scattering being a function of the number and mass of atoms that lie in the electron path. Since most of the constituent elements in biological materials are of low mass, the contrast of these materials is weak. It can be greatly enhanced by "staining" with the salts of various heavy metals (e.g., lead, tungsten, uranium). These may be either fixed on the specimen (*positive stain-*

*ing*) or used to increase the electron opacity of the surrounding field (*negative staining*). Negative staining is particularly valuable for the examination of very small structures such as virus particles, protein molecules, and bacterial flagella (e.g., see Figure 7.2, p. 149). However, cells (even of very small microorganisms) are too thick to be examined satisfactorily in whole mounts. Observation of their internal fine structure requires that they be fixed, dehydrated, embedded in a plastic, and sectioned. Ultrathin sections (not more than 50 nm thick) are then positively stained with heavy metal salts, and mounted for examination.

Two other preparatory techniques, *metal shadowing* and *freeze-fracturing*, are frequently used for the observation of biological specimens with the transmission electron microscope. In metal shadowing the dried specimen is exposed at an acute angle to a directed stream of a heavy metal (platinum, palladium, or gold), thus producing an image that reveals the three-dimensional structure of the object (see Figure 8.10(b), p. 188). In the process of freeze-fracturing, the specimen is frozen, and the frozen block is fractured with a knife, exposing various surfaces on and within the specimen. The fractured surface is shadowed at an acute angle with a heavy metal, and a supporting layer of carbon is evaporated onto the metal surface. The shadowed specimen is then destroyed by chemical treatment, and the replica is examined (see Figure 3.3, p. 47).

### the scanning electron microscope
The light microscope and the transmission electron microscope are fundamentally similar in operation: a broad beam of electromagnetic energy (light or a beam of electrons) is passed through the specimen, and a magnified image of it is formed by passing the emergent beam through lenses (glass or electromagnetic). In both instruments, the principle of refraction of electromagnetic energy by lenses is used to form a magnified image of the specimen. The recently developed *scanning electron microscope* utilizes a totally different principle of image formation, that of electronic amplification of signals generated by irradiating the surface of the specimen with a very narrow beam of electrons. Such irradiation causes low energy (secondary) electrons to be ejected from the specimen; these can be collected on a positively-charged plate (an *anode*) thereby generating an electric signal that is proportional to the number of electrons striking the anode. Since this number of electrons depends on the number ejected (in turn, a function of angle of a particular region of the surface with respect to the electron beam) and on the number reabsorbed by surrounding protuberances on the surface of the specimen, the electric signal can be used to generate an image of the specimen (Figure 2.6). By use of a scanning generator, the electron beam is caused to traverse the specimen in a raster pattern. The signal generated by secondary electrons striking the anode is amplified and used to modulate the intensity of a spot scanning a cathode ray tube (essentially the

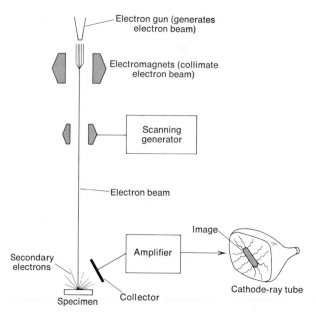

**Figure 2.6**
Schematic representation of a scanning electron microscope.

same as the picture tube of a television receiver) in precise register with the scanning pattern of the electron beam. Hence a magnified image of the surface topography of the specimen is presented on the cathode ray tube. The depth of focus of this instrument is several millimeters; and its range of effective magnification extends from about 20 × to 20,000 ×. An example of the type of image obtained is shown in Figure 11.15 (p. 264).

# THE
# NATURE
# OF THE
# MICROBIAL WORLD

The term *microorganism* does not have the precise taxonomic significance of such terms as *vertebrate* or *angiosperm,* each of which defines a restricted biological group, and all members of which share numerous common structural and functional properties. In contrast, any organism of microscopic dimensions is by definition a microorganism; and microorganisms occur in a wide diversity of taxonomic groups, some of which (e.g., the algae) also contain members far too large to be assigned to this category. In this chapter, two major taxonomic groups that consist either in whole or in part of microorganisms—the procaryotes and the protists—will be distinguished.

## the common properties of biological systems

Most organisms share a common physical structure, being organized into microscopic subunits termed *cells.* Cellular organisms share a *common chemical composition,* their most distinctive chemical attribute being the presence of three classes of complex macromolecules: deoxyribonucleic acid (DNA), ribonucleic acid (RNA), and protein. All cells are enclosed by a thin membrane, the *cytoplasmic membrane,* which retains within its boundary the various molecules, large and small, necessary for the maintenance of biological function, and which at the same time regulates the passage of solutes between the interior of the cell and its external environment. Cells never arise *de novo:* they are always derived from preexisting cells, by the process of growth and cell division.

These generalizations apply to all living objects with the exception of viruses (see Chapter 7).

## patterns of cellular organization

The simplest cellular organisms consist of a single cell. Since cells are always of microscopic dimensions, such *unicellular organisms* are necessarily small, and thus fall in the general category of microorganisms. Unicellularity is widespread, though not universal, in the microbial groups known as *bacteria, protozoa,* and *algae:* it likewise occurs, though more rarely, in *fungi.* The very considerable differences that exist among the various groups of microorganisms are expressed solely in terms of differences with respect to the *size, form,* and *internal structure* of the cell: sketches of a few unicellular organisms, all drawn to the same scale, are shown in Figure 3.1.

A more complex mode of organization is *multicellularity.* Although a multicellular organism arises initially from a single cell, it consists in the mature state of many cells, attached to one another in a characteristic fashion which determines the gross external form of the organism. Multicellular organisms composed of a small number of cells may still be of microscopic dimensions; many examples exist among bacteria and algae. Such organisms are usually composed of similar cells, arranged in the form of a thread or filament. However, when the number of cells composing the organism is larger, the organism acquires a certain degree of structural complexity, simply from the manner in which the constituent cells are arranged. The best illustrations of such simple multicellular organization occur among the larger algae (for example, marine algae), which often have a characteristically plantlike form, even though there is little or no specialization of the component cells. Form is derived by the specific pattern in which the like structural units are arranged.

In metazoan animals and vascular plants, multicellular organization leads to a much higher degree of intrinsic structural complexity, as a consequence of the *differentiation of distinct cell types* during the development of the mature organism. This leads, through cell division, to the emergence of distinct *tissue regions,* each composed of a special type of cell; a further level of internal complexity may be attained by the association of

**Figure 3.1**

Drawings of several unicellular microorganisms on the same relative scale: (a) an amoeba; (b) a large bacterium; (c) a yeast; (d) a flagellate alga; (e) a small bacterium (×1,000).

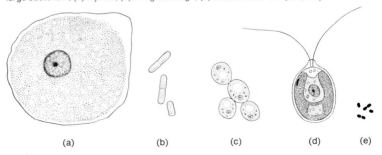

(a)          (b)          (c)          (d)          (e)

different cell types into functional units known as *organs*. The structural complexity of a vascular plant or a metazoan animal thus proves, upon microscopic analysis, to be vastly greater than that of large but undifferentiated multicellular organisms such as the marine algae.

In a few biological groups, biological organization assumes a third form, known as *coenocytic structure*, which at first sight seems to contradict the axiom that organisms are composed of cells. A coenocytic organism is not composed of cellular subunits, separated from one another by their bounding membranes; instead, the cytoplasm is continuous throughout the individual organism, which grows in size without undergoing cell division. This type of organization is characteristic of most fungi, and also occurs in many algae.

### the problem of primary
### divisions of organisms

It is a judgment of common sense, as old as mankind, that our planet is populated by two different kinds of organisms, plants and animals. Early in the history of biology this prescientific opinion became formalized in scientific terms: biologists recognized two primary kingdoms of organisms, the *Plantae* and the *Animalia*. The members of the two kingdoms appeared to be readily distinguishable by a whole series of characters, both structural and functional. This traditional bipartite division was in fact a satisfactory one, as long as biologists had to take into account only the more highly differentiated groups of multicellular organisms.

When exploration of the microbial world got under way in the eighteenth and nineteenth centuries, there seemed no reason to doubt that these simple organisms could be distributed between the plant and animal kingdoms. However, as knowledge of the properties of the various microbial groups deepened, it became apparent that at this biological level a division of the living world into two kingdoms could not really be maintained on a logical and consistent basis. Some groups (notably the flagellates and the slime molds) were claimed both by botanists as plants and by zoologists as animals. The problem is easy enough to understand in evolutionary terms. The major microbial groups can be regarded as the descendants of very ancient evolutionary lines, which antedated the emergence of the two great lines that eventually led to the development of plants and animals. Hence, most microbial groups cannot be pigeonholed in terms of the properties that define these two more advanced evolutionary groups.

## eucaryotes and procaryotes

About 1950 the development of the electron microscope and of associated preparative techniques for biological materials made it possible to examine the structure of cells with a degree of resolution many times greater than that previously possible by the use of the light microscope. Within a few years many hitherto unperceived features of cellular fine structure were revealed. This led to the recognition of a profoundly important dichotomy among the various groups of organisms with respect to *the internal architecture of the cell:* two radically different kinds of cell exist in the contemporary living world (Figure 3.2). The more complex *eucaryotic cell* is the unit of structure in plants, metazoan animals, protozoa, fungi, and algae. Despite the extraordinary diversity of the eucaryotic cell, its basic architecture always has many common denominators. The less complex *procaryotic cell* is the unit of structure in the bacteria (including the organisms formerly known as blue-green algae). The placement of the so-called blue-green "algae" with other algal groups can thus no longer be justified, and they will be treated in this book as one group of photosynthetic bacteria, the blue-green bacteria.

The flood of new information about cellular fine structure provided by electron microscopy has been paralleled by an equally

**Figure 3.2**

Electron micrographs illustrating the differences between the eucaryotic and procaryotic cell. (a) A protist, *Labyrinthula* which has a relatively undifferentiated eucaryotic cell structure. (b) A bacterium, *Bacillus subtilis,* which has a typical procaryotic cell structure. The *Labyrinthula* cell lacks a wall, but is surrounded by a loose, extracellular slime matrix (sm). Other recognizable structures include the endoplasmic reticulum (er); Golgi bodies (g); mitochondria (m); a nucleus (n) surrounded by a nuclear membrane (nm); 80S ribosomes* (r); large lipid droplets (ld); and the cell membrane (cm). The dividing cell of *Bacillus subtilis* is surrounded by a relatively dense wall (cw), enclosing the cell membrane (cm). Within the cell, the nucleoplasm (n) is distinguishable by its fibrillar structure from the cytoplasm, densely filled with 70S ribosomes (r). Note the absence of internal unit membrane systems. (a) Courtesy of David Porter; (b) courtesy of C. F. Robinow. (*Note: The size of ribosomes is frequently expressed in "Svedberg units" (S), a measure of how rapidly they sediment in a high-speed centrifuge; the larger the ribosome, the faster it sediments and the higher the S value.)

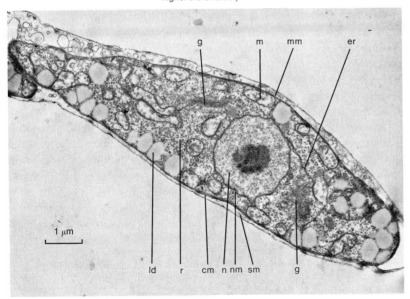

(a)

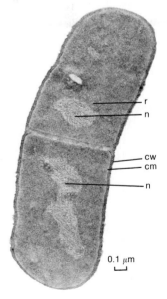

(b)

rapid growth of knowledge about cellular function, in large part a result of the rise of molecular biology. This has made it evident that the structural differences between eucaryotic and procaryotic cells are expressive of highly important differences in the way that universal cell functions are accomplished: notably the transmission and expression of genetic information, the performance of energy metabolism, and the entry and exit of materials. It is evident that the line of demarcation between eucaryotic and procaryotic cellular organisms is the largest and most profound single evolutionary discontinuity in the contemporary biological world. This discontinuity splits the microbial world into two major groups: the bacteria (procaryotic) and the *protists,* relatively simple and usually microscopic eucaryotes consisting of the algae, fungi, and protozoa. In the following pages the principal features of organization and function that distinguish eucaryotic from procaryotic cells will be summarized.

## organization and function in the eucaryotic cell

As already mentioned, all cells are bounded by a surface membrane known as the *cytoplasmic membrane.* This can be resolved by electron microscopy as a triple-layered structure with a total width of about 8 nm.* Membranes possessing this fine structure are termed *unit membranes.* A characteristic property of the eucaryotic cell is the presence within it of a *multiplicity of unit membrane systems,* many of which are both structurally and topologically distinct from the cytoplasmic membrane. These internal unit membrane systems serve to segregate many of the functional components of the eucaryotic cell into specialized and partly isolated regions.

The most complex internal membrane system, topologically speaking, is the *endoplasmic reticulum* (ER). It consists of an irregular network of interconnected membrane-delimited channels that traverses much of the interior of the cell and is in direct contact with two other major cell components: the nucleus and some of the cytoplasmic ribosomes. Part of the ER surrounds the nucleus, being expanded to form the *nuclear membrane,* which has a distinctive structure, perforated by numerous pores about 40 nm in diameter (Figure 3.3). In other regions of the ER, known as the *rough endoplasmic reticulum,* the surfaces of the membranes are coated with small particles termed *ribosomes* (Figure 3.4), which function in protein synthesis.

Another characteristic membranous organelle is the Golgi apparatus (Figure 3.5), which consists of a densely packed mass of flattened sacs and vessels of varying size. Its manifold functions include

---

*The nanometer (nm) is now the standard unit of length used to describe biological fine structure. It equals $10^{-3}$ micrometer ($\mu$m), or $10^{-7}$ cm. The Angstrom unit (Å), frequently used in older descriptions of biological fine structure, equals $10^{-1}$ nm.

**Figure 3.3**
An electron micrograph of a freeze-fractured preparation of the resting nucleus of a mouse cell. The plane of fracture has passed in part through the nuclear membrane and reveals the nuclear pores both on the inner surface of the membrane (npi) and on its outer surface (npo) (×24,000). Courtesy of Dr. L. G. Chevance, Institut Pasteur.

the packaging of material, both proteinaceous and nonproteinaceous, synthesized in the endoplasmic reticulum, and their transport to other regions of the cell, or to the cell surface, where Golgi vesicles may coalesce with the cytoplasmic membrane and liberate their contents to the exterior of the cell, a process known as *exocytosis*.

The machinery responsible for the performance of respiration and—in photosynthetic eucaryotes—of photosynthesis is segregated in two functionally distinct classes of membrane-bounded organelles, mitochondria and chloroplasts (Figure 3.6). Each type of organelle contains an internal membrane system of characteristic structure and function. The internal membranes of the mitochondrion (*cristae*) house the respiratory electron transport system; the internal membranes of the chloroplast (*thylacoids*) house the photosynthetic pigments and electron transport system, as well as the photochemical reaction centers (see Chapter 4).

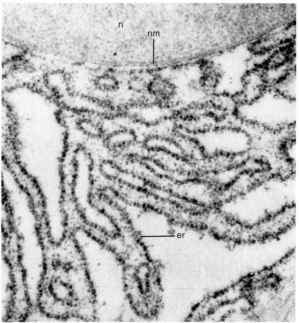

**Figure 3.4**
Electron micrograph of a thin section of a rabbit cell, showing a portion of the cytoplasm filled with rough endoplasmic reticulum (er); the field also includes a portion of the nucleus (n), surrounded by the nuclear membrane (nm) (×35,000). Courtesy of Dr. L. G. Chevance, Institut Pasteur.

**Figure 3.5**
The Golgi apparatus as seen in an electron micrograph of a thin section of *Euglena gracilis* (×28,000). Two adjacent Golgi bodies have been sectioned in different planes. At left, vertical section through the membrane stack. At right, section parallel to the stack. Courtesy of Gordon F. Leedale.

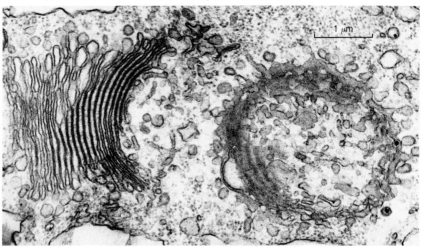

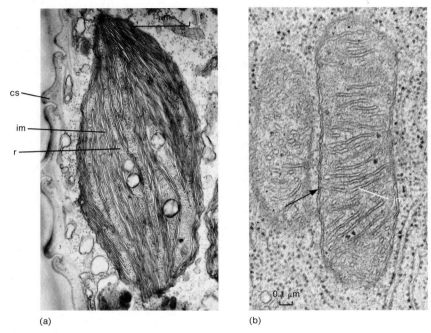

(a)                                                    (b)

**Figure 3.6**
The structure of chloroplasts and mitochondria are revealed in electron micrographs of thin sections of eucaryotic cells. (a) Chloroplast of the unicellular alga *Euglena* ( × 21,200). The internal membranes (im) are arranged in irregular parallel groups and run in the long axis of the chloroplast. Ribosomes (r) are scattered between the lamellae. The chloroplast lies just below the sculptured cell surface (cs). From G. F. Leedale, B. J. D. Meeuse, and E. G. Pringsheim, "Structure and Physiology of *Euglena spirogyra*," *Arch. Mikrobiol.* **50**, 68 (1965). (b) Mitochondria in a mammary gland cell of the mouse ( × 56,100). Numerous flattened internal membranes (im) arise by invagination from the inner enclosing membrane of the organelle (arrow). Courtesy of Dorothy Pitelka.

## the eucaryotic genome

In a eucaryotic cell the nucleus is the principal but rarely the sole repository of genetic information. A quantitatively minor but functionally very important part of the total cellular genome is located in mitochondria and (in photosynthetic organisms) in chloroplasts. The organellar DNA determines some (but by no means all) of the properties of the specific organelle with which it is associated.

The genetic information contained in the eucaryotic nucleus is dispersed over a limited number of distinct structural elements known as *chromosomes*. Each chromosome is a threadlike structure containing DNA and a special class of basic proteins known as *histones*. In the nondividing or *interphase* nucleus, each chromosome is greatly elongated and is only 20 to 30 nm in width; it therefore cannot be resolved by light microscopy.

The replication and partition of the genetic information carried on the chromosomes involves a complex cyclic process known as *mitosis*. Replication of chromosomal DNA occurs prior to the onset of mitosis, in the interphase nucleus. As mitosis begins, the chromosomes coil into compact structures visualizable by light microscopy: the total number and form of the chromosomes are fundamental characters of each eucaryotic species. The shortening of the chromosomes is accompanied by the rapid assembly in the nuclear region of a bipolar, spindle-shaped microtubular structure known as the *mitotic apparatus* (Figure 3.7). Its formation is usually accompanied by the disintegration of the nuclear membrane. The chromosomes become aligned in the equatorial region of the spindle, each splitting longitudinally into two identical daughter structures, the *chromatids*. One set of chromatids then moves to each pole of the mitotic apparatus, two daughter nuclei being reorganized at the poles, with an accompanying disassembly of the spindle. Cell division normally takes place during the terminal phase of mitosis, the plane of division following the equatorial plane of the spindle.

REPLICATION, TRANSCRIPTION, AND TRANSLATION OF THE ORGANELLAR GENOMES
The DNA in chloroplasts and mitochondria exists as small circular double-helical molecules, not associated with histones (Figure 3.8). The organellar genetic material is thus located in structures that appear to be very similar to procaryotic chromosomes (see below), although of considerably smaller size. Chloroplasts and mitochondria likewise contain the machinery of protein synthesis, including organelle-specific ribosomes, which are smaller than the 80S cytoplasmic ribosomes and resemble in size the 70S ribosomes of procaryotes. Organellar protein synthesis can be inhibited by chloramphenicol and certain other antibiotics, which are likewise inhibitors of this process in procaryotes but do not affect eucaryotic cytoplasmic protein synthesis. In a number of important respects, accordingly, chloroplasts and mitochondria show fundamental resemblances to procaryotic cells. Mitochondria also possess a property characteristic of cells, but not of other cell constituents: they are formed by division of preexisting organelles. This has also been demonstrated for many types of chloroplasts.

### eucaryotic sexual processes

Cellular fusion is the first step in the process of *sexual reproduction*. The two cells that participate are known as *gametes* and the resulting fusion cell as a *zygote*. In all eucaryotic organisms, gametic fusion is followed by nuclear fusion, with the result that the zygote nucleus contains *two complete sets of chromosomes*, one derived from each gametic nucleus.

Sexual reproduction is common in the life cycle of plants and animals. In vertebrates and many invertebrates, it is the *only* method for the production of a new individual. Plants can also be propagated asexually (e.g., by cuttings), and asexual modes of reproduction exist in many

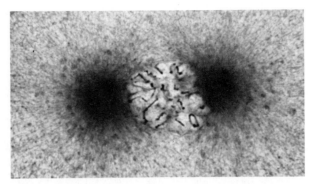

(a)

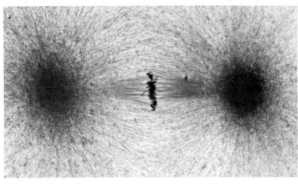

(b)

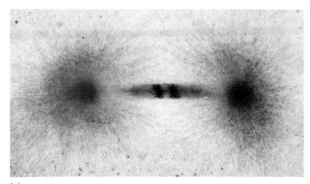

(c)

**Figure 3.7**
Photomicrographs illustrating three successive phases of a mitotic nuclear division. (a) Early organization of the spindle; the nuclear membrane has disappeared, and the chromosomes are already visible in the region of the organizing spindle. (b) The spindle is now fully developed, and the chromosomes are regularly aligned in its equatorial plane; this stage is often referred to as the *metaphase of mitosis*. (c) Separation of the two daughter sets of chromosomes has occurred, and each set is being withdrawn toward one pole of the spindle.

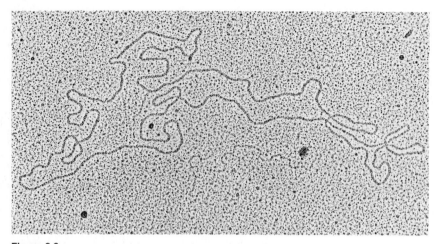

**Figure 3.8**
Electron micrograph of a circular double helical molecule of mitochondrial DNA, isolated from the yeast, *Saccharomyces cerevisiae*. The total length of the molecule is 9.2 nm (×160,000). Courtesy of J. Lazowska and P. Slonimski.

groups of invertebrates. Among the protists, sexual reproduction is rarely an obligatory event in the life cycle. Many of these organisms completely lack a sexual stage in their life cycles, and even in species in which sexuality does exist, sexual reproduction may occur infrequently, the formation of new individuals taking place principally by asexual means (for example, by binary fission or the formation of spores).

Sexual fusion results in a *doubling of the number of chromosomes,* since the nuclei of the gametes, each containing $N$ chromosomes, fuse to form the nucleus of the zygote, which consequently contains $2N$ chromosomes. Hence, in passing from one sexual generation to the next, there must at some stage be a *halving of the number of chromosomes* if the chromosome content of the nucleus is not to increase indefinitely. In fact, the halving of the chromosome number is a universal accompaniment of sexuality. It is brought about by a special process of nuclear division termed *meiosis*. Meiosis is a process comprised of two sequential specialized nuclear divisions: in the first homologous chromosomes are separated and in the second sister chromatids separate. The result is four daughter nuclei, each with half the chromosome number ($N$) of the parent ($2N$). In animals, *meiosis takes place immediately prior to the formation of gametes*. In other words, each individual of the species has $2N$ chromosomes in its cells through most of the life cycle. Such an organism is termed *diploid*. This state of affairs is, however, by no means universal among sexually reproducing eucaryotic organisms. In many protists, *meiosis takes place immediately after zygote formation,* with the consequence that the organisms have $N$ chromosomes through most of the life cycle. Such organisms are termed *haploid*. In many algae and plants, as well as in some fungi and protozoa, there is a well-marked *alternation of haploid and diploid generations*. In this type of life cycle, the diploid zygote gives rise to a diploid individual, which forms, by meiosis, haploid *asexual* reproductive cells. Each such haploid cell gives rise to a haploid individual, which eventually forms haploid gametes; gametic fusion, with the formation of a diploid zygote once again, completes the cycle.

### endocytosis and exocytosis

Although small molecules in solution can enter the eucaryotic cell by passage through the cytoplasmic membrane, the entry of other materials can occur by a second quite distinct mechanism: bulk transport of small droplets, enclosed by an infolding of the cytoplasmic membrane to form a membrane-enclosed vacuole. The most familiar example of this phenomenon is the *phagocytosis* of bacteria or other small solid objects by phagotrophic protozoa, or by the phagocytic cells of metazoan animals. Droplets of liquid can enter the eucaryotic cell in a similar fashion, this process being termed *pinocytosis*. Phagocytosis and pinocytosis are known collectively as *endocytosis*, a distinctively eucaryotic process of fundamental biological importance, which initiates both *intracellular digestion* (hydrolysis of biological macromolecules) and the *establishment of endosymbionts*.

One of the products formed in the Golgi apparatus is a membrane-bounded vesicle known as the *lysosome*. Lysosomes contain an extensive array of hydrolytic enzymes capable of breaking down most classes of biological macromolecules. These enzymes do not normally act upon the constituents of the cell in which they are formed, since they are segregated within the lysosomal membrane. However, the lysosomes can fuse with vacuoles formed through endocytosis, thus permitting hydrolysis of the materials (or cells) contained in these vacuoles: the soluble hydrolytic products then diffuse into the surrounding cytoplasm (Figure 3.9). This process of intracellular digestion is the major feeding mechanism in many protozoa and primitive invertebrate animals. Throughout the animal world it plays

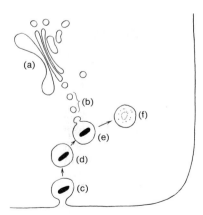

**Figure 3.9**
A diagrammatic representation of the events of intracellular digestion: (a) Golgi apparatus; (b) lysosomes produced from the Golgi apparatus; (c) phagocytic capture of a food particle (a bacterium) at the surface of the cell, during which the particle is almost completely surrounded by the cell membrane; (d) newly formed food vacuole; (e) coalescence of the food vacuole with a lysosome; (f) digestion of the vacuolar contents by hydrolytic enzymes released from the lysosome. Modified from N. Novikoff, E. Essner, and N. Quintana, *Federation Proc.* 23, 1011 (1964).

a protective role by enabling the body to destroy potentially dangerous microorganisms that have entered its fluids or tissues. Indeed, this protective function has become the primary role of phagocytosis in vertebrates, where digestion of food materials takes place within the alimentary tract, outside the body tissues.

Droplets or solid particles of material synthesized within the eucaryotic cell can also pass to the exterior by the converse mechanism, known as *exocytosis*. Here also the Golgi apparatus plays a key role, since materials destined for exocytosis are packaged initially in Golgi vesicles. The secretion of enzymes and of hormones by specialized animal cells occurs in this manner, and in algae it has been shown that formation of the cell wall involves the exocytosis of small fragments of the wall fabric, synthesized endogenously and transported to the cell surface in Golgi vesicles.

### microtubular systems

An element of structure that has many functions in the eucaryotic cell is the *microtubule*, an extremely thin cylinder, some 20 to 30 nm in diameter and of indefinite length. The walls of microtubules are composed of globular protein subunits that can assemble in regular array.

Microtubules provide the structural framework of the mitotic spindle; they also appear to play a role in the establishment and maintenance of the shape of many types of eucaryotic cells. In addition, a regular longitudinal array of microtubules occurs within the eucaryotic locomotor organelles, known as *cilia* or *flagella*. The cilium or flagellum is enclosed by an extension of the cytoplasmic membrane and contains a set of nine outer pairs of radially arranged microtubules, in turn surrounding an inner central pair (Figure 3.10). The central microtubules arise from a plate near the surface of the cell, whereas the outer pairs originate from a cylindrical body or *centriole,* which is composed of nine triple rows of microtubules.

In some eucaryotes the centrioles are also associated with the formation of the microtubular system of the mitotic apparatus, being located at the two poles of the spindle.

### directed intracellular movement
### in the eucaryotic cell

Examination of many types of eucaryotic cells in the living state reveals that the cytoplasm is frequently in active movement, a phenomenon known as *cytoplasmic streaming*. Furthermore, it is evident that intracellular movements are often closely directed: the light-induced orientation of chloroplasts, the localized concentration in cells of mitochondria, the roles played by the Golgi apparatus in the intracellular transport of packaged materials, and the movement of chromosomes during mitosis are only some

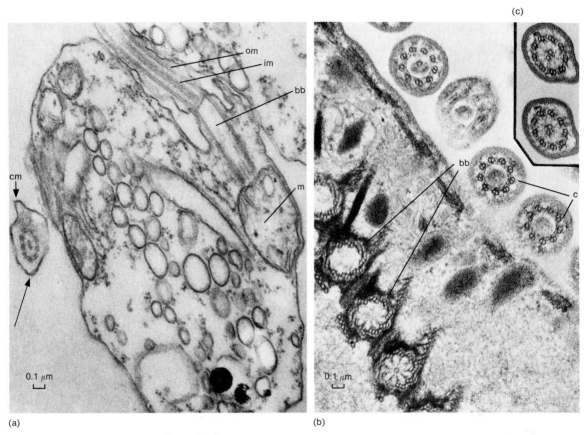

**Figure 3.10**
The fine structure of eucaryotic flagella and cilia, as revealed by electron micrographs of thin sections. (a) Longitudinal section through the cell of *Bodo*, a nonphotosynthetic flagellate (×38,800): centriole (bb); outer microtubules (om); inner microtubules (im). Underlying the centriole is a specialized mitochondrion (m). At left (arrow), transverse section of a flagellum external to the cell. Note enclosure by an extension of the cell membrane (cm). (b) Section through the body surface of a ciliate, *Didinium* (×51,800). Within the cell (lower left), centrioles (bb) have been sectioned transversely; their walls are composed of nine triple rows of microtubules. Just above the cell surface, several cilia (c) have been sectioned transversely; note the nine outer pairs of microtubules and the absence of the inner pair of microtubules. (c) Insert at upper right: section through two cilia at a point some distance from the cell surface. Note the inner pair of microtubules, the nine outer pairs, and the enclosing membrane. Courtesy of Dorothy Pitelka.

of the phenomena that reveal the precision with which the relative positions of the cell components can be controlled.

In many protists that have cells not enclosed by walls, directed cytoplasmic streaming can be used as a means of cellular translocation over a solid substrate: this is the characteristic mode of movement in ameboid cells.

56

## organization and function in
## the procaryotic cell

One of the most striking structural features of the procaryotic cell is the *absence of internal compartmentalization by unit membrane systems* (see Figure 3.2). The *cytoplasmic membrane is, in the great majority of procaryotes, the only unit membrane system of the cell.* However, the topology is often complex, membranous infoldings penetrating deeply into the cytoplasm. The blue-green bacteria provide the sole known exception to the rule that there is only one unit membrane system in the procaryotic cell. In these organisms the photosynthetic apparatus is located on a series of membranous flattened sacs or thylakoids similar in structure and function to the thylakoids within a chloroplast. However, in blue-green bacteria the thylakoids are not segregated within an organelle but are dispersed throughout the cytoplasm (Figure 3.11).

**Figure 3.11**
Electron micrograph of a thin section of a unicellular blue-green bacterium. The cell is enclosed by a cell wall (cw) and cell membrane (cm). The thylacoids (th) which bear the photosynthetic apparatus are located in the cytoplasm, and are thicker than the cell membrane, since they are composed of two closely appressed membranes (×53,000). Courtesy of Dr. G. Cohen-Bazire.

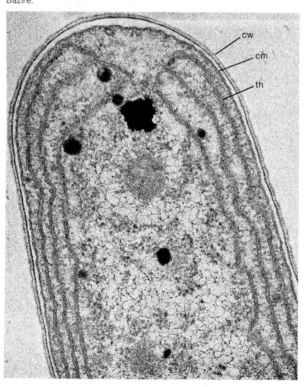

Electron microscopy of most procaryotes reveals only two structurally distinguishable internal regions in the cell: cytoplasm and nucleoplasm (see Figure 3.2). The cytoplasm has a finely granular appearance as a result of its content of ribosomes, each about 10 nm in diameter. These are always of the so-called 70S type, smaller than the cytoplasmic ribosomes of eucaryotes, but similar in size to eucaryotic organellar ribosomes. The nucleoplasm is of irregular contour, sharply segregated from the cytoplasm, even though a bounding membrane never separates the two regions.

A *cell wall* encloses the cells of most procaryotes; only the *Mycoplasma* group is devoid of this structure. The wall serves as the principal determinant of cell shape, and in most bacterial groups serves to protect the cells from lysis by high internal osmotic pressure. With very few exceptions, the procaryotic cell wall contains a distinctive macromolecular polymer termed *peptidoglycan*, not synthesized by any eucaryote. Despite the near universality of peptidoglycan, there is considerable ultrastructural and chemical variability of the cell wall among procaryotes. It is nevertheless possible to discern two major assemblages of organisms on the basis of the gross organization of cell-wall constituents. One group possesses a thin peptidoglycan layer overlain by a unit membrane that in appearance is identical to the cytoplasmic membrane [Figure 3.12(b)]. These bacteria are termed *Gram-negative*, because, as discovered in 1885 by Christian Gram, they fail to retain a crystal violet-iodine stain complex when washed with organic solvents like ethanol or acetone. The other group has a much thicker peptidoglycan layer and no outer membrane [Figure 3.12(a)]. This group is termed *Gram-positive*, as the stain complex cannot be washed from the cells by organic solvents. The chemical or physical basis of dye retention by Gram-positive bacteria is not known; however, it is clearly dependent on the cell-wall structure. Enzymatic digestion of the Gram-positive cell wall renders the cell no longer capable of dye retention.

**Figure 3.12**
Electron micrographs of sections of the surface layers of (a) a Gram-positive bacterium and (b) a Gram-negative bacterium, illustrating the differences in cell wall profiles: c, cytoplasm; cm, cell membrane; p, peptidoglycan; and om, outer membrane.

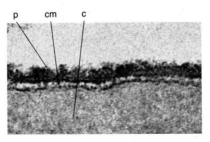

(a)

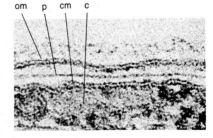

(b)

58

THE NATURE OF THE MICROBIAL WORLD

### functions of the cytoplasmic membrane

In many procaryotes, the cytoplasmic membrane performs a role in energy metabolism, a role that it never plays in the eucaryotic cell. Among aerobic bacteria, the respiratory electron transport system is incorporated into the cell membrane. In eucaryotes this part of the machinery of respiration is incorporated into the inner membrane system of the mitochondrion. In both the purple and green bacteria, the centers of photosynthetic activity are likewise incorporated into the cytoplasmic membrane. Both in purple bacteria and in aerobic bacteria that have high rates of respiration, the topology of the membrane is often very complex: it is extensively infolded, lamellar or vesicular intrusions penetrating deep into the cytoplasm. The greatly increased total area of the membrane thus permits the cell to accommodate many centers of respiratory (or photosynthetic) function.

There are strong indications that the cytoplasmic membrane also contains specific attachment sites for the DNA of the procaryotic cell, membrane growth being the mechanism responsible for the separation of genomes following completion of their replication. This is, of course, another function never played by the cytoplasmic membrane of the eucaryotic cell in which the separation of genomes is accomplished by mitosis.

The procaryotic membrane is incapable of mediating endocytosis or exocytosis.

### the procaryotic genome

The genetic information of a procaryotic cell is carried in the nucleoplasm on the structure termed the *bacterial chromosome*. It consists of a single circular DNA molecule, not associated with histones. The bacterial chromosome is consequently not structurally homologous with the nuclear chromosomes of the eucaryotic cell, but rather with the organellar DNA present in the eucaryotic mitochondria and chloroplasts.

Many bacteria can also harbor small (*ca* 0.01 to 5 percent of the size of the chromosome) double-helical circular molecules of DNA termed *plasmids*. These genetic elements are capable of replicating in synchrony with the cell, although errors in replication or segregation may occur either spontaneously or in response to laboratory manipulation such that the plasmid is lost from some cells of the population. Some plasmids, although demonstrable by physical methods, are of unknown function to the cell and are called *cryptic* plasmids.

A number of plasmids mediate their own transfer (and frequently the transfer of other plasmids or of chromosomal material) from their host cell to another cell. The sex factor of *Escherichia coli* (termed "F") is the classic example of this type, and is described in more detail in Chapter 8.

A very large class of plasmids confers on the host resistance to one or more antibacterial substances. These plasmids include the group termed *R factors,* which are found in a variety of Gram-negative bacteria. Most of these R factors are transmissible from cell to cell, and have in fact been shown to be composed of two elements: a nontransmissible plasmid encoding the resistance, and a second factor termed *RTF* for *resistance transfer factor.* Although the two elements are capable of autonomous existence and replication, they are also capable of joining covalently, and it is presumably in this state that they are transferred. Under the intense selection of antibiotic prophylaxis and therapy, R factors have proliferated to the point where multiply-resistant pathogenic bacteria are becoming a major medical problem.

Another class of plasmids are the *degradative plasmids,* found principally in *Pseudomonas* but which probably occur in other genera. These plasmids, most of which are transmissible, encode enzymes that degrade a variety of rather unusual compounds, such as camphor, toluene, or octane. Some of the nutritional versatility of the pseudomonads (Chapter 11) thus reflects the possession by some strains of such plasmids.

Although plasmids may be essential to cell survival and multiplication under certain specific environmental conditions, no bacterial strains have been isolated from nature that contain plasmids that are necessary to the cell under all conditions.

### ploidy

Most procaryotes (probably all) normally exist and reproduce by asexual means in the haploid state. Consequently, persistent diploidy, which is characteristic of many groups of eucaryotes and which has had profound evolutionary consequences, plays no role in the evolution of procaryotes. A partial diploid state can arise transiently in procaryotes, as a result of genetic transfer, but full diploidy is rarely attained, as a consequence of the special mechanisms of genetic transfer characteristic of procaryotes (see Chapter 8).

### procaryotic movement

Directed internal cytoplasmic movements which are so conspicuous in most eucaryotic cells do not occur in procaryotic cells. Consequently, ameboid locomotion, characteristic of many eucaryotic protists that lack cell walls, does not occur among procaryotes without walls, namely the members of the *Mycoplasma* group. Many procaryotes that possess cell walls exhibit active movement. One kind of active movement, *gliding motility,* is manifested only when the cell is in contact with a solid substrate; it is not known to be mediated by specific locomotor organelles. Gliding motility is characteristic of many blue-green bacteria, and also occurs in certain groups of nonphotosynthetic bacteria.

The second kind of active movement, *swimming motility,* occurs in cells suspended in a liquid medium and is brought about by locomotor organelles termed *bacterial flagella.* The bacterial flagellum is completely different in fine structure and organization from the eucaryotic flagellum or cilium. It is a proteinaceous filament of molecular dimensions (approximately 12 to 18 nm in diameter), which extends out through the wall and membrane from an anchoring structure located just below the membrane (see Chapter 6). A compound structure known as the *axial filament,* comprised of two sets of bacterial flagella that lie within the cell wall, is the organelle responsible for cellular movement in the group of procaryotes known as *spirochetes.*

### intracytoplasmic organelles of procaryotes

The only intracytoplasmic organelles of procaryotes that bear a structural resemblance to components of a eucaryotic cell are the thylacoids—flattened sacs composed of a unit membrane system—which house the photosynthetic apparatus in one group of photosynthetic procaryotes, the blue-green bacteria. They appear to be homologous in function and structure to the thylacoids of chloroplasts. However, a few other types of distinctively procaryotic organelles do occur in some groups of procaryotes, although none of these structures is of wide distribution. All these organelles are characterized by the *absence of unit membranes:* they are enclosed by a single-layered membrane with a thickness of only some 2 to 3 nm. They include the *chlorobium vesicles* (Chapter 10), which house the light-harvesting photosynthetic pigments of green bacteria and the *gas vesicles* (Chapter 6), which confer bouyancy on the cells of a diversity of aquatic procaryotes.

## the differences between procaryotes and eucaryotes: a summing up

The numerous and profound divergences with respect both to organization and to function between eucaryotic and procaryotic cells have been revealed gradually and have been fully recognized only recently. It is now evident that many cellular functions are mediated in markedly different ways by the two kinds of cells. This fact raises some major evolutionary questions concerning the primary origins and relationships between the two kinds of cells. Does the relatively simple procaryotic cell represent an early stage in the evolution of the more complex eucaryotic cell, or did the two kinds of cells have independent evolutionary origins? The apparent resemblances between procaryotes and two classes of eucaryotic organelles, chloroplasts and mitochondria, suggest another fascinating evolutionary possibility that has been much discussed. These two classes of organelles might have been derived from free-living procaryotes endowed respectively with respiratory and photosynthetic function, which entered primitive eucaryotes and gradually became so closely integrated with the host cell that they ultimately became incapable of independent existence.

Some of the major differences between the eucaryotic and the procaryotic cell are summarized in Table 3.1.

**Table 3.1**
Differences between Procaryotes and Eucaryotes

| CELLULAR PROPERTY | EUCARYOTES | PROCARYOTES |
|---|---|---|
| Intracellular unit membrane systems (endoplasmic reticulum, Golgi, lysosomes, mitochondria, chloroplasts, nuclear membrane) | + | − |
| Number of chromosomes | >1 | 1 |
| Chromosomes contain histones | + | − |
| Nuclear division by mitosis | + | − |
| Means of genetic recombination: | | |
| Fusion of gametes to produce complete diploids | + or − | − |
| Unidirectional transfer of DNA to produce partial diploids | − | + or − |
| Ribosomes | 80S* | 70S |
| Microtubules | + | − |
| Peptidoglycan | − | + (rarely −) |
| Endocytosis | + | − |
| Cytoplasmic streaming and ameboid movement | + | − |

*Organelles contain additional procaryote-type ribosomes.

# MICROBIAL METABOLISM

The sum total of all the chemical transformations that occur in cells is termed *metabolism*. In the case of microorganisms, the vast majority of these transformations are involved in growth of the cell; i.e., they comprise reaction sequences whereby substrates available from the medium are converted to cell material. Microbial cells, like all cells, are composed largely of high molecular weight substances, termed *biopolymers*,* which include *proteins, nucleic acids, polysaccharides,* and *complex lipids.*

Each biopolymer is composed of a number of chemically distinct monomeric subunits (see p. 93). Thus the total number of chemical constituents that are synthesized during growth is large, as is the number of chemical reactions involved in their synthesis. Around 2,000 distinct chemical reactions comprise the metabolism of a microorganism. Almost all of these reactions proceed only when catalyzed by catalytically-active proteins, termed *enzymes,* which are quite specific in their action. With very few exceptions, each reaction of metabolism is catalyzed by a distinct enzyme. Thus the synthesis of enzymes constitutes a significant portion of metabolism.

The chemical details of the many reactions of metabolism are complex, but their consequences fit simply (in the case of nonphotosynthetic organisms) into two categories: (1) those reactions leading directly to the synthesis of biopolymers, the portion of metabolism termed *biosynthesis,*† and (2) those reactions by which exogenous substrates are degraded, the portion of metabolism termed *catabolism.*

The processes of catabolism and biosynthesis are interrelated. Catabolic reactions are catalyzed by enzymes, one of the products of biosynthesis, and chemical intermediates of catabolic pathways enter

*Also termed macromolecules.

†Sometimes termed *anabolism.*

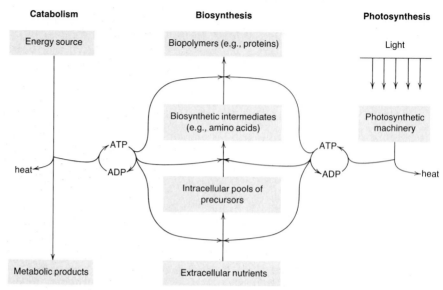

**Figure 4.1**
A schematic representation of the role of ATP in coupling catabolism and photosynthesis to biosynthesis. ATP is synthesized from adenosine diphosphate (ADP) by catabolism or photosynthesis; reactions of biosynthesis utilize ATP and produce ADP as a product.

into biosynthesis. Less obviously, the processes are related energetically. Many of the reactions of biosynthesis require an activation of their reactants to an increased energy state; i.e., an input of energy is required for biosynthesis. Hence, an important part of the metabolism of any organism involves the mobilization of chemical energy to drive biosynthesis. The usual form of mobilized energy is the highly reactive compound adenosine triphosphate (ATP). Certain reactions of catabolism generate ATP for use in biosynthesis.

Although the processes of biosynthesis are nearly uniform among organisms, the mechanisms of generating ATP are diverse, particularly among procaryotes. Catabolism, as has been discussed, involves the generation of ATP from chemical energy of organic or inorganic compounds. Alternatively, by the process of *photosynthesis,* light energy can be utilized to generate ATP.

The coupling between the processes of ATP generation (photosynthesis and catabolism) and biosynthesis is schematized in Figure 4.1.

## some thermodynamic considerations

The second law of thermodynamics states that only part of the energy released in a chemical reaction, called *free energy* ($\Delta G$), is available for the performance of work; the rest is lost through the increase in entropy (randomness of the system). Chemical reactions occur only if the free-energy change, $\Delta G$, is negative.

Enzymes, like all catalysts, have no effect on the $\Delta G$ of a reaction; instead they merely increase the rate of reaction. All biological reactions must therefore be thermodynamically favorable in order to proceed; catalysis by enzymes allows them to proceed at significant rates.

### the role of ATP in biosynthesis

The chemical structure of ATP is schematized in Figure 4.2. The two bonds indicated by the symbol $\sim$ are high-energy bonds;* i.e., they are particularly reactive. Hence, ATP is able to donate phosphate groups to a number of metabolic intermediates, thereby converting them to activated forms. The phosphorylated intermediate can participate in biosynthetic reactions that are thermodynamically favorable ($-\Delta G$), while the comparable reaction of which the unphosphorylated form as a reactant would be thermodynamically unfavorable ($+\Delta G$). Thus, ATP generation is required in order for biosynthetic pathways to function.

OTHER COMPOUNDS WITH HIGH-ENERGY BONDS. We have emphasized the primary role of ATP in trapping a portion of the free energy made available through catabolic and photosynthetic reactions, and its role in driving biosynthetic reactions. Although ATP is directly involved in the majority of such reactions, a number of other highly reactive metabolites that also contain high-energy bonds drive specific steps in certain pathways of biosynthesis. All these high-energy compounds can be formed at the expense of one or more of the high-energy bonds of ATP, but sometimes they are formed directly in catabolic reactions. These compounds include triphosphate compounds whose structure and mode of action are similar to ATP and a set of compounds with completely different structures, i.e., acyl coenzyme A in which organic acids are linked to a particular coenzyme by a high-energy bond.

---

*The term *high-energy bond* should not be confused with the term *bond energy* that is used by the physical chemist to denote the energy *required to break a bond between two atoms.*

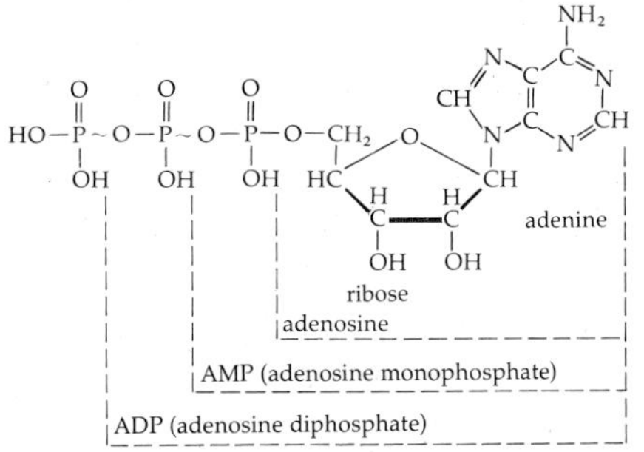

**Figure 4.2**
The structure of ATP (adenosine triphosphate), showing the various components of the molecule.

## the role of pyridine nucleotides in metabolism

Like all oxidations, the biological oxidation of organic intermediates of metabolism (*metabolites*) is the removal of electrons. In most cases, each step of oxidation of a metabolite involves the removal of two electrons and thus the simultaneous loss of two protons ($H^+$); this is equivalent to the removal of two hydrogen atoms (H) and is called *dehydrogenation*. Conversely, the *reduction* of a metabolite involves the addition of two electrons and two protons and can therefore be considered a *hydrogenation*. For example, the oxidation of lactic acid to pyruvic acid and the reduction of pyruvic acid to lactic acid can be expressed:

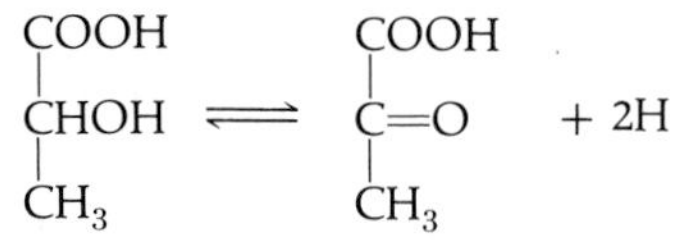

COOH          COOH
|             |
CHOH  ⇌   C=O      + 2H
|             |
CH₃           CH₃
lactic acid    pyruvic acid

The compounds that most often mediate biological oxidations and reductions (i.e., that serve as acceptors for hydrogen atoms released by dehydrogenation reactions, and as donors of hydrogen atoms required for hydrogenation reactions) are two pyridine nucleotides: *nicotinamide adenine dinucleotide* (NAD) and *nicotinamide adenine dinucleotide phosphate* (NADP), the structures of which are schematized in Figure 4.3.

Both of these pyridine nucleotides can readily undergo reversible oxidation and reduction. The oxidized form of the pyridine nucleotides carries one hydrogen atom less than the reduced form; in addition, it has a positive charge on the nitrogen atom, which enables it to accept a second electron upon reduction (Figure 4.3). The reversible oxidation-reduction of NAD and NADP can thus be symbolized:

$$NAD^+ + 2H \rightleftharpoons NADH + H^+$$

**Figure 4.3**
Reversible oxidation and reduction of pyridine nucleotides. In the case of NAD, R represents an adenine, two phosphate, and two ribose groups. An additional phosphate group is included in NADP.

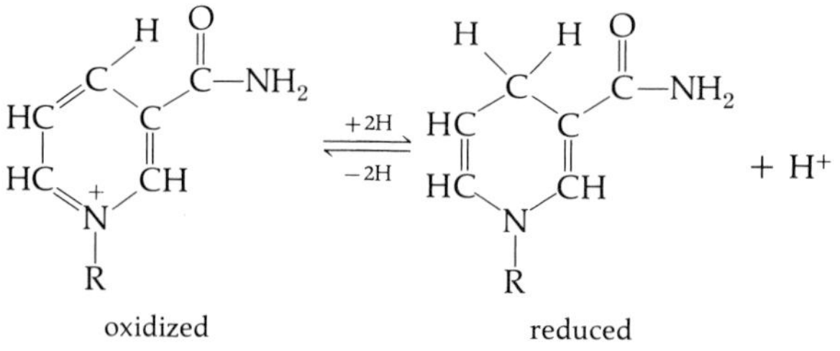

oxidized                    reduced

and

$$NADP^+ + 2H \rightleftharpoons NADPH + H^+$$

The pyridine nucleotides function in metabolism in the *transfer of reducing power*. For example, they couple the oxidation of one metabolic intermediate with the reduction of another (e.g., NAD couples the oxidation of glyceraldehyde phosphate with the reduction of pyruvic acid in lactic acid bacteria).

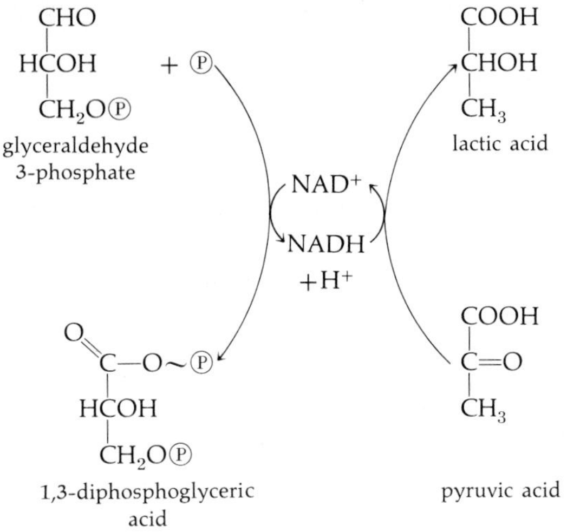

glyceraldehyde 3-phosphate

lactic acid

1,3-diphosphoglyceric acid

pyruvic acid

# the biochemical mechanisms of ATP synthesis

In cellular metabolism, ATP is generated by two fundamentally different mechanisms: *substrate-level phosphorylation* and *electron transport*.

### substrate-level phosphorylation

In substrate-level phosphorylation, ATP is formed from ADP by direct transfer of a high energy phosphate group from an intermediate of a catabolic pathway. The following reactions of the glyco-lytic pathway (see pp. 71–72) serve as an example:

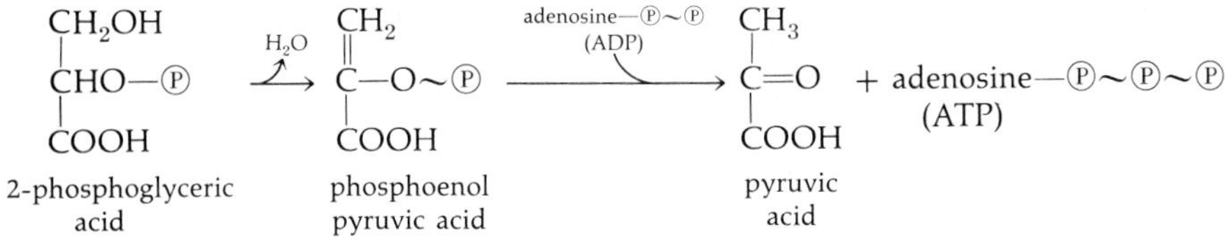

2-phosphoglyceric acid

phosphoenol pyruvic acid

pyruvic acid

As a consequence of the removal of a molecule of water, the low-energy ester linkage of phosphate* in 2-phosphoglyceric acid is converted to the high-energy enol linkage in phosphoenol pyruvic acid. This high-energy linked phosphate group can then be transferred to ADP, the consequence of which is the generation of a molecule of ATP.

### electron transport

ATP is also generated by a more complex process as a consequence of transport of electrons between certain carrier molecules with fixed orientation in membranes. Collectively these carrier molecules comprise an *electron transport chain*. The components of the chain are capable of carrying electrons by undergoing reversible oxidation and reduction; a metabolite that transfers electrons to the chain is termed a *primary electron donor*, and one that accepts electrons from the end of the chain is termed a *terminal electron acceptor*. The functioning of a typical electron transport chain is schematized in Figure 4.4.

Although the complexity and components of electron transport chains vary, they have certain common features: (1) They are composed of a number of distinct electron carriers, physically associated with the lipid component of membranes. (2) Each component of the chain must be capable of being reduced by the reduced form of the previous component and oxidized by the oxidized form of the subsequent component (i.e., there must be a gradient of susceptibility to oxidation in an electron transport chain). (3) As a consequence of passage of electrons through the chain, ATP is generated from ADP.

*Here and subsequently, the symbol $\textcircled{P}$ indicates an orthophosphate group, *i.e.*,

$$-O-\overset{\overset{\textstyle O}{\|}}{\underset{\underset{\textstyle OH}{|}}{P}}-O-$$

**Figure 4.4**

A schematic representation of the functioning of an electron transport chain. The subscript, *red*, indicates the reduced form of the molecule, and *ox*, the oxidized form. In each transfer of electrons the donor molecule becomes oxidized as the acceptor is reduced.

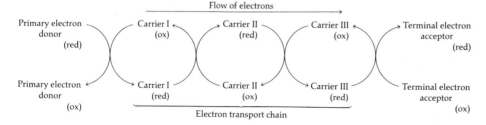

THE MAJOR COMPONENTS OF ELECTRON TRANSPORT CHAINS.  Two classes of electron carriers are proteins with firmly bound prosthetic groups: the *flavoproteins* and the *cytochromes*. A third class consists of nonprotein carriers of relatively low molecular weight, the *quinones*.

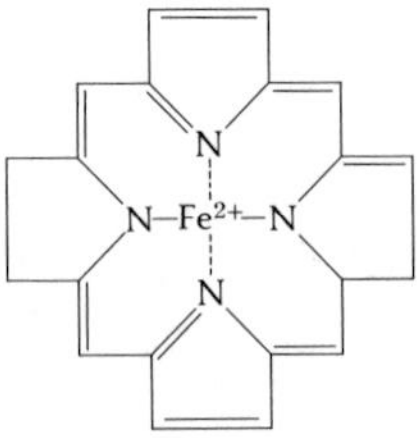

Flavoproteins have a yellow-colored prosthetic group, derived biosynthetically from the vitamin riboflavin (vitamin $B_2$).

The cytochromes belong to a class of *heme proteins*, their prosthetic group being a tetrapyrrole with an atom of iron chelated within the ring system (Figure 4.5). Electrons are carried by the iron, which is oxidized and reduced:

**Figure 4.5**
Basic structure of a tetrapyrrole showing the site of chelation of iron.

$$Fe^{2+} \xrightleftharpoons[e^-]{e^-} Fe^{3+}$$

Cytochromes comprise a group of compounds that differ with respect to their protein moiety, the chemical modifications of their tetrapyrrole prosthetic group, and the manner by which the prosthetic group and protein are attached. The cytochromes are classified on the basis of the latter two properties, the classes being distinguished by a terminal letter (e.g., cytochrome a, cytochrome b, or cytochrome c). Bacteria contain a remarkable diversity of cytochromes, in contrast to the essential uniformity of those in mitochondria. This diversity is taxonomically significant. Although very delicate and precise optical measurements are necessary for the complete characterization of cytochrome content of a bacterium, a simple biochemical test, the *oxidase test,* distinguishes two major groups of bacteria. The results of this test correlate strongly with the presence (oxidase-positive) or absence (oxidase-negative) of cytochrome c.

The cytochromes, as their name implies, absorb visible light, particularly when in the reduced state, and frequently confer a slight pink color to a mass of cells.

A diversity of quinones occur in bacterial electron transport chains. They, like the flavoproteins, transfer electrons in association with a proton, *i.e.,* in the form of a hydrogen atom (Figure 4.6).

ATP GENERATION BY ELECTRON TRANSPORT.  In spite of its fundamental biological importance, the mechanism by which ATP is generated as a consequence of passage of electrons through a transport chain is still a subject of some controversy, but the *chemiosmotic hypothesis* (Figure 4.7) is now the favored explanation.

It states that the electron transport chain is oriented across the membrane, so that oxidation of the electron carriers is accompanied by translocation of protons out of the procaryote cell. This is accomplished in two steps and is dependent on the fact that certain compo-

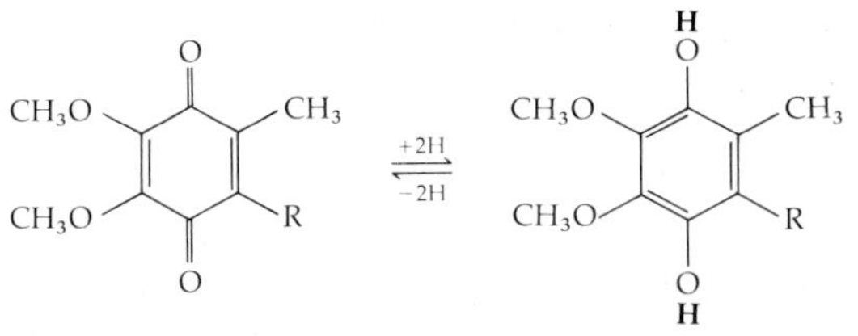

$$CH_3O \cdots \overset{O}{\cdots} CH_3 \quad \underset{-2H}{\overset{+2H}{\rightleftharpoons}} \quad CH_3O \cdots \overset{H}{\underset{O}{\cdots}} CH_3$$

**Figure 4.6**
The basic structure of a quinone, showing how it is reversibly oxidized and reduced during electron transport. R represents a side chain of relatively low molecular weight; its structure varies in different organisms.

nents of an electron transport chain, e.g., quinones, carry hydrogen atoms (an electron associated with a proton); while others, e.g., cytochromes, carry free electrons. Thus a hydrogen-carrying component oriented in an outward direction, followed by an electron-carrying component with an inward orientation, results in the outward translocation of a proton (Figure 4.7). The chemiosmotic hypothesis further states that membranes are impermeable to protons. Hence, as a consequence of electron transport, a difference in pH is created between the outside and inside of the membrane. Protons can only reenter the cell at specific sites. At these sites, specific proteins (ATPase) are located, which catalyze the following reaction:

$$ADP^{3-} + ⑰^{2-} + H^+ = ATP^{4-} + H_2O$$

Thus the pH gradient generated by electron transport drives the generation of ATP.

**Figure 4.7**
A schematic representation of ATP generation according to the chemiosmotic hypothesis.

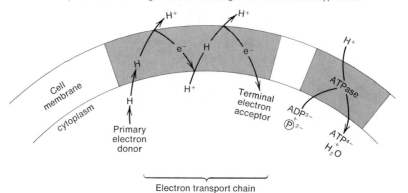

Electron transport chain

# the biochemistry of catabolism

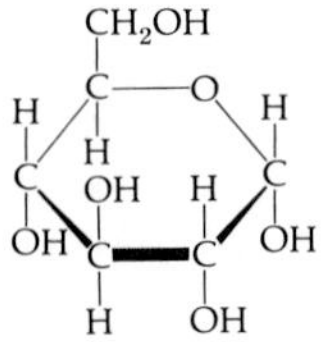

Regardless of mechanism, the generation of ATP by heterotrophs is dependent on the catabolism of organic compounds. Although microorganisms are able to attack a seemingly endless variety of organic compounds (probably each naturally occurring one is metabolized by some microorganism), the biochemical reactions by which they are metabolized to generate ATP are fewer than might be expected because the pathways of their catabolism are convergent. For example, the naturally abundant 6-carbon sugar glucose (Figure 4.8) is an intermediate of many pathways through which simple and complex sugars are metabolized.

**Figure 4.8**
The structure of glucose.

## the pathways of catabolism of glucose

Two of the major pathways by which glucose is metabolized are the *Embden-Meyerhof pathway* (also called the *glycolytic pathway*) and the *Entner-Doudoroff pathway* (Figure 4.9). The first occurs in many organisms, including both procaryotes and eucaryotes; the second is restricted to certain groups of procaryotes.

Both of these pathways involve the cleavage of glucose, eventually yielding two molecules of another key intermediate of metabolism, pyruvic acid,

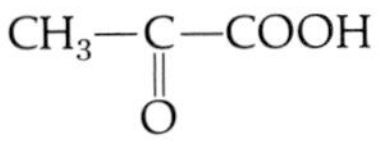

$$CH_3—\overset{\displaystyle}{\underset{\displaystyle O}{C}}—COOH$$

and in both, glucose is activated by a series of reactions prior to its cleavage into two 3-carbon moieties.

In the case of the Embden-Meyerhof pathway, this activation requires the expenditure of two molecules of ATP to produce an activated compound, fructose-1, 6-diphosphate, which is then split to yield two 3-carbon sugars each containing a phosphate (i.e., triosephosphates). In the subsequent conversion of each triosephosphate to pyruvic acid, another phosphate group is incorporated directly from inorganic phosphate. The linkages of both of these phosphate groups become converted to high-energy bonds, and by substrate level phosphorylation, each yields one molecule of ATP. The net yield, therefore, is two molecules of ATP per molecule of glucose.

In the Entner-Doudoroff pathway only a single molecule of ATP is expended to generate the activated form, which in this case is the phosphorylated 6-carbon sugar acid, 2-keto-3-deoxy-6-phosphogluconic acid (KDPG). Cleavage of KDPG yields one molecule of pyruvic acid directly and one molecule of triosephosphate. Only the latter generates ATP

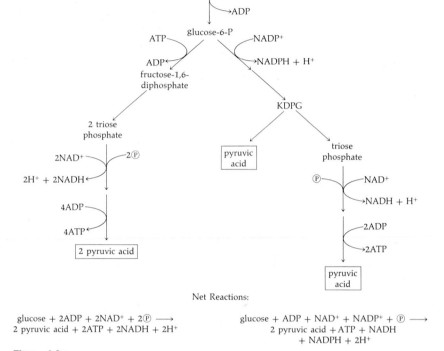

**Figure 4.9**
Schematic outline of two pathways of glucose degradation. KDPG is 2-ketodeoxygluconate.

as it is converted to pyruvic acid by the same series of reactions that occur in the Embden-Meyerhof pathway; i.e., it yields two molecules of ATP. The net yield by the Entner-Doudoroff pathway, therefore, is only one molecule of ATP per molecule of glucose.

Metabolism of a molecule of glucose by both the Embden-Meyerhof and the Entner-Doudoroff pathways generates two molecules of reduced pyridine nucleotide, which can be reoxidized by an electron transport chain, thereby generating ATP, or by some other metabolite.

### the pathways of catabolism of pyruvic acid

OXIDATIVE METABOLISM OF PYRUVIC ACID: THE TRICARBOXYLIC ACID CYCLE. Pyruvic acid, the product of the Embden-Meyerhof and the Entner-Doudoroff pathways is the starting point of a vast number of other metabolic pathways specific to certain microbial groups; in addition it is oxidized through a pathway known as the *tricarboxylic acid (TCA) cycle* or the *Krebs cycle* (Figure 4.10). This cycle is the major route of ATP generation in aerobes because it produces reduced pyridine nucleotides that donate electrons to an electron transport chain; it also generates certain metabolic intermediates that are required in biosynthetic pathways, so that even anaerobes in which little or no ATP is generated by electron transport possess most of the enzymes of this cycle. The TCA cycle effects the complete oxidation of one molecule of acetic acid (which is derived from pyruvic acid) to $CO_2$ with concomitant reduction of pyridine nucleotides.

REDUCTIVE METABOLISM OF PYRUVIC ACID. In addition to being oxidized by the ubiquitous tricarboxylic acid cycle, pyruvic acid can be reduced to a variety of end products by pathways characteristic of specific groups of microorganisms. Some of these pathways are summarized in Figure 4.11. Being reductive, they can occur in the absence of air and hence comprise a portion of anaerobic modes of catabolism collectively termed *fermentations* (see p. 74).

**Figure 4.10**
A schematic representation of the TCA cycle by which pyruvic acid is oxidized.

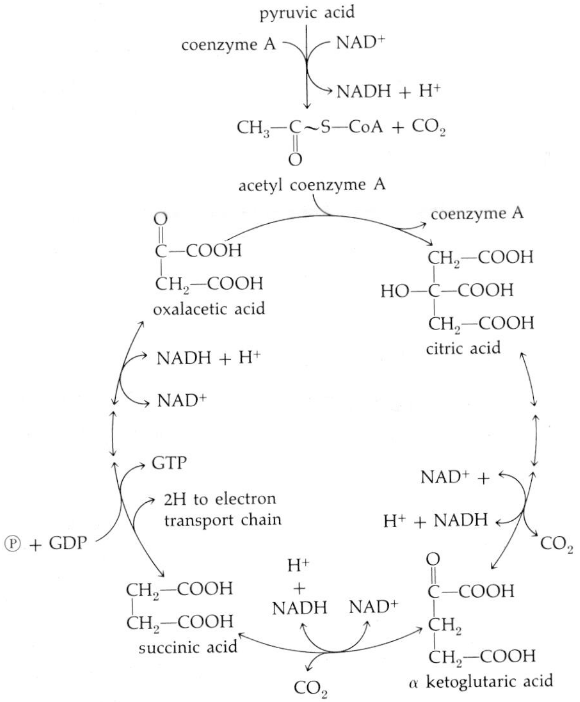

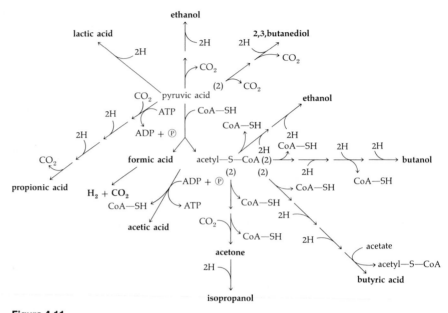

**Figure 4.11**
Some products of the reductive metabolism of pyruvic acid. The symbol (2) indicates that two molecules of the compound condense to form the first intermediate.

## modes of ATP-generating metabolism

The two biochemical mechanisms of generation of ATP, substrate level phosphorylation and electron transport, are distributed among three fundamentally different ATP-generating processes or *modes of metabolism*. These are: *fermentation, respiration,* and *photosynthesis.*

### fermentation

In terms of mechanism, fermentation is the simplest mode of ATP-generating metabolism. It is a metabolic process in which *organic compounds serve both as electron donors* (becoming oxidized) *and electron acceptors* (becoming reduced); *the electrons are not passed through an electron transport chain*. Rather, electrons are passed to an oxidized pyridine nucleotide, thereby reducing it; the reduced pyridine nucleotide then donates electrons directly to the electron acceptor. The organic compounds that serve as electron donors and acceptors are usually two different metabolites derived from a single fermentable substrate (such as a sugar). Fermentations often give rise to a mixture of end products; some may be more oxidized and some may be more reduced than the primary substrate, but their average oxidation level must be identical to that of the substrate. This can be readily seen in the case of alcoholic fermentation of glucose (Table 4.1). *Carbohydrates* are the principal sub-

**Table 4.1**
Alcoholic Fermentation of Glucose

|  | SUBSTRATE | END PRODUCTS | |
| --- | --- | --- | --- |
| Overall reaction | $C_6H_{12}O_6$ (glucose) $\longrightarrow$ | $2CO_2$ | $+\ 2C_2H_5OH$ (ethanol) |
| Number of carbons | 6 | 2 | 4 |
| Oxidation state of carbon | 0 | $+4$ | $-2$ |
| Oxidative balance | $6 \times 0$ = | $2 \times (+4) + 4 \times (-2)$ | |

strates of fermentation. Among the bacteria, some compounds belonging to the other chemical classes can also be fermented: organic acids, amino acids, purines, and pyrimidines.

Pyridine nucleotides reduced in one step of a fermentation are subsequently oxidized in another. This general principle is illustrated by two fermentations: the *alcoholic fermentation* (typical of the anaerobic metabolism of yeasts) and the *homolactic fermentation* (typical of the metabolism of certain lactic acid bacteria). Both of these fermentations (Figure 4.12) are slight modifications of the Embden-Meyerhof pathway; the two molecules of pyridine nucleotide ($NAD^+$) reduced by this pathway are reoxidized by reactions involving the subsequent metabolism of pyruvic acid. In the case of the homolactic acid fermentation, this oxidation occurs as a direct consequence of the reduction of pyruvic acid to lactic acid. In the case of the alcoholic fermentation, pyruvic acid is first decarboxylated to form acetaldehyde; the oxidation of NADH occurs concomitantly with the reduction of acetaldehyde to form ethanol. Since no ATP is generated by the reductive

**Figure 4.12**
The reoxidation of NADH by reduction of pyruvic acid in the homolactic and alcoholic fermentations.

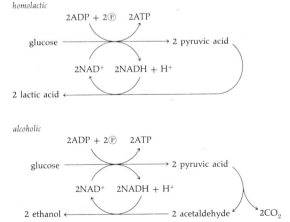

catabolism of pyruvic acid, the overall yield of ATP from the alcoholic and homolactic fermentations is the same as from the Embden-Meyerhof pathway alone, namely, two molecules of ATP per molecule of glucose. Catabolism of pyruvic acid in these fermentations serves only to reoxidize NADH.

The substrates and end products of bacterial fermentations and the pathway by which they are formed are group specific. All fermentations can proceed under strictly anaerobic conditions. Many of the organisms that generate ATP by fermentation are strict anaerobes (the physiological basis of their inability to grow in the presence of air will be considered in Chapter 5). Others are facultative anaerobes, being able to grow either in the presence or absence of air. As a rule, facultative anaerobes change their mode of ATP generation on exposure to air: the presence of molecular oxygen induces a metabolic shift from fermentation to respiration (p. 27). However, in a few facultatively anaerobic groups of bacteria, the presence of oxygen does not modify the mode of ATP-generating metabolism. Fermentation continues even in the presence of air.

All the reductive pathways shown in Figure 4.11 oxidize NADH, and thereby regenerate NAD, which is required for both the Embden-Meyerhof and Entner-Doudoroff pathways. Thus either pathway combined with one or more of the reductive pathways constitutes a fermentation. Indeed all the compounds shown in boldface type in Figure 4.11 are end products of one or more types of fermentation.

### respiration

Respiration can be defined as an *ATP-generating process that can occur in the dark and in which electrons are passed through an electron transport chain.* Both the primary electron donors and terminal electron acceptors can be either organic or inorganic compounds. Although ATP generation by electron transport defines respiration, the process often also includes the generation of ATP by substrate level phosphorylation during the primary biochemical processing of the carbon source. For example, respiration of glucose often involves catabolism by the Embden-Meyerhof pathway, which generates ATP by substrate level phosphorylation, followed by catabolism of pyruvate by the TCA cycle, which generates electrons that are passed through an electron transport chain. Usually the terminal electron acceptor is molecular oxygen. However, in the *anaerobic respirations*, characteristic of a few bacteria, an inorganic compound other than oxygen serves as the terminal electron acceptor; the compounds that can so act include sulfates, nitrates, and certain organic compounds. To distinguish oxygen-linked respiratory processes from these anaerobic respirations, it is useful to qualify the former as *aerobic respiration*.

Many microorganisms that perform aerobic respiration are strict aerobes. Some, however, are facultative anaerobes because they can also generate ATP either by fermentation (as stated above), or by anaerobic respiration with nitrate or organic compounds as the terminal electron acceptor. The bacteria that perform anaerobic respiration utilizing sulfate as the terminal electron acceptor are strict anaerobes.

Many kinds of organic compounds can be degraded by respiration. As stated earlier, probably all naturally occurring organic compounds can be respired by some microorganism. Respiration of organic compounds usually results in their complete oxidation to $CO_2$. The free-energy change (and hence yield of ATP) for the complete respiration of an organic compound is very much greater than for its fermentation. For example, the complete respiration of one mole of glucose liberates about 20 times as much energy as is derived by its fermentation.

Certain bacteria belonging to a number of special physiological groups and known collectively as *chemoautotrophs* are capable of respiring inorganic compounds; i.e., inorganic compounds serve as primary electron donors. Usually the terminal electron acceptor in such respirations is $O_2$. The inorganic substrates that can serve as energy sources (electron donors) include $H_2$, $CO_2$, $NH_3$, $NO_2^-$, $Fe^{2+}$, and certain reduced sulfur compounds. In this mode of respiratory metabolism, the sole function of substrate oxidation is to provide ATP and reduced pyridine nucleotides; all the organic constituents of the cell are synthesized from $CO_2$. Oxidation of $H_2$ is characterized by direct transfer of electrons to $NAD^+$. However, such direct transfer of electrons to $NAD^+$ from other inorganic substrates is thermodynamically unfavorable (the $\Delta G$ of the reaction is positive). In the latter case, the cell's requirement for reduced pyridine nucleotide is met by *reverse electron transport:* ATP generated by passage of electrons through one electron chain is used to drive electrons back through another to reduce $NAD^+$ (Figure 4.13).

### photosynthesis

The third (and mechanistically most complex) mode of ATP generation is *photosynthesis*, which can be defined as *the use of light as a source of energy*. Historically, the term *photosynthesis* was used to describe the overall metabolism of plants, algae, and blue-green bacteria, represented by

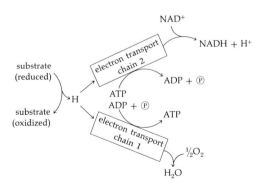

**Figure 4.13**
Schematic representation of ATP generation and $NAD^+$ reduction by certain chemoautotrophs. ATP generation by the passage of electrons through one chain (1) is utilized to accomplish the otherwise thermodynamically unfavorable reduction of $NAD^+$ (2).

the following reaction:

$$CO_2 + H_2O \xrightarrow{\text{light}} (CH_2O) + O_2$$

where the term $(CH_2O)$ represents organic compounds at the average oxidation state found in cells.

This reaction does not describe a process by which ATP is generated, but rather its biosynthetic consequence: the light-mediated conversion of $CO_2$ to organic cell materials. ATP is generated by the light energy absorbed by the photosynthetic pigment system; this process is called *photophosphorylation*. It is mechanistically analogous to respiration in that ATP is formed by electron transport. Substrate-level phosphorylations, which provide all the ATP in fermentations and some of the ATP in respirations, do not occur in photosynthetic metabolism.

Since most photosynthetic organisms use the highly oxidized compound $CO_2$ as their principal carbon source, they have a considerable requirement for reducing power. The majority of photosynthetic organisms (plants, algae, and blue-green bacteria) use water as an ultimate source of reductant; the concomitant oxidation of water leads to the formation of $O_2$ as a photosynthetic product. This kind of photosynthesis is termed *oxygenic photosynthesis*.

Certain photosynthetic procaryotes (purple and green bacteria) are unable to use water as an ultimate reductant, and their photosynthetic metabolism is never accompanied by the formation of $O_2$. Instead, other reduced inorganic compounds (e.g., $H_2S$, $H_2$) are used by some of these organisms as reductants. Some purple and green bacteria use organic compounds in place of $CO_2$ as the principal carbon source; in these groups, the net requirement for reducing power becomes negligible. These modes of photosynthesis are termed collectively *anoxygenic photosynthesis*.

Molecular oxygen ($O_2$) is not a reactant of the ATP-generating reactions in any of these forms of photosynthesis. Consequently, all photosyntheses can, in principle, occur under strictly anaerobic conditions. However, all organisms that carry out oxygenic photosynthesis are aerobes in the limited sense that they must be able to tolerate oxygen. The fundamentally anaerobic nature of photosynthesis is evident among the organisms that carry out anoxygenic photosynthesis. Most of these organisms are strict anaerobes; in the minority that are facultative aerobes, photosynthetic generation of ATP is suppressed by the presence of oxygen, being replaced by the respiratory generation of ATP.

The site of photosynthetic energy conversion is a membrane system containing pigments, electron carriers, lipids, and proteins. This is termed the *photosynthetic apparatus*. In eucaryotic organisms, it is contained within the chloroplast. Among procaryotes, the location within the cell of the photosynthetic apparatus is different in the three major photosynthetic groups (see Chapter 10 for further discussion).

LIGHT ABSORPTION. Radiant energy is always transferred in discrete packets known as *photons;* the energy content of a photon is inversely related to its wavelength. When radiant energy is absorbed by matter, its possible effects are a function of the energy content of the photon and hence the wavelength of the radiation. Infrared light of wavelengths longer than 1,200 nm has an energy content so small that the absorbed energy is immediately converted to heat; it cannot mediate chemical change. So-called *ionizing* radiations of very short wavelength (X rays, $\alpha$ particles, cosmic rays) have such a high energy content that molecules in their path are immediately ionized. Between these two extremes, radiations ranging in wavelengths from 200 to 1,200 nm (ultraviolet, visible, and near-infrared light) have an energy content such that their absorption is capable of producing a chemical change in the absorbing molecule; these wavelengths can provide energy for photosynthesis.

However, solar radiation is profoundly modified by its passage through the atmosphere, which effectively filters out much of the shorter, high-energy radiation. At sea level about 75 percent of the total energy of sunlight is contained in the light of wavelengths between 400 and 1,100 nm (the visible and near-infrared portions of the spectrum), and it is within these limits that the pigments responsible for light capture in photosynthesis have their effective absorption bands.

THE PHOTOSYNTHETIC APPARATUS. The photosynthetic apparatus of all organisms capable of carrying out photosynthesis consists of three essential components:

1. *An antenna of light-harvesting pigments.* These pigments can include chlorophylls, carotenoids, and (in some groups) phycobiliproteins (see p. 219). The particular light-harvesting pigments that comprise the antenna system are group specific, and their cumulative light-absorptive properties determine the range of wavelength of light over which photosynthesis occurs.
2. *A photosynthetic reaction center.* Light energy absorbed by the antenna causes an energy conversion at the photosynthetic reaction center, which contains chlorophyll molecules in a special state. Light energy causes an ejection of an electron from a molecule of chlorophyll.
3. *An electron transport chain.* Electrons generated at the reaction center are passed through an electron transport chain, thereby generating ATP.

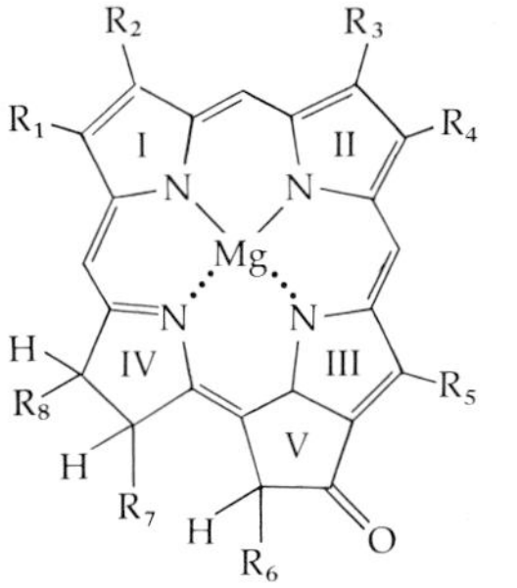

Chlorophyll (Figure 4.14) plays two roles in photosynthesis: as a light-harvesting pigment and as the photosynthetic reaction center. Carotenoids and phycobiliproteins function only as light-harvesting pigments.

The process of photosynthetic generation of ATP is initiated when a molecule of reaction center chlorophyll absorbs energy; the consequence of this event is the ejection of an electron producing oxidized chlorophyll:

$$\text{chlorophyll} + \text{light energy} \longrightarrow \text{chlorophyll}^{\oplus} + e^-$$

**Figure 4.14**
The molecular ground plan of the chlorophylls. The tetrapyrrolic nucleus (rings I, II, III, and IV) has the same derivation as that of the hemes, but it is chelated with magnesium. In chlorophylls, one or more of the pyrrole rings are reduced; in this diagram, ring IV is shown reduced, as is characteristic of chlorophyll a; $R_1$, $R_2$, and so forth, designate aliphatic side chains attached to the tetrapyrrolic nucleus. The presence of ring V, the pentanone ring, and the substitution of $R_7$ by a long-chain alcohol are characteristic features of the chlorophylls that do not occur in hemes.

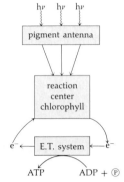

**Figure 4.15**
Functions of three components of the photosynthetic apparatus in cyclic photophosphorylation. Light energy (h$\nu$) is absorbed by the antenna of light-harvesting pigments (Pigment antenna) and transferred (arrows) to the action center chlorophyll causing an electron (e$^-$) to be ejected which flows through the photosynthetic electron transport system (E. T. system) back to oxidized reaction center chlorophyll (see text). By this transport of electrons ATP is generated from ADP.

The ejected electron (e$^-$) passes along the photosynthetic electron transport chain, thereby generating ATP. If the terminal electron acceptor of the chain is a molecule of oxidized chlorophyll (chlorophyll$^\oplus$) formed by the primary photochemical event, the process of ATP generation is termed *cyclic photophosphorylation* because the electrons flow through a closed circuit, the flow being driven by light energy (see Figure 4.15).

Alternatively, the electron ejected by light energy from reaction center chlorophyll can be used to reduce pyridine nucleotides. In this case, the flow of electrons becomes noncyclic because chlorophyll$^\oplus$ is reduced to its original state by electrons derived from $H_2O$ (in the case of blue-green bacteria or plants) via the photosynthetic electron transport chain. Since this noncyclic transfer of electrons, viz.,

$$H_2O \longrightarrow \text{electron transport chain} \longrightarrow \text{chlorophyll} \longrightarrow \text{pyridine nucleotide}$$

also generates ATP, the process is termed *noncyclic photophosphorylation*.

THE DIFFERENCE BETWEEN OXYGENIC ("PLANT") AND ANOXYGENIC ("BACTERIAL") PHOTO-SYNTHESIS. The difference between oxygenic and anoxygenic photosynthesis is the nature of the electron donor and the nature of the process by which it is oxidized. In the case of oxygenic photosynthesis, the transfer of electrons from $H_2O$ does not proceed spontaneously; on the contrary, it requires energy. The photosynthetic apparatus of organisms capable of oxygenic photosynthesis contains *two distinct kinds of photochemical reaction centers*. One kind (Type I) mediates the reactions we have already discussed. The other (Type II) mediates a photochemical removal of electrons from water. The coupling of the two light-induced reactions characteristic of oxygenic photosynthesis is schematized in Figure 4.16.

All current evidence suggests that the inability of purple and green bacteria to bring about a photosynthetic production of oxygen is a consequence of the absence of Type II reaction centers from the anoxygenic photosynthetic apparatus.

The mechanism of reduction of pyridine nucleotides in anoxygenic phototrophs is unclear. The participation of electron donors (e.g., $H_2S$ or $H_2$), which are more readily oxidized than $H_2O$, makes the direct reduction of chlorophyll in Type I reaction centers thermodynamically feasible. Alternatively, reverse electron transport driven by photochemically generated ATP may accomplish this reduction.

### modes of ATP-generating metabolism: a summary

The essential features of the various modes of ATP-generating metabolism are summarized in Table 4.2.

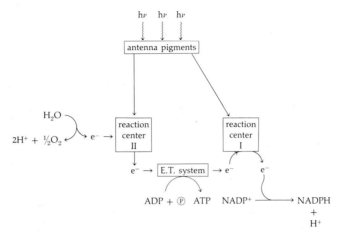

**Figure 4.16**
Schematic representation of noncyclic photophosphorylation in oxygenic
photosynthesis by blue-green bacteria. Light energy ($h\nu$) is absorbed by the
antenna of light-harvesting pigments and transferred (arrows) to reaction
centers I and II. Reaction center II catalyzes the photo oxidation of $H_2O$ to $O_2$.
The electrons ($e^-$) from this oxidation flow through the photosynthetic electron
transport system (E. T. System) to oxidized chlorophyll ($^+$) generated in reaction center I. Electrons generated by the photooxidation of chlorophyll in
reaction center I reduce pyridine nucleotide (NADP). For simplicity, the final
reaction of $NADP^+$ reduction is not balanced in this scheme. In fact, two
electrons and two protons (generated from water at photosystem II) are required for the reduction of $NADP^+$ to $NADPH + H^+$.

**Table 4.2**
Summary of Modes of ATP Generation

| MODE OF ATP-GENERATING METABOLISM | ELECTRON DONOR | ELECTRON ACCEPTOR | BIOCHEMICAL MECHANISM OF ATP SYNTHESIS |
|---|---|---|---|
| Fermentation | An organic compound | An organic compound | Substrate level phosphorylation |
| Respiration | | | |
|   Aerobic: | | | |
|     By chemoheterotrophs | Organic compounds | $O_2$ | Electron transport and some substrate level phosphorylation |
|     By chemoautotrophs | Inorganic compounds[a] | $O_2$ | Electron transport |
|   Anaerobic | Organic compounds[b] | Inorganic compounds[c] | Electron transport and some substrate level phosphorylation |
| Photosynthesis (cyclic photophosphorylation) | Reaction center chlorophyll | Oxidized reaction center chlorophyll | Electron transport |

[a] $H_2$, CO, $NH_3$, $NO_2^-$, $Fe^{2+}$, or inorganic sulfur compounds ($H_2S$, S, or $S_2O_3^{2-}$).
[b] In one case an inorganic sulfur compound.
[c] $NO_3^-$, $NO_2^-$, $SO_4^{2-}$, and rarely organic compounds.

# biosynthesis

Despite their mechanistic diversity, all the metabolic pathways discussed thus far have the same common function: the provision of ATP and reduced pyridine nucleotides. In this sense, there is a fundamental unity underlying the superficial diversity of catabolic metabolism. This unity of biochemistry, a concept first emphasized by the microbiologist A. J. Kluyver in 1926, becomes even more evident when we analyze the ways in which ATP is employed in biosynthesis. In all cells the major end products of biosynthesis are proteins and nucleic acids, and the biochemical reactions leading to their formation show little variation from group to group among procaryotes and even between procaryotes and eucaryotes. There is, accordingly, a *central core of biosynthetic reactions that are similar in all organisms.* A greater degree of diversity occurs in the synthesis of certain other classes of cell constituents, in particular polysaccharides and lipids; the chemical composition of these substances is often group specific.

The sum total of biosynthesis is the formation of a new cell from the compounds available in the environment. Since most of the organic matter of the cell consists of *biopolymers* that belong to four classes (nucleic acids, proteins, polysaccharides, and complex lipids), their formation constitutes the major portion of biosynthesis. Each of these classes of biopolymers is defined by the type of precursors that are polymerized to form it: nucleotides in the case of nucleic acids, amino acids in the case of proteins, and simply sugars (monosaccharides) in the case of polysaccharides. Complex lipids are more variable and heterogeneous in composition; their precursors include fatty acids, polyalcohols, simple sugars, amines, and amino acids. As shown in Table 4.3, approximately 60 different kinds of precursors are required to synthesize the four major classes of biopolymers.

**Table 4.3**
Classes of Macromolecules of the Cell and Their
Component Precursors

| MACRO-MOLECULE | CHEMICAL NATURE OF PRECURSOR | NUMBER OF KINDS OF PRECURSOR |
|---|---|---|
| Nucleic acids | | |
| RNA | Ribonucleotides | 4 |
| DNA | Deoxyribonucleotides | 4 |
| Proteins | Amino acids | 20 |
| Polysaccharides | Monosaccharides | ~15[a] |
| Complex lipids | Variable | ~20[a] |

[a]The number of building blocks in any particular representative of these macromolecules is usually much smaller.

In addition to biopolymers the cell must synthesize about 20 coenzymes* and electron carriers.

In all, about 150 different small molecules are required to produce a new cell. These small molecules are composed of the major bioelements: carbon, hydrogen, oxygen, sulfur, nitrogen, and phosphorus. The source of the major bioelements may be organic compounds or inorganic ones, e.g., $CO_2$, $SO_4^{2-}$, $NH_4^+$, $N_2$, $NO_3^-$, and $PO_4^{3-}$.

We will consider the process of biosynthesis in three phases: (1) assimilation of bioelements from inorganic sources, (2) synthesis of small precursor molecules, and (3) their polymerization into biopolymers.

## assimilation of the major bioelements from inorganic sources

Many microorganisms are able to assimilate the major bioelements from inorganic sources. The use of $CO_2$ as a source of carbon is characteristic of the autotrophs. Ammonia can be used as a sole nitrogen source by many microorganisms belonging to all nutritional categories; a lesser number of these can likewise utilize nitrate. The ability to use dinitrogen ($N_2$) as a nitrogen source (by the process of nitrogen fixation) is restricted to a relatively few procaryotes. Most microorganisms can use sulfate as a source of sulfur; all can use phosphate as a source of phosphorus.

### assimilation of $CO_2$

Chemoheterotrophs incorporate some $CO_2$ as the carboxyl group of oxalacetate, one of the intermediates of the TCA cycle, but in such organisms this contributes only a minor fraction of cellular carbon; most is supplied from the organic carbon source. Autotrophs, for which $CO_2$ serves as the total source of cellular carbon, fix $CO_2$ by a different reaction (see Figure 4.17):

ribulose diphosphate + $CO_2$ $\longrightarrow$ 2 glyceric acid phosphate

The primary product of such $CO_2$ fixation is glyceric acid phosphate, from which all other organic molecules of the cell are synthesized. However, $CO_2$ fixation is dependent on a supply of the other substrate (the 5-carbon sugar ribulose diphosphate) of the fixation reaction. Consequently, most of the glyceric acid phosphate must be utilized to regenerate ribulose diphosphate. Thus, the process of $CO_2$ fixation is cyclic, each turn of the cycle resulting in

---

*Coenzymes are small molecules that act with enzymes in catalyzing certain reactions. $NAD^+$, $NADP^+$, and coenzyme A are examples of coenzymes.

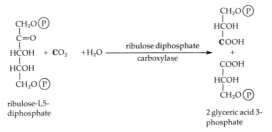

**Figure 4.17**
The $CO_2$-fixing reaction of autotrophs. The reaction is catalyzed by *ribulose diphosphate carboxylase*. The phosphorylated pentose, ribulose-1,5-diphosphate, accepts 1 molecule of $CO_2$ and is simultaneously cleaved yielding 2 molecules of glyceric acid-3-phosphate; the carboxyl group of one glyceric acid-3-phosphate molecule is thus derived from $CO_2$.

the fixation of one molecule of $CO_2$. Intermediates of the cycle are drawn off and enter various biosynthetic pathways.

This mechanism of $CO_2$ fixation was first elucidated in a green alga by M. Calvin, A. Benson, and J. Bassham and is sometimes called the *Calvin cycle*. Subsequent work has shown it to be nearly universal among autotrophs.

The Calvin cycle is complex, sharing certain reactions with other metabolic pathways. Only two reactions are specific to the cycle: the $CO_2$ fixation reaction itself and the reaction that generates the $CO_2$ acceptor (ribulose-1,5-diphosphate) from its immediate precursor, ribulose-5-phosphate.

The Calvin cycle can be considered to consist of three phases: (i) the unique reaction of $CO_2$ fixation, (ii) the reactions through which the fixed $CO_2$ is reduced to the oxidation state of triosephosphate, and (iii) the reactions that regenerate ribulose diphosphate, the compound that accepts $CO_2$ in the fixation reaction. These various phases of the Calvin cycle and the points at which intermediates are withdrawn for biosynthesis are shown in Figure 4.18.

### assimilation of inorganic forms of nitrogen

The nitrogen atom of ammonia is at the same oxidation level as the nitrogen atoms of organic cell constituents. Its assimilation, therefore, does not necessitate oxidation or reduction. It is incorporated directly by three reactions: one forms the amino group of glutamic acid, and the two others form the amido groups of asparagine and glutamine (Figure 4.19).

All three products of $NH_3$ assimilation, being amino acids, are direct precursors of protein, and one (asparagine) serves

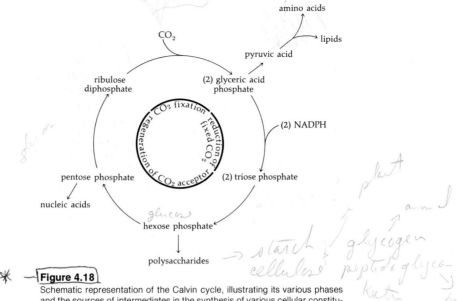

**Figure 4.18**

Schematic representation of the Calvin cycle, illustrating its various phases and the sources of intermediates in the synthesis of various cellular constituents.

**Figure 4.19**

The cellular reactions by which ammonia is assimilated. $\alpha$-ketoglutaric acid is an intermediate of the TCA cycle; aspartic acid is formed from glutamic acid and another intermediate of the TCA cycle (oxalacetic acid). The formation of glutamic acid is reductive and therefore requires the participation of NADPH; the other two reactions (forming amido groups) require the participation of ATP.

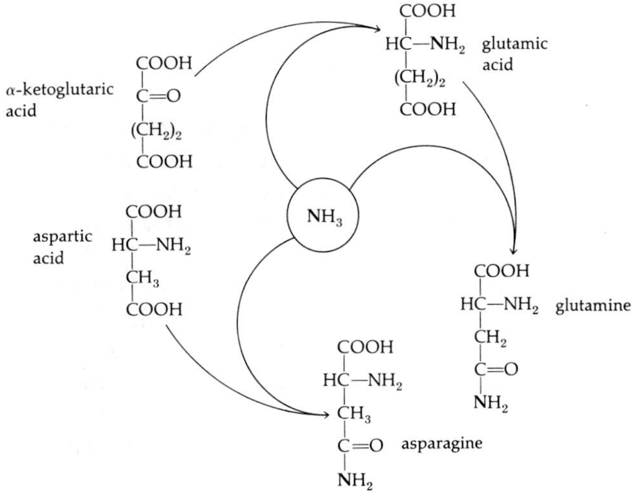

exclusively in this role. However glutamic acid and glutamine play the additional roles of nitrogen donors in the synthesis of all other nitrogenous compounds in the cell.

Nitrate ion ($NO_3^-$) can be used by many microorganisms as a source of ammonia, and, hence, by the reactions just discussed, as a total source of cellular nitrogen. The valence of $NO_3^-$ is $+5$; consequently, its assimilation involves reduction, which is mediated in two steps by enzymes called *nitrate reductase* and *nitrite reductase*:

$$NO_3^- \xrightarrow[\text{reductase}]{\text{nitrate}} NO_2^- \xrightarrow[\text{reductase}]{\text{nitrite}} NH_3$$

As was discussed previously, nitrate is also reduced when in certain microorganisms it serves as a terminal electron acceptor for anaerobic respiration. This process leads to the formation of the gaseous compounds nitrous oxide ($N_2O$) and dinitrogen ($N_2$) rather than ammonia, and hence does not result in assimilation into cellular components. To distinguish this process from *assimilatory nitrate reduction* that produces ammonia, it is called *dissimilatory nitrate reduction*. Dissimilatory nitrate reduction is restricted to certain members of a few procaryotic groups.

Gaseous dinitrogen ($N_2$) can also be reduced to ammonia and hence assimilated by a limited number of procaryotes. This process, called *nitrogen fixation*, constitutes the major route by which atmospheric nitrogen enters the biosphere and is of enormous ecological and agricultural importance (see Chapter 15).

The enzyme responsible for $N_2$ fixation, called *nitrogenase*, contains molybdenum and requires ATP as well as a source of electrons. It is extremely sensitive to oxygen; hence all organisms that fix $N_2$ are either anaerobes or aerobes that possess mechanisms for protecting nitrogenase from $O_2$.

## assimilation of sulfate

The great majority of microorganisms can fulfill their sulfur requirements from sulfate. Sulfate, with a valence of $+6$, is reduced to sulfide (valence $-2$) prior to its incorporation into cellular organic compounds. Chemically, this is equivalent to the reduction of sulfate by the sulfate-reducing bacteria that use it as the terminal electron acceptor in anaerobic respiration. The enzymatic mechanisms are different, however; the reduction of sulfate for use as a sulfur source is termed *assimilatory sulfate reduction* (by analogy with assimilatory nitrate reduction) to distinguish it from *dissimilatory sulfate reduction*, the use of sulfate as a terminal electron acceptor.

Assimilatory reduction of sulfate requires the expenditure of high-energy bonds of ATP; dissimilatory sulfate reduction is associated with the generation of ATP by anaerobic respiration.

# the synthesis of small molecules—precursors of biopolymers

The various precursors of biopolymers listed in Table 4.3 are synthesized by sequences of distinct metabolic reactions. A detailed understanding of most of these reactions and of the biosynthetic pathways they constitute has now been achieved as a consequence of biochemical investigations conducted during the last 30 years.

As examples of such pathways we shall outline the synthesis of purines, pyrimidines, and amino acids.

### the synthesis of purines and pyrimidines

The precursors of nucleic acids are purine and pyrimidine nucleoside triphosphates, all of which have the same general structure. A purine or pyrimidine base is attached through a nitrogen atom to a pentose; this combination is called a *nucleoside*. Phosphate groups are attached to the 5' position of the nucleoside (to distinguish between the base and pentose moieties of a nucleoside, positions on the pentose are assigned a prime following the number). The general structure of nucleoside triphosphates is shown in Figure 4.20. The names and structures of specific nucleosides are shown in Figure 4.21. Nucleotides are symbolized by letters, A, G, U, C, or T, to indicate the purine or pyrimidine base they contain; MP, DP, or TP indicates whether they are mono-, di- or triphosphates. Deoxynucleotides are indicated by a "d" (e.g., CDP symbolizes cytidine diphosphate, and dGTP symbolizes 2'deoxyguanosine triphosphate). The two purine (dATP and

**Figure 4.20**
The general structure of nucleoside triphosphates. High-energy (anhydride) phosphate bonds are symbolized by a wavy line (∼); low-energy (ester) phosphate bonds are symbolized by a straight line (−).

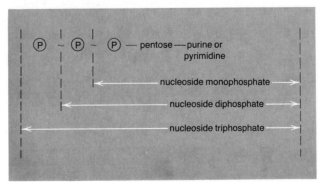

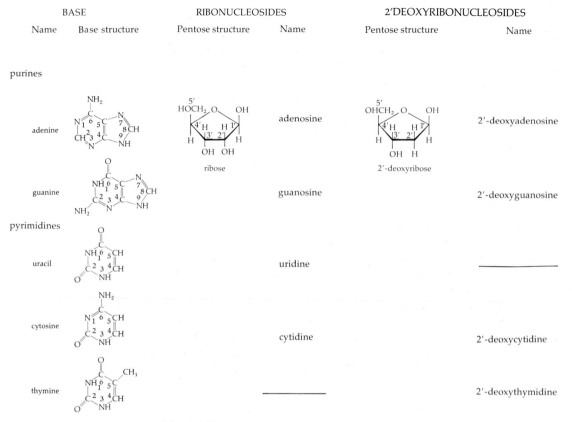

**Figure 4.21**

Names and composition of bases and nucleosides. Purines at the 9 position, and pyrimidines at the 3 position, are attached to the 1 position of pentoses to form nucleosides.

dGTP) and two pyrimidine (dCTP and dTTP) nucleoside triphosphates containing deoxyribose are the specific precursors of DNA; the two purine (ATP and GTP) and two pyrimidine (CTP and UTP) nucleoside triphosphates containing ribose are specific precursors of RNA.

The general outline of biosynthesis of purine and pyrimidine nucleoside triphosphates is schematized in Figure 4.22.

The four deoxyribonucleoside triphosphate precursors of DNA (dATP, dGTP, dCTP, and dTTP) are synthesized from ribonucleoside phosphates. Three of them (dATP, dGTP, and dCTP) are formed by reduction of the corresponding ribonucleotides by a single enzyme. The products of reduction, the deoxynucleoside diphosphates (dADP, dGDP, and dCTP), are converted to triphosphates by the same enzyme that converts ribonucleoside diphosphates to triphosphates.

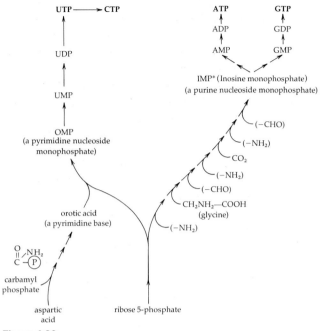

**Figure 4.22**
The general outline of the pathways of synthesis of purine and pyrimidine ribonucleoside triphosphates. Parentheses indicate the addition of the indicated functional group.

The fourth precursor of DNA, dTTP, is synthesized by an independent route.

### the synthesis of amino acids

Twenty amino acids are required for the biosyn-
thesis of proteins. Only one, histidine, has a completely isolated biosynthetic
origin. The other 19 are derived through branched pathways from a relatively
small number of intermediates of catabolism; in these terms, amino acids can
be grouped into five families (Table 4.4).

### interconnections between catabolic
### and biosynthetic pathways

The interconnections between catabolic and biosynthetic pathways can be summarized in the form of a metabolic map
(Figure 4.23). This makes more evident a point, already alluded to, that the
special pathways of biosynthesis are interconnected through reaction sequences (e.g., the reactions of the Embden-Meyerhof pathway and the tricarboxylic acid cycle), which also play important roles in the ATP-generating
metabolism of organic substrates.

**Table 4.4**
Biosynthetic Derivations of Amino Acids[a]

| PRECURSOR | AMINO ACIDS | FAMILY |
|---|---|---|

**α-Ketoglutarate** → glutamate ($COOH$—$CHNH_2$—$(CH_2)_2$—$COOH$)

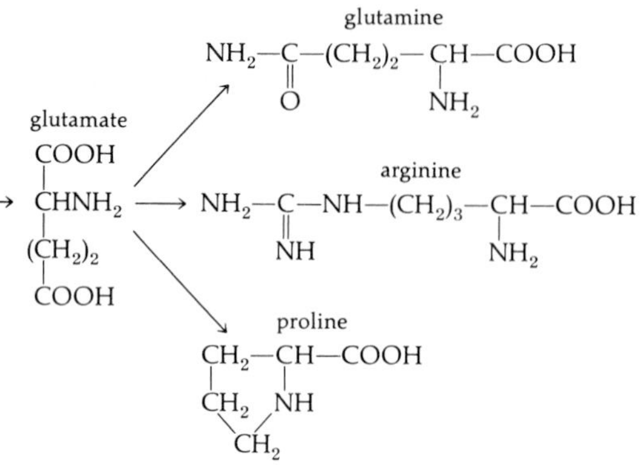

glutamine
$NH_2$—$C$(=$O$)—$(CH_2)_2$—$CH(NH_2)$—$COOH$

arginine
$NH_2$—$C$(=$NH$)—$NH$—$(CH_2)_3$—$CH(NH_2)$—$COOH$

proline
$CH_2$—$CH$—$COOH$ / $CH_2$—$NH$ / $CH_2$

**glutamate family**

**Oxalacetate** → aspartate ($COOH$—$CHNH_2$—$CH_2$—$COOH$)

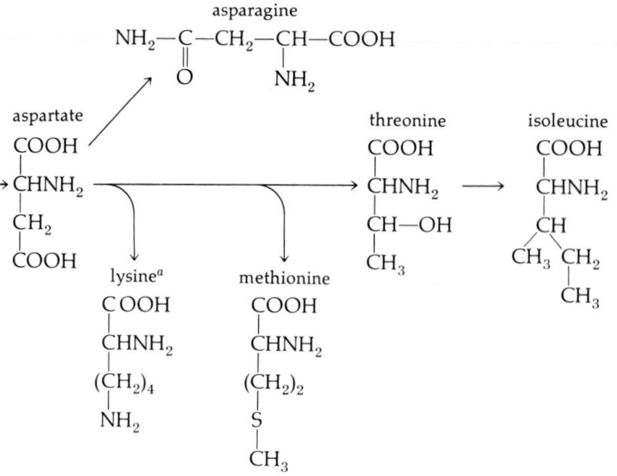

asparagine
$NH_2$—$C$(=$O$)—$CH_2$—$CH(NH_2)$—$COOH$

lysine[a]
$COOH$—$CHNH_2$—$(CH_2)_4$—$NH_2$

methionine
$COOH$—$CHNH_2$—$(CH_2)_2$—$S$—$CH_3$

threonine
$COOH$—$CHNH_2$—$CH$—$OH$—$CH_3$

isoleucine
$COOH$—$CHNH_2$—$CH$—$CH_3$ $CH_2$—$CH_3$

**aspartate family**

**Phosphoenol pyruvate + Erythrose-4-phosphate**

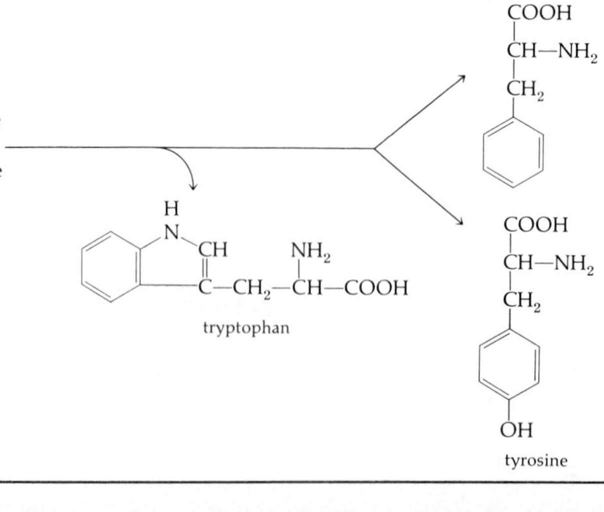

phenylalanine
$COOH$—$CH$—$NH_2$—$CH_2$—(phenyl ring)

tryptophan
(indole ring)—$C$—$CH_2$—$CH(NH_2)$—$COOH$

tyrosine
$COOH$—$CH$—$NH_2$—$CH_2$—(phenol ring)—$OH$

**aromatic family**

**Table 4.4 (cont.)**

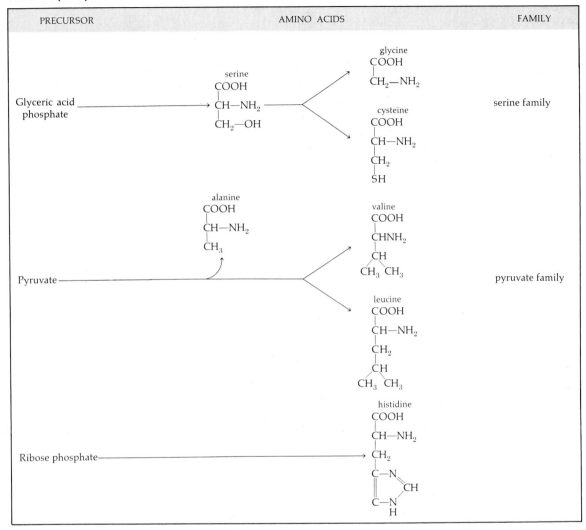

| PRECURSOR | AMINO ACIDS | FAMILY |
|---|---|---|

*a* In certain algae and fungi, lysine is synthesized from β-ketoglutarate.

# the synthesis of biopolymers: general principles

Biopolymers are composed of subunits (monomers) linked together by bonds that are characteristic of each class of molecule (Figure 4.24). The subunits of all biopolymers can be liberated in free form by hydrolysis. Thus, the biosynthesis of biopolymers involves the joining of subunits through reactions that are, in a formal chemical sense, the reverse of hydrolysis: namely, *dehydration*.

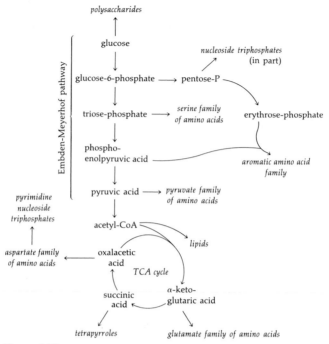

**Figure 4.23**
The relationship between catabolic and biosynthetic pathways.

Biopolymers can be hydrolyzed to their subunits by either chemical or enzymatic means. Thus, their biosynthesis by simple dehydration is thermodynamically unfavorable; in the aqueous intracellular environment, breakdown through hydrolysis predominates over synthesis through dehydration. The net synthesis of all biopolymers is therefore accomplished by a preliminary *chemical activation* of the monomer. Such activation requires the expenditure of ATP, and involves the attachment of the monomer to a carrier molecule. Polymerization then occurs by transfer of the monomer from the carrier to the growing polymer chain by a thermodynamically favorable reaction. The activated forms of monomers of the major classes of biopolymers are shown in Table 4.5.

### the general plan of synthesis of nucleic acids and proteins
A bacterial cell can synthesize several thousand different kinds of proteins, each containing, on the average, approximately 200 amino acid residues linked together in a definite sequence. The information required to direct the synthesis of these proteins is encoded by the sequence of nucleoside phosphates (nucleotides) in the cell's complement of DNA, most of which is in the form of a double-stranded circular molecule, the bacterial chromosome. By the process of *replication* the chromosome is pre-

| Type of polymer | Designation of bond | Structure of bond |
|---|---|---|
| protein | peptide | 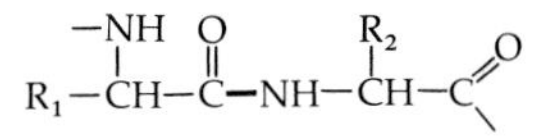 |
| polysaccharide | glycoside | 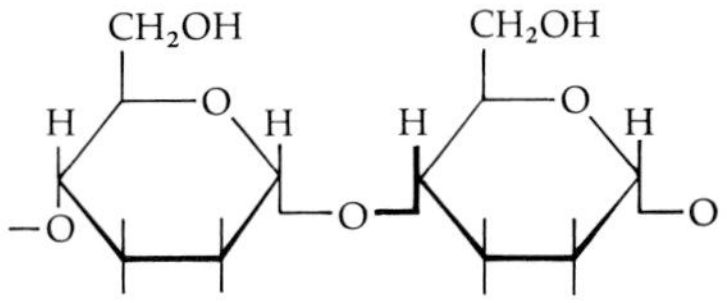 |
| nucleic acid | phosphodiester | 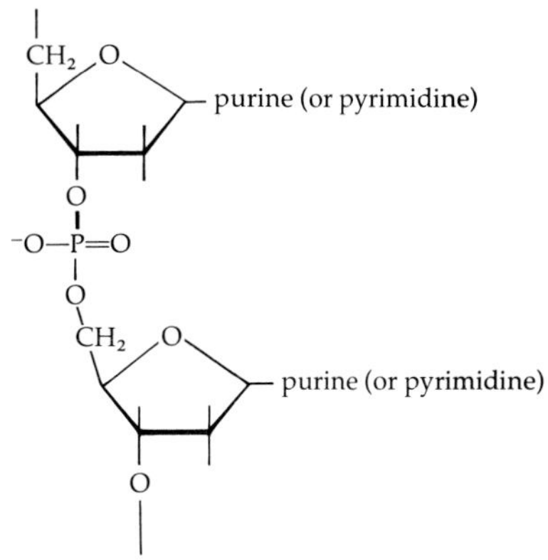 |

**Figure 4.24**
Nature of the bonds that link together the subunits in the major classes of biological polymers.

cisely duplicated, thus assuring that progeny cells receive information enabling them to synthesize the same proteins.

The process by which the information encoded in the chromosome directs the order of polymerization of amino acids into proteins occurs in two steps: transcription and translation (see Figure 4.25).

TRANSCRIPTION. The information content of one of the strands of DNA is transcribed into RNA; i.e., the DNA strand serves as a template upon which a single strand of RNA is polymerized, the length of which corresponds to from

**Table 4.5**
Bipolymers, Their Monomeric Constituents, and the Activated Forms From Which They Are Polymerized

| BIPOLYMER | CONSTITUENT MONOMER[a] | ACTIVATED FORM OF MONOMER |
|---|---|---|
| Protein | Amino acids | Aminoacyl-tRNA |
| Nucleic acid | Nucleoside monophosphphates | Nucleoside triphosphate |
| Polysaccharide | Sugars | Sugar-nucleoside monophosphate |

[a] Products formed by hydrolysis.

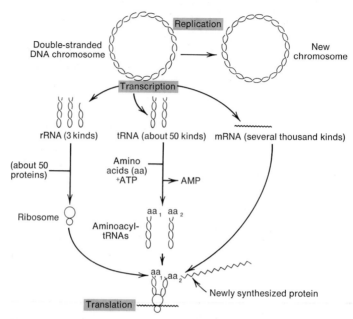

**Figure 4.25**
The general plan of synthesis of nucleic acids and proteins.

one to several genes on the bacterial chromosome. One class of these RNA molecules, termed *messenger RNA* (mRNA), carries the information encoded in the DNA to the protein-synthesis machinery.

TRANSLATION.   Protein synthesis takes place on ribonucleoprotein particles called *ribosomes* [composed of ribosomal RNA (rRNA) and protein], which attach themselves to the molecule of mRNA. The information carried by the mRNA molecules is translated into protein molecules by a special class of RNA molecules called *transfer RNA* (tRNA). These molecules are multifunctional: they are able to bind to the ribosome, to be attached to specific amino acids, and to recognize specific nucleotide sequences of the mRNA. Each molecular species of tRNA recognizes a specific sequence of three nucleotides (a *codon*) on the mRNA molecule, and can be attached to a specific amino acid. Thus, the various amino acids are brought by their cognate tRNA molecules to the ribosome, where they are polymerized into protein in the sequence encoded by the mRNA.

Within the cell, translation of a molecule of mRNA begins before its synthesis is complete, and it is brought about by the simultaneous functioning of a large number of ribosomes distributed over the length of the mRNA molecule. A molecule of mRNA to which a number of ribosomes is attached is called a *polysome*. Excellent electron micrographs have been made that show the concurrent nature of transcription and translation as well as the formation of polysomes (Figure 4.26).

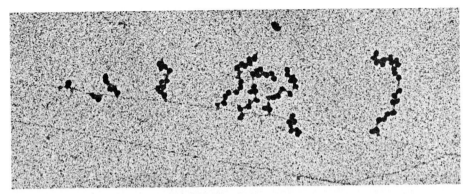

**Figure 4.26**
Photomicrograph of the simultaneous transcription and translation of a fragment of the chromosome of *E. coli*. The central horizontal line is DNA. The more wavy lines extending from it are molecules of mRNA to which a number of ribosomes are attached (polysomes). The gradual increase in length of the mRNA from left to right of the picture indicates that transcription was proceeding in that direction, and that it began near the left edge of the picture. (×62,350) From B. Hamkalo and O. Miller, Jr., "Electronmicroscopy of Genetic Material," *Ann. Revs. Biochem.* **42**, 379 (1973).

THE GENETIC CODE. The *genetic code,* representing the correspondence between codons on mRNA and the amino acid that they cause to be incorporated into protein can now be written in its entirety. Since three nucleotides encode the insertion of each amino acid in the peptide chain, the code is said to be *triplet.* The same code (Table 4.6) applies to all living organisms. The "breaking of

**Table 4.6**
The Genetic Code

| FIRST LETTER | SECOND LETTERS | | | | | | | |
|---|---|---|---|---|---|---|---|---|
| | U | | C | | A | | G | |
| U | UUU | phe[a] | UCU | ser | UAU | tyr | UGU | cys |
| | UUC | phe | UCC | ser | UAC | tyr | UGC | cys |
| | UUA | leu | UCA | ser | UAA | (none)[b] | UGA | (none)[b] |
| | UUG | leu | UCG | ser | UAG | (none)[b] | UGG | try |
| C | CUU | leu | CCU | pro | CAU | his | CGU | arg |
| | CUC | leu | CCC | pro | CAC | his | CGC | arg |
| | CUA | leu | CCA | pro | CAA | glu-N | CGA | arg |
| | CUG | leu | CCG | pro | CAG | glu-N | CGG | arg |
| A | AUU | ileu | ACU | thr | AAU | asp-N | AGU | ser |
| | AUC | ileu | ACC | thr | AAC | asp-N | AGC | ser |
| | AUA | ileu | ACA | thr | AAA | lys | AGA | arg |
| | AUG | met | ACG | thr | AAG | lys | AGG | arg |
| G | GUU | val | GCU | ala | GAU | asp | GGU | gly |
| | CUC | val | GCC | ala | GAC | asp | GGC | gly |
| | GUA | val | GCA | ala | GAA | glu | GGA | gly |
| | GUG | val | GCG | ala | GAG | glu | GGG | gly |

[a] Amino acids are abbreviated as the first three letters in each case, except for glutamine (glu-N), asparagine (asp-N), and isoleucine (ileu).
[b] The codons UAA, UAG, and UGA are nonsense codons (see Chapter 8).

the genetic code," representing the work of a number of groups of investigators over the remarkably short period of about five years, ranks as a major achievement of biology and, indeed, of science in general.

## the synthesis of DNA

The structure of the DNA molecule, elucidated by J. D. Watson and F. H. C. Crick in 1953, immediately suggests how it can be accurately replicated. They showed it to be a double helix, each strand of which consists of 2'-deoxyribonucleotide molecules linked together by phosphodiester bonds between the 3' hydroxyl group of one and the 5' hydroxyl group of the next. The purine and pyrimidine bases (attached to the 1 position of deoxyribose) project toward the center of the molecule, holding the two strands together by hydrogen bonding between specific purine-pyrimidine pairs. Guanine is paired with cytosine (G—C) and adenine is paired with thymine (A—T) (Figure 4.27). The entire molecule can thus be described as a linear sequence of *nucleotide pairs*; the order of these pairs constitutes the genetic message that contains all the information necessary to determine the structures and functions of the cell.

The polymerization of DNA is catalyzed by enzymes called *DNA polymerases*. In addition to the four deoxynucleoside triphosphates (dATP, dGTP, dCTP, and dTTP), which are substrates for the reaction, two molecules of single-stranded DNA are required: one is the *template* to which the substrate deoxynucleoside triphosphate molecules pair according to the rules of hydrogen bonding (G with C and A with T) (Figure 4.28); the second is the *primer* to which the nucleotides are attached as a consequence of polymerization (Figure 4.28). DNA synthesis proceeds by the sequential formation of phosphoester bonds between the $\alpha$-phosphate of the deoxynucleoside triphosphate and the terminal 3'-hydroxy group of the primer, with the release of one pyrophosphate ($\circled{P}$—$\circled{P}$) molecule. The template DNA determines the sequence of addition of nucleotides to the primer DNA molecule.

### replication of the chromosome

Although the process of polymerization of DNA is simple and thoroughly understood, the replication of the intact double-stranded bacterial chromosome is much more complicated, and many questions remain to be answered. As shown in Figure 4.28, DNA polymerases require a single-stranded template; thus by still incompletely understood mechanisms, the double-stranded chromosome must be separated into single strands in the region in which replication starts. The requirement for initial primer is probably met by synthesis of a short complementary piece of RNA on each of the separated strands. From the point of initiation, replication occurs in both directions around the chromosome.

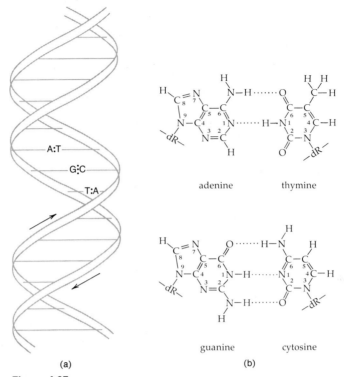

**Figure 4.27**

(a) Schematic representation of the DNA double helix. The outer ribbons represent the two deoxyribosephosphate strands. The parallel lines between them represent the pairs of purine and pyrimidine bases held together by hydrogen bonds. Specific examples of such bonding is shown in the center section, each dot between the pairs of bases representing a single hydrogen bond. The direction of the arrows correspond to the 3' to 5' direction of the phosphodiester bonds between adjacent molecules of 2'-deoxyribose. After J. Mandelstam and K. McQuillen, *Biochemistry of Bacterial Growth*, 2nd ed. New York: Wiley, 1973. (b) The pairing of adenine with thymine and guanine with cytosine by hydrogen bonding. The symbol—dR— represents the deoxyribose moieties of the sugar-phosphate backbones of the double helix. Hydrogen bonds are shown as dotted lines.

# the synthesis of proteins

The products of transcription, mRNA, tRNA, and rRNA, all function in protein synthesis. rRNA is a component of *ribosomes*, which are the sites of protein synthesis. The functional form of the procaryotic ribosome, which has a sedimentation velocity of 70S, can be dissociated reversibly into a 30S and a 50S subunit by changing the concentration of $Mg^+_2$ in the suspending buffer. The molecular composition of procaryotic ribosomes is shown is Figure 4.29.

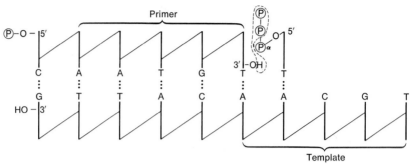

**Figure 4.28**
Synthesis of DNA showing the regions which serve as primer and template. Vertical lines representing 2'-deoxyribose molecules are linked by phosphodiester bonds (diagonal lines). The purine and pyrimidine bases (A, C, G and T) which project toward the center of the molecule are held together by hydrogen bonding (dotted lines). A molecule of dTTP is shown in the upper right corner; it has become hydrogen bonded to A on the template strand. DNA polymerase will catalyze the formation of a phosphodiester bond between the $\alpha$-phosphate of dTTP and the 3'-hydroxyl of the primer strand with the elimination of a molecule of pyrophosphate ($\text{(P)}-\text{(P)}$) which is shown within the circled region.

**Figure 4.29**
The composition of procaryotic ribosomes.

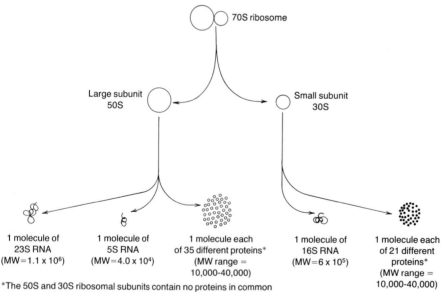

## amino acid activation

The activated forms of amino acids that are polymerized to form proteins are aminoacyl-tRNAs. They are synthesized in two steps catalyzed by a group of enzymes, *aminoacyl-tRNA synthetases*. There are 20 such enzymes, each specific for a particular amino acid and particular molecules of tRNA.*

## initiation of translation

The 30S ribosomal subunit, along with a special aminoacyl-tRNA (a methionyl-tRNA, to the amino group of which a formyl residue is attached), combine at a specific site on a molecule of mRNA, forming the *initiation complex*. Then, a 50S ribosomal subunit attaches to the complex to form the 70S ribosome, which mediates polymerization. Three accessory proteins (termed *initiation factors*) take part in the reaction; the expenditure of energy in the form of one high-energy bond of GTP is also required.

## polymerization of amino acids

The addition of each amino acid during the polymerization process involves the same sequence of events (Figure 4.30).

1. *Recognition.* A molecule of aminoacyl-tRNA attaches to a particular region of the 70S ribosome, termed the *A* (or aminoacyl) *site*, which exposes a sequence of three bases (a *codon*) on the mRNA molecule. The bases must be complementary to three bases (an *anticodon*) on the distal end of the aminoacyl-tRNA molecule, in order for polymerization to proceed. Since the *anticodon* of tRNA corresponds to the amino acid that can be attached to the opposite end of the molecule, the codon in the A site determines which amino acid will be added to the growing peptide that is attached to another molecule of tRNA located at another site on the ribosome, the *P* (or peptidyl) *site*. The recognition step requires the participation of two more accessory proteins (termed *elongation factors*) and the conversion of 1 molecule of GTP to GDP.

2. *Peptidyl transfer.* The peptide at the P site is then transferred to the amino acid attached to the tRNA molecule at the A site, forming a new peptide bond and thereby lengthening the peptide chain by one aminoacyl residue. Peptidyl transfer is catalyzed by the 50S ribosomal subunit itself; no accessory proteins are required.

3. *Translocation.* Following peptidyl transfer, the peptide-bearing tRNA moves to the P site and the mRNA moves with it. The free tRNA molecule is thus displaced and the next codon is brought into the A site. Translocation requires an additional elongation factor and the hydrolysis of another molecule of GTP.

4. *Chain termination.* By repetition of the recognition, peptidyl transfer, and translocation steps, successive aminoacyl residues are added to the peptide chain in the order encoded by the sequence of codons in the mRNA molecule. The process continues until a codon is reached (UAG, UAA, or UGA), which causes the release

*The cell contains more than 20 different kinds of tRNA molecules because, in certain cases, several different tRNA molecules can accept the same amino acid.

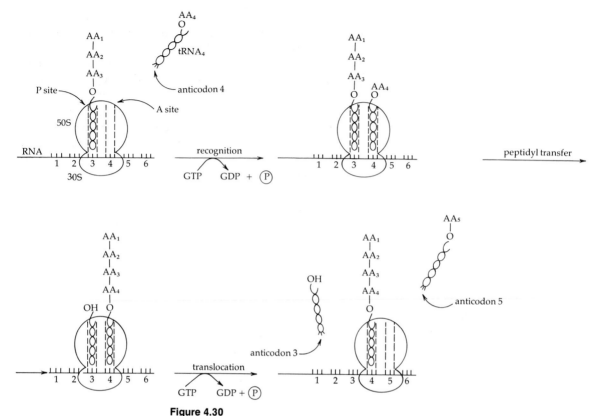

**Figure 4.30**
The sequence of events in the lengthening of the peptide chain. Amino acids (AA) are numbered by their order of addition to the peptide; numbered trios of lines symbolize the codons on the mRNA molecule.

of the completed peptide from the 70S ribosome. This process requires the intervention of a protein *release factor*. Then, the 70S ribosome dissociates into its 30S and 50S subunits.

### the secondary, tertiary, and quaternary structure of proteins

Even before the nascent polypeptide chain detaches from the ribosome, it begins to fold into a compact, three-dimensional mass. First, parts of the polypeptide become coiled into a regular, helical structure called an *α-helix*: this is designated as the *secondary structure* of the protein. Next, the entire molecule, including those regions that have the α-helical configuration, folds on itself to assume a specific three-dimensional shape, called the *tertiary structure* of the protein.

Both the secondary and tertiary configurations of a protein are determined by its primary structure (the sequence of amino acids in the polypeptide); a polypeptide with a given primary structure will ultimately assume one particular secondary and tertiary structure that represents its most stable state.

The quaternary structure of a protein is formed by the noncovalent association of several polypeptides that might be identical or completely different. Glutamic dehydrogenase, for example, is composed of between 24 and 30 identical subunits, each with a molecular weight of about 40,000. In contrast, carbamylphosphate synthase is composed of two subunits that are dissimilar, one with a molecular weight of 110,000 and the other of 40,000.

# MICROBIAL GROWTH

# the definition of growth

In any biological system, growth can be defined as *the orderly increase of all chemical components*. Increase of mass might not really reflect growth because the cells could be simply increasing their content of storage products such as glycogen or poly-$\beta$-hydroxybutyrate (see Chapter 6). In an adequate medium to which they have become fully adapted, bacteria are in a state of *balanced growth*. During a period of balanced growth, a doubling of the biomass is accompanied by a doubling of all other measurable properties of the population, e.g., protein, RNA, DNA, and intracellular water. In other words, cultures undergoing balanced growth maintain a constant chemical composition. The phenomenon of balanced growth simplifies the task of measuring the rate of growth of a bacterial culture; since the rate of increase of *all* components of the population is the same, measurements of *any* component suffice to determine the growth rate.

# the mathematical nature and expression of growth

A bacterial culture undergoing balanced growth mimics a first-order autocatalytic chemical reaction; i.e., the rate of increase in bacteria is proportional to the number or mass of bacteria present.

$$\text{rate of increase of bacteria} = \mu(\text{number of bacteria}) \qquad [5.1]$$

The constant of proportionality, $\mu$, is an index of the rate of growth and is called the *growth rate constant*. Since we assume

103

growth to be balanced, $\mu$ also relates the rate of increase of any given cellular component to the amount of that cellular component, or in mathematical terms,

$$\frac{dN}{dt} = \mu N \qquad [5.2]$$

where $N$ is the number of cells/ml, or the mass of cells/ml, or the amount of any cellular component/ml; $t$ is time; and $\mu$ is the growth rate constant. This equation, in fact, accurately describes the growth of most unicellular organisms. Other (nondifferential) forms of this equation are more useful in practice. Upon integration, Eq. (5.2) yields:

$$\ln N - \ln N_0 = \mu(t - t_0) \qquad [5.3]$$

and on converting natural logarithms to logarithms to the base 10,

$$\log_{10} N - \log_{10} N_0 = \frac{\mu}{2.303}(t - t_0) \qquad [5.4]$$

where the values of $N$ and $N_0$ correspond to the amount of any bacterial component of the culture at times $t$ and $t_0$, respectively. By measuring $N$ and $N_0$, one can compute the value of $\mu$, the growth rate constant of the culture. Thus, if the culture contains $10^4$ cells/ml at $t_0$ and $10^8$ cells/ml 4 hours later, the specific growth rate of the culture is:

$$\mu = \frac{(8 - 4)\, 2.303}{4} = 2.3 \text{ hours}^{-1} \qquad [5.5]$$

The value of $\mu$ suffices to define the rate of growth of a culture. However, certain other parameters are also commonly used. One is the mean doubling time, or generation time ($g$), defined as the time required for all components of the culture to increase by a factor of 2.* The relationship between $g$ and $\mu$ can be derived from Eq. 5.3, since if the time interval considered ($t - t_0$) is equal to $g$, then $N$ will be twice $N_0$. Making these substitutions, one obtains:

$$\mu = \frac{\ln 2}{g} = \frac{0.693}{g} \qquad [5.6]$$

In the case of the example we have chosen, the mean doubling time, $g$, of the culture *is $g$ = 0.693/2.3 = 0.3 hour or 18 minutes. Generation times for many bacteria under common laboratory conditions range from about 20 minutes to several hours, although many other bacteria, particularly autotrophic ones, grow considerably more slowly.
We have considered growth as a chemical process in which rate of increase is an exponential function of time. If one observes the actual process of division, the same conclusion would be drawn. Most

*Sometimes the reciprocal of doubling time is used as an index of growth rate. This index ($k$) is numerically similar to $\mu$ and can be confusing; its use should be avoided.

bacteria divide by a process known as *binary fission*,* in which a cell grows in mass and size, then divides into two equal *daughter cells*. These two cells repeat the process, taking about the same period of time for division, to produce four cells, which then produce eight cells, etc. From this observation, we can conclude that the number of cells in a culture ($N$) is related to the number initially present ($N_0$) after $n$ generations, by the equation:

$$N = N_0 2^n \tag{5.7}$$

$$N/N_0 = 2^n \tag{5.8}$$

Converting to logarithmic form generates:

$$\log_2 N - \log_2 N_0 = n \tag{5.9}$$

Since $n = (t - t_0)/g$ and $g = 0.693/\mu$,

$$\log_2 N - \log_2 N_0 = (t - t_0)\mu/0.693 \tag{5.10}$$

Converting to $\log_{10}$ yields Eq. (5.4) again:

$$\log N - \log N_0 = (t - t_0)\mu/2.303$$

## the growth curve

Equation (5.4) predicts a straight-line relationship between the logarithm of cell number (or any other measurable property of the population) and time [Figure 5.1(a)] with a slope equal to $\mu/2.303$ and an ordinate intercept of log $N_0$ (Figure 5.1). Populations of bacteria growing in a manner that obeys these equations are said to be in the *exponential phase* of growth.

Microbial populations seldom maintain exponential growth at high rates for long. The reason is obvious if one considers the consequences of exponential growth. After 48 hours of exponential growth, a single bacterium with a doubling time of 20 minutes would produce a progeny of $2.2 \times 10^{31}$ grams, or roughly 4,000 times the weight of the earth.

The growth of bacterial populations is normally limited either by the exhaustion of available nutrients or by the accumulation of toxic products of metabolism. As a consequence, the rate of growth declines and growth eventually stops. At this point a culture is said to be in the *stationary phase* (Figure 5.2). The transition between the exponential phase and the stationary phase involves a period of *unbalanced growth*, during which the

*Two less common modes of bacterial growth are *multiple fission* and *budding*. In multiple fission, the cell grows to a number of times its original size, then divides into many daughter cells. In budding, the cell produces a *bud*, an outgrowth of the cell that may be directly on the cell surface or at the tip of an extension of the cell called a *hypha*. The bud enlarges until it is roughly the size of the *mother cell*, at which point division occurs. Both of these modes of growth obey the kinetics developed in this chapter if the number of cells whose growth is being measured is reasonably large.

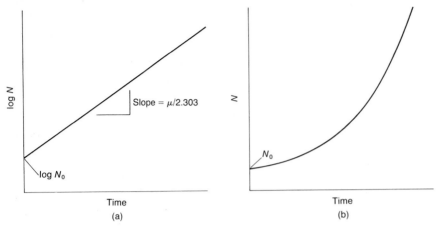

**Figure 5.1**

Comparison of methods of plotting growth data. Plotting the logarithm of cell density (number of cells/ml, $N$) of a culture undergoing balanced growth as a function of time yields a straight line (a); the slope of the line is the growth rate constant ($\mu$) divided by 2.303, and the intercept is log $N_0$. Plotting the cell density directly as a function of time yields an exponential curve (b).

various cellular components are synthesized at unequal rates. Consequently, cells in the stationary phase have a chemical composition that is different from that of cells in the exponential phase. The cellular composition of cells in the stationary phase depends on the specific growth-limiting factor. Despite this, certain generalizations hold: cells in the stationary phase are small relative to cells in the exponential phase (since cell division continues for a period after increase in mass has stopped), and they are more resistant to adverse physical (heat, cold, radiation) and chemical agents.

**Figure 5.2**

Generalized growth curve of a bacterial culture.

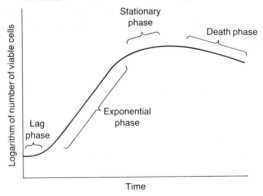

the death phase
Bacterial cells held in a nongrowing state eventually die. Death results from a number of factors; an important one is depletion of the cellular reserves of energy. Like growth, death is an exponential function, and hence in a logarithmic plot (Figure 5.2) the death phase is a linear decrease in number of *viable* cells with time. The death rate of bacteria is highly variable, being dependent on the environment as well as on the particular organism.

the lag phase
Cells transferred from a culture in the stationary phase to a fresh medium of the same composition undergo a change of chemical composition before they are capable of initiating growth. This period of adjustment, called the *lag phase* (Figure 5.2), is extremely variable in duration; in general, its length is directly related to the duration of the preceding stationary phase.

# the measurement of growth

To follow the course of growth, it is necessary to make quantitative measurements. Because exponential growth is usually balanced growth, any property of the biomass can be measured to determine growth rate. As a matter of convenience, the properties measured are usually cell mass or cell number.

measurement of cell mass
The only *direct* way to measure cell mass is to determine the dry weight of cell material in a fixed volume of culture by removing the cells from the medium, drying them, and then weighing them. Such determinations are time-consuming and relatively insensitive.
The method of choice for measuring the cell mass of unicellular microorganisms is an optical one: the determination of the amount of light scattered by a suspension of cells. This technique is based on the fact that small particles scatter light proportionally, within certain limits, to their concentration. When a beam of light is passed through a suspension of bacteria, the reduction in the amount of light transmitted as a consequence of scattering is thus a measure of the cell density. Such measurements are usually made in a spectrophotometer. This instrument reads in *absorbancy* ($A$) units; absorbancy is defined as the logarithm of the ratio of intensity of light striking the suspension ($I_0$) to that transmitted by the suspension ($I$):

$$A = \log \frac{I_0}{I}$$

[5.11]

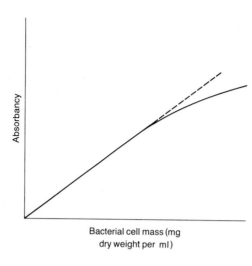

**Figure 5.3**
The relationship between absorbancy of a suspension of bacteria and bacterial cell mass. Note that proportionality is strict only at low values of absorbancy and deviates from strict proportionality (dashed line) at higher absorbancy values.

The instrument is convenient for estimating cell density, and when calibrated against bacterial suspensions of known density (Figure 5.3), it becomes an accurate and rapid way to estimate the dry weight of bacteria per unit volume of culture.

### measurement of cell number

The number of unicellular organisms in a suspension can be determined microscopically by counting the individual cells in an accurately determined very small volume. Such counting is usually done with special microscope slides known as *counting chambers*. These are ruled with squares of known area and are so constructed that a film of liquid of known depth can be introduced between the slide and the cover slip. Consequently, the volume of liquid overlying each square is accurately known. Such a direct count is known as the *total cell count*. It includes both viable and nonviable cells, since, at least in the case of bacteria, these cannot be distinguished by microscopic examination.

The enumeration of viable microorganisms can also be made by *plate count*, because single viable cells separated from one another in space by dispersion on or in an agar medium give rise through growth to separate, macroscopically visible colonies. Hence, by preparing appropriate dilutions of a bacterial population and using them to seed an appropriate medium, one can ascertain the number of viable cells in the initial population by counting the number of colonies that develop after incubation of the plates, and multiplying this figure by the dilution factor. This method of enumeration is often termed a *viable count*: in contrast to direct microscopic enumeration, it measures only those cells that are capable of growth on the plating medium used. The viable count is by far the most sensitive method of estimating bacterial number, since even a single viable cell in a suspension can be detected.

# the efficiency of growth: growth yields

The net amount of growth of a bacterial culture is the difference between the cell mass (number of cells) used as an inoculum and the cell mass (number of cells) present in the culture when it enters the stationary phase. When growth is limited by a particular nutrient, there is a fixed linear relationship between the concentration of that limiting nutrient initially present in the medium and the net growth that results, as shown in Figure 5.4. The mass of cells produced per unit of limiting nutrient is, accordingly, a constant, the *growth yield* ($Y$). The value of $Y$ can be calculated from single measurements of total growth by the equation:

$$Y = \frac{X - X_0}{C} \qquad [5.12]$$

where $X$ is the dry weight/ml of cells present when culture enters the stationary phase, $X_0$ is the dry weight/ml of cells immediately after inoculation, and $C$ is the concentration of limiting nutrient.

The growth yield can be measured for any required nutrient and once determined, can then be used to calculate the concentration of that nutrient in an unknown mixture simply by measuring how much growth a sample of the unknown mixture supports when added to a medium complete in all respects except for the limiting nutrient. Such a determination is called a *bioassay* (see Chapter 18).

In the case of a chemoheterotrophic bacterium, the growth yield measured in terms of organic substrate utilized becomes an index of the efficiency of conversion of substrate into bacterial mass. The data

**Figure 5.4**

The relationship between total growth of an aerobic bacterium (*Pseudomonas sp.*) and the initial concentration of the limiting nutrient (fructose). The experiments were done in a synthetic medium with fructose as the sole source of carbon and energy. The slope of the line is the growth yield ($Y$) of the bacterium on fructose (see text).

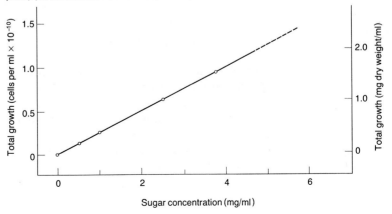

shown in Figure 5.4 were obtained with an obligately aerobic chemohetero-trophic pseudomonad growing in a synthetic medium with fructose as the sole source of carbon and energy. Inspection of the graph reveals a growth yield of approximately 0.4. Considering that the carbon content of fructose and cell material is 40 and 50 percent, respectively, the fraction of fructose carbon converted to cell carbon can be calculated to be about 0.5. Accordingly, this microorganism uses about one-half the carbon of fructose to make cells and oxidizes the other one-half to $CO_2$. Analogous experiments with other aerobic chemoheterotrophs utilizing sugars as the sole source of carbon reveal that the efficiency of conversion of carbon from the sugars to cellular carbon varies between about 20 and 50 percent.

# effect of nutrient concentration on growth rate

In many respects, the bacterial growth process can be likened to a chemical reaction in which the components of the medium (the reactants) produce more cells (the product of the reaction), a process catalyzed by the bacterial population. The velocity of chemical reactions is determined by the concentration of reactants, but as we have seen, bacterial growth rate remains constant until the medium is almost exhausted of the limiting nutrient. This seeming paradox is explained by the action of permeases, which are capable of maintaining saturating intracellular concentrations of nutrients over a wide range of external concentrations (Chapter 6). Nevertheless, at extremely low concentrations of external nutrients, the permease systems are no longer able to maintain saturating intracellular concentrations; the growth rate falls and becomes roughly proportional to concentration of the limiting nutrient (Figure 5.5).

# continuous culture of microorganisms

Cultures of the type so far discussed are called *batch cultures;* nutrients are not renewed, and hence growth remains exponential for only a few generations. Microbial populations can be maintained in a state of exponential growth over a long period of time by using a system of continuous culture (Figure 5.6). The growth chamber is connected to a reservoir of sterile medium. Once growth has been initiated, fresh medium is continuously supplied from the reservoir. The volume of liquid in the growth chamber is maintained constant by allowing the excess volume to be removed continuously through a siphon overflow.

If the fresh medium enters at a constant rate, the density of bacteria in the growth chamber remains constant after an initial

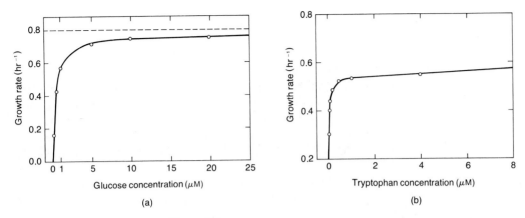

**Figure 5.5**
The effect of nutrient concentration on the specific growth rate of *E. coli*. (a) Effect of glucose concentration, and (b) effect of tryptophan concentration (for a tryptophan requiring mutant). From T. E. Shehata and A. G. Marr, "Effect of Nutrient Concentration on the Growth of *Escherichia coli*," *J. Bacteriol.* **107,** 210 (1971).

**Figure 5.6**
Simplified diagram of a continuous culture system.

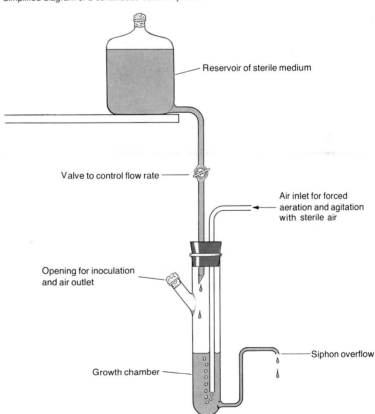

period of adjustment. In other words, the bacteria in the growth chamber grows just fast enough to replace those lost through the siphon overflow. If the rate of entry of fresh medium is changed, another adjustment period occurs, followed by maintenance of a constant population at a new density; the growth rate changes to match the new rate of loss of cells through the overflow. A continuous culture system responds in this manner to a wide variation in the rate of addition of fresh medium. Regardless of the rate of inflow of medium, bacteria cannot grow faster than they would in batch culture.

The question posed by the observation that culture densities in continuous culture systems remain constant is the following: How does the rate of addition of fresh medium to the culture vessel determine the growth rate of the culture? The explanation lies in the fact that the rate of growth of bacteria in continuous culture devices is always limited by the concentration of one nutrient. Consequently, the rate of addition of fresh medium determines the rate of growth of the culture: the system is self-regulating. Consider a continuous culture device that is operating at a constant rate of addition of fresh medium. After inoculation, the culture will at first grow at maximum rate. As the culture density increases, the rate of utilization of nutrients will increase until the depletion of one nutrient begins to limit the growth rate (Figure 5.5). As long as the growth rate exceeds the rate of loss through the siphon overflow, density will continue to increase, the steady-state concentration of limiting nutrient in the growth vessel will continue to decrease, and as a consequence the growth rate will decrease until the rate of increase of cells through growth will just equal the rate of loss of cells through the overflow. Were the growth rate transiently to become lower than the rate of loss of cells, cell density would decrease, limiting nutrient concentration would increase, and the growth rate would increase until the balance between the growth rate and the loss of cells is again reached.

## regulation of growth and metabolism

The overall metabolic activity of a growing microorganism reflects the simultaneous operation of a large number of interconnected pathways, both ATP-yielding and biosynthetic. Each specific pathway comprises a number of individual reactions catalyzed by specific enzymes. The product of these many reactions is a new cell. Successful competition in nature demands that the growth process be both rapid and efficient. The cell must therefore possess appropriate means of balancing the rates of the constituent reactions in each metabolic pathway as well as the overall rates of flow through the various pathways. The cell is also able to make qualitative adjustments in its metabolic machinery in response to changing needs imposed by environmental changes. Microorganisms have evolved a variety of *regulatory mechanisms* that accomplish this.

The fact that microbial metabolism is highly regulated can be inferred from several observations. One concerns the changes in macromolecular composition of bacteria that occur in response to the kinds of nutrients available to the cells. The first systematic study of this aspect of regulation was made almost 20 years ago by O. Maaløe and his associates with enteric bacteria, principally *Salmonella typhimurium*. Enteric bacteria can synthesize all cellular components from a single carbon source and inorganic salts, and they can utilize a number of organic compounds as carbon sources. Certain carbon sources such as acetate are metabolized slowly and support growth at a low rate; others, such as glucose, are metabolized more rapidly and support more rapid growth. If precursors of macromolecules (e.g., amino acids) are added to the medium, growth is even more rapid. Thus, by changing the composition of the medium while maintaining temperature constant, one can obtain growth of *S. typhimurium* with doubling times that vary between 20 minutes and several hours (at 37°C). Furthermore, the size and composition of the cells vary systematically with growth rate (Table 5.1). Cells growing at high rates are richer in RNA, poorer in DNA, and larger than cells growing at lower rates. Changes in cellular composition depend on growth rate alone, provided that the temperature is not changed.

The variation of the macromolecular composition of cells with growth rate can be interpreted as follows: a cell growing at a high rate must synthesize protein much more rapidly than does a slowly growing cell. This higher rate of protein synthesis requires that the cell contain more ribosomes, since the rate of protein synthesis per ribosome is constant. The ability of a bacterium to modulate its content of ribosomes is of great importance for the maintenance of high growth rates under changing environmental conditions. An insufficient complement of ribosomes would clearly restrict growth rate; an excess of ribosomes would also do so because the cell would be engaged in nonproductive synthesis of ribosomes.

The biosynthesis and degradation of small molecules by bacteria are similarly subject to close regulation, as shown by the following observations on enteric bacteria:

**Table 5.1**
Size and Composition of *Salmonella typhimurium* Growing Exponentially at Various Rates

| DOUBLING TIME (MIN) | AVERAGE WEIGHT OF A CELL (pg)[b] | CONTENT OF | | 70s RIBOSOMES (NUMBER/CELL) |
|---|---|---|---|---|
| | | DNA (%)[a] | RNA (%)[a] | |
| 25 | 0.77 | 3.0 | 31 | 69,800 |
| 50 | 0.32 | 3.5 | 22 | 16,300 |
| 100 | 0.21 | 3.7 | 18 | 7,100 |
| 300 | 0.16 | 4.0 | 12 | 2,000 |

[a] Calculated as a percentage of the dry weight of the cell.
[b] One *picogram* (pg) is equal to $10^{-12}$ g.

1. When growing in synthetic media containing a single organic compound as source of energy, bacteria synthesize all the monomeric precursors (e.g., amino acids) of macromolecules at rates that are precisely coordinated with the rates of macromolecule synthesis.
2. The endogenous synthesis of any one of these monomeric precursors is immediately arrested when the same compound is added to the medium, provided that the exogenously furnished monomer can enter the cell.
3. Formation of the enzymes that mediate biosynthesis of the monomers in question is also arrested when the compound is added to the medium.
4. Bacteria frequently synthesize enzymes responsible for the dissimilation of certain organic substrates only if the compounds in question are present in the medium.
5. When presented with two organic substrates, a bacterium often first synthesizes the enzymes required to dissimilate the compound that supports the more rapid growth; only after this compound has been completely utilized are the enzymes required to dissimilate the second compound synthesized.

## the biochemical basis of regulation

Two different regulatory mechanisms operate in the cell: the regulation of *enzyme synthesis*, and the regulation of *enzyme activity*. Both are mediated by compounds of low molecular weight, which are either formed in the cell as intermediary metabolites or enter it from the environment. Both regulatory mechanisms involve the operation of a special class of proteins, called *allosteric proteins*.* The fundamental importance of allosteric proteins as the key components of regulatory systems was first perceived by J. Monod in 1963.

Allosteric proteins are proteins whose *properties change* if certain specific small molecules, *effectors*, are bound to them. Hence, *allosteric proteins are mediators of metabolic change that is directed by changes in concentration of the small effector molecules.*

There are two classes of allosteric proteins: *allosteric enzymes*, whose activities are either enhanced or inhibited when combined with their effectors, and *regulatory allosteric proteins*, devoid of catalytic activity, which modulate the synthesis of specific enzymes.

Regulatory allosteric proteins attach to the bacterial chromosome near the specific structural genes that they control. This attachment can be modified by the binding of small effector molecules to the regulatory proteins, thereby changing the rate at which specific messenger RNAs are synthesized.

---

*The word allosteric means *differently shaped,* and it alludes to the fact that the effectors that regulate the activity of an allosteric enzyme have a structure different from that of the substrate of the enzyme.

## the regulation of enzymatic activity

The most thoroughly studied allosteric proteins are the allosteric enzymes, exemplified by aspartic transcarbamylase (ATCase). ATCase catalyzes the first reaction in the pathway of biosynthesis of pyrimidines (Figure 5.7). Its activity is inhibited by an end product of the pathway, cytidine triphosphate (CTP). Thus, elevated intracellular concentrations of CTP progressively inhibit the functioning of ATCase and consequently the formation of more CTP until its concentration decreases to an optimal level. The precise mechanism of such allosteric inhibition remains unexplained, but it evidently involves a conformational change in the enzyme. When the concentration in the cell of the end product (effector) of a given biosynthetic pathway rises, the catalytic activity of the allosteric enzyme with which it combines is reduced. Since the activity of this enzyme in turn controls the rate of biosynthesis of the end product (effector), the formation of the latter is also reduced and its intracellular concentration begins to fall. Therefore, allosteric inhibition also decreases. Through this device of *feedback regulation,* termed end-product inhibition, the intracellular concentrations of biosynthetic intermediates are very closely controlled. Typically, the enzyme that mediates the first reaction of a given biosynthetic pathway is the specific target of inhibition by the end product (or products) of that pathway. It is evident that when the first enzyme of a specific pathway is the target of regulation, neither the end product nor the intervening intermediates in its formation can accumulate in the cell. By this means, the primary rate of carbon flow into all biosynthetic pathways is controlled.

**Figure 5.7**

The allosteric control of the first step in pyrimidine biosynthesis (condensation of carbamyl phosphate and aspartic acid to form carbamyl aspartic acid). The enzyme responsible, aspartic transcarbamylase, is allosterically inhibited (bold arrow) by cytidine triphosphate, the eventual product of the biosynthetic sequence.

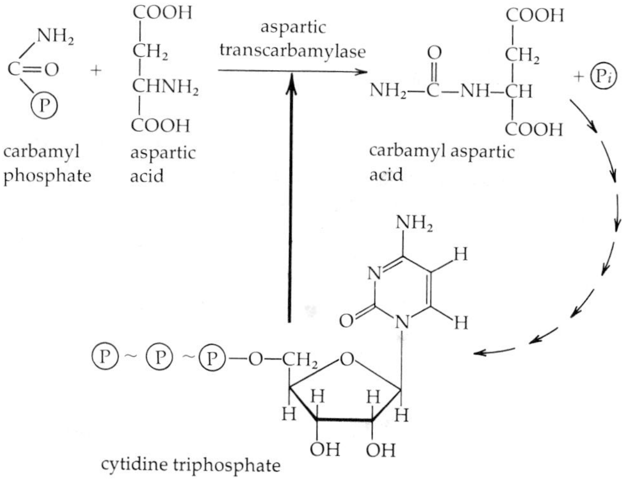

# regulation of enzyme synthesis

End-product inhibition mediated by allosteric enzymes is in large part sufficient to assure that all biosynthetic and catabolic pathways operate in balance with one another. However, when the product of a pathway is not required, the enzymes that catalyze the reactions of that pathway become superfluous. The regulation of microbial metabolism also includes mechanisms that modulate the enzymatic composition of the cell; this regulation is effected at the level of gene expression, and its genetic aspects are considered in Chapter 8.

## induction of enzyme synthesis

Many bacteria can use a wide range of different organic compounds as carbon and energy sources, but at any given time, only one of these compounds may be present in the environment. Although the genetic information necessary to synthesize the relevant enzymes is always present, its phenotypic expression is environmentally determined, a given enzyme being synthesized in response to the presence of its substrate.

Induction of enzyme synthesis is mediated by noncatalytic allosteric proteins. These substances are the products of specific regulatory genes; they frequently control enzyme synthesis in a negative manner by binding to the bacterial chromosome at a site near the structural genes that determine enzyme synthesis, thus preventing their transcription. They are termed *repressors*. When a repressor is bound to its specific allosteric effector, called an *inducer*, its ability to block transcription is lost, and specific enzyme synthesis is initiated.

This kind of regulation of enzyme synthesis was first elucidated by J. Monod and his collaborators through study of the induction of the enzymes that govern the dissimilation of lactose by *Escherichia coli*. Although *E. coli* contains the enzymes necessary for the metabolism of glucose under all growth conditions (such enzymes are termed *constitutive*), glucose-grown cells contain only barely detectable levels of the enzymes necessary for the initial steps in the metabolism of lactose. These enzymes are, however, present at high levels in lactose-grown cells. Since lactose induces their formation, they are termed *inducible enzymes*.

Alternatively, the binding of the regulatory allosteric protein to the bacterial chromosome adjacent to the structural genes may be *required* for transcription. In such positive control systems the regulatory protein is termed an *activator*. In some systems (e.g., induction of enzymes that metabolize arabinose) both positive and negative control mechanisms are combined. In the absence of the inducer the regulatory protein is a *repressor*; when bound to the inducer it is an *activator*.

## catabolite repression

About 35 years ago, J. Monod discovered a phenomenon known as *diauxy*. He noted that a culture of *E. coli* in a medium containing pairs of compounds as carbon sources (e.g., glucose and lactose) underwent two distinct growth cycles, characterized by two exponential phases of growth separated by a distinct lag phase (Figure 5.8). Glucose is utilized during the first growth cycle and lactose is utilized during the second. The enzymes necessary for the metabolism of lactose are not synthesized (even though the inducer is always present) until the glucose in the medium has been exhausted. While glucose is being metabolized, the induction of enzymes that metabolize lactose is prevented. Growth curves of similar form are obtained in media containing glucose and various other carbon sources that are metabolized by inducible enzymes. Subsequent work has shown that all rapidly metabolizable energy sources repress the formation of enzymes necessary for the dissimilation of energy sources that are more slowly attacked, and the phenomenon is now known as *catabolite repression*. As a result of catabolite repression, the cell first utilizes the substrate that supports the most rapid rate of growth.

The extent of catabolite repression is roughly proportional to the intracellular concentration of ATP. A single allosteric protein, termed CAP (*catabolite activator protein*), is responsible for the regulation of all enzymes under this form of control. Its effector is a cyclic nucleotide, 3′,5′-cyclic adenosine monophosphate (cyclic AMP), which is synthesized from ATP. By mechanisms that are still incompletely understood, the intracellular concentration of cyclic AMP varies inversely with ATP supply. Hence, when the cell is growing on rapidly metabolizable substrates, intracellular concentrations of cyclic AMP are low; when the cell is growing on slowly metabolizable substrates, the concentrations are high. Cyclic AMP has been found in most bacteria so far examined. Since it is not an intermediate of any known metabolic pathway, its only physiological role in bacteria is pre-

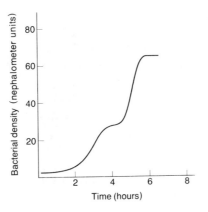

**Figure 5.8**

Diauxy of *E. coli* in a mineral salts medium initially containing equal quantities of glucose and lactose as sources of carbon. The transient cessation of growth after about 4 hours reflects the complete utilization of glucose. After J. Monod, *La Croissance des Cultures Bacteriennes*. Paris: Hermann, 1942.

sumably a regulatory one.* Binding of cyclic AMP to CAP causes CAP to undergo an allosteric transition that allows it to bind to the chromosome near the genes encoding catabolite-repressible enzymes and, through such binding, to stimulate transcription of these genes. When unbound to cyclic AMP, the CAP protein is inactive. Hence, both the CAP protein and relatively high internal concentrations of cyclic AMP are necessary for the synthesis of enzymes under control of catabolite repression.

### end-product repression

In many cases the addition of a compound that is the end product of a biosynthetic pathway (e.g., an amino acid) to the growth medium of a bacterium results in immediate slowing or arrest of the *synthesis* of the enzymes of that particular pathway. This is called *end-product repression*.

Thus, biosynthetic pathways are subject to two types of feedback regulation: end-product inhibition, which regulates activity of the first enzyme of the pathway, and end-product repression, which regulates enzyme synthesis. Although the effect of end-product repression is immediate in regulating the rate of enzyme synthesis, were it to act alone, the pathway would continue to function until preexisting enzymes were diluted to low levels as a consequence of further growth; end-product inhibition, however, brings about an immediate cessation of the operation of the pathway. Thus, end-product repression and end-product inhibition are complementary mechanisms, which in combination achieve highly efficient regulation of biosynthetic pathways. Most biosynthetic pathways are regulated by end-product repression, and usually all enzymes of each specific pathway are regulated by end-product repression.

End-product repression of the enzymes catalyzing the biosynthetic pathway for tryptophan in *E. coli* has been explored in considerable detail. A specific gene (*trpR*) governs the synthesis of an allosteric protein called the *tryptophan repressor*, whose only function is to regulate the biosynthesis of the enzymes of this pathway. In its free state the repressor is inactive, but when bound to tryptophan (the *corepressor*), it undergoes an allosteric change that enables it to bind to a region of the chromosome near the structural genes, thus preventing their transcription. Mutant strains that are unable to produce the repressor become insensitive to the presence of the end product; they produce high levels of the tryptophan biosynthetic enzymes under all conditions (i.e., they are *constitutive* for these enzymes).

### regulatory mechanisms: a summary

All known regulatory mechanisms are mediated through allosteric proteins, the activity of which is altered by the binding of small molecules. Hence, allosteric proteins serve as sensitive detectors of the intracellular concentration of key metabolites and modulate the overall met-

*Cyclic AMP also acts as a regulator in eucaryotes where it has been shown to play a role not solely in enzyme expression, but also in cellular differentiation.

abolic activity of the cell in such a manner as to maximize the rate of growth, by ensuring the conversion of nutrients into cell material with maximal efficiency. These mechanisms are summarized and compared in Table 5.2.

# effects of the environment on the growth of bacteria

### effects of solutes

Most bacteria do not need to regulate their internal osmolarity with precision because they are enclosed by a cell wall capable of withstanding considerable internal osmotic pressure.* Bacteria always maintain their osmolarity well above that of the medium. If the internal osmotic pressure of the cell falls below the external osmotic pressure, water leaves the cell and the volume of the cytoplasm decreases with accompanying damage to the membrane. In Gram-positive bacteria, this causes the cell membrane to pull away from the wall; the cell is said to be *plasmolyzed*. Gram-negative bacteria do not undergo plasmolysis, since the wall retracts with the membrane; this also damages the membrane.

Bacteria vary widely in their osmotic requirements. Some are able to grow in very dilute solutions, and some in solutions saturated with sodium chloride. Microorganisms that can grow in solutions of high osmolarity are called *osmophiles*. Most natural environments of high osmolarity contain high concentrations of salts, particularly sodium chloride. Microorganisms that grow in this type of environment are called *halophiles*. Bacteria can be divided into four broad categories in terms of their salt tolerance: *nonhalophiles, marine organisms, moderate halophiles,* and *extreme halophiles* (Table 5.3). Some halophiles, for example *Pediococcus halophilus*, can tolerate high concentrations of salt in the growth medium, but they can also grow in media without added NaCl. Other bacteria, including marine bacteria and certain moderate halophiles, as well as all extreme halophiles, require NaCl for growth. The tolerance of environments of high osmolarity and the specific requirement for NaCl are distinct phenomena, each of which has a specific biochemical basis.

* When a *solution* of any substance (*solute*) is separated from a *solute-free solvent* by a membrane that is freely permeable to solvent molecules, but not to molecules of the solute, the solvent tends to be drawn through the membrane into the solution, thus diluting it. Movement of the solvent across the membrane can be prevented by applying a certain hydrostatic pressure to the solution. This pressure is defined as *osmotic pressure*. A difference in osmotic pressure also exists between two solutions containing different concentrations of any solute.

The osmotic pressure exerted by any solution can be defined in terms of *osmolarity*. An osmolar solution is one that contains one *osmole* per liter of solutes, i.e., a 1.0 molal solution of an ideal nonelectrolyte. If the solute is an electrolyte, its osmolarity is dependent on the degree of its dissociation, since both ions and undissociated molecules contribute to osmolarity. Consequently, the osmolarity and the molarity of a solution of an electrolyte are grossly different. If both the molarity and the dissociation constant of a solution of an electrolyte are known, its osmolarity can be calculated with some degree of approximation, as the sum of the moles of undissociated solute and the mole equivalents of ions.

**Table 5.2**
**Key Regulatory Mechanisms Operative in Bacteria**

| MECHANISM | ALLOSTERIC PROTEIN | EFFECTOR | ACTIVITY OF ALLOSTERIC PROTEIN | | PHYSIOLOGICAL RESULT |
|---|---|---|---|---|---|
| | | | ALONE | BOUND TO EFFECTOR | |
| End-product inhibition (feedback control of pyrimidine synthesis)[a] | First enzyme of a pathway (aspartic transcarbamylase) | End product of pathway (CTP) | Catalyzes first step of pathway | Has decreased catalytic activity | Regulates biosynthesis of small molecules (CTP) |
| Enzyme induction: negative control (induction of β-galactosidase) | The repressor (product of lacI gene) | The inducer (lactose) | Binds to chromosome; prevents enzyme synthesis | Cannot bind to chromosome; enzyme synthesis occurs | Enzymes synthesized only when substrate is present in medium |
| Enzyme induction: positive control (induction of enzymes that metabolize arabinose) | Repressor-activator (product of araC gene) | The inducer (arabinose) | Repressor form; binds to chromosome and prevents enzyme synthesis | Activator form: binds to chromosome and permits enzyme synthesis | Enzymes synthesized only when substrate is present in medium |
| Catabolite repression (glucose repression of β-galactosidase synthesis) | CAP protein | Cyclic AMP | Cannot bind to chromosome; enzyme synthesis prevented | Binds to chromosome and stimulates enzyme synthesis | Allows cell to use most favorable source of carbon; regulates catabolic rate |
| End-product repression (regulation of enzymes necessary for synthesis of tryptophan) | Repressor (product of trpR gene) | End product of the pathway (tryptophan) | Cannot bind to chromosome; permits enzyme synthesis | Binds to chromosome and prevents enzyme synthesis | Regulates synthesis of biosynthetic enzymes (enzymes of tryptophan pathway) |

[a]Note that specific examples are shown in parentheses, for each of the key regulatory mechanisms.

**Table 5.3**
Osmotic Tolerance of Certain Bacteria

| PHYSIOLOGICAL CLASS | REPRESENTATIVE ORGANISMS | APPROXIMATE RANGE OF NaCl CONCENTRATION TOLERATED FOR GROWTH (g/100 ml) |
|---|---|---|
| Nonhalophiles | *Spirillum serpens* | 0.0–1 |
| | *Escherichia coli* | 0.0–4 |
| Marine forms | *Alteromonas haloplanktes* | 0.2–5 |
| | *Pseudomonas marina* | 0.1–5 |
| Moderate halophiles | *Micrococcus halodenitrificans* | 2.3–20.5 |
| | *Vibrio costicolus* | 2.3–20.5 |
| | *Pediococcus halophilus* | 0.0–20 |
| Extreme halophiles | *Halobacterium salinarium* | 12–36 (saturated) |
| | *Sarcina morrhuae* | 5–36 (saturated) |

*Note:* Ranges of tolerated salt concentrations are only approximate; they vary with the strain and with the presence of other ions in the medium.

### the requirement for Na+ in bacteria

In most nonhalophilic bacteria it has not been possible to demonstrate a specific $Na^+$ requirement. In view of the extreme experimental difficulty of preparing a medium that is rigorously free of this very abundant ion, the possibility that nonhalophiles might require a very low concentration of $Na^+$ cannot be excluded.

In contrast, bacteria of marine origin and halophiles normally require $Na^+$ for growth at concentrations so high that their absolute dependence on this cation can be demonstrated experimentally without difficulty, even if the basal medium employed has not been prepared from specially purified ($Na^+$-free) ingredients.

In the latter organisms, $Na^+$ probably plays a number of indispensable roles. It is an essential component of certain transport mechanisms and confers stability on certain cellular constituents, including enzymes and cell walls.

## effect of temperature on microbial growth

Figure 5.9 shows a plot of the rate of growth of *E. coli* as a function of temperature. The abrupt fall in growth rate at high temperatures is caused by the thermal denaturation of proteins and possibly of such cell structures as membranes. The *maximum temperature for growth* is the temperature at which these destructive reactions become overwhelming. This temperature is usually only a few degrees higher than the temperature at which growth rate is maximal (the *optimum temperature*).

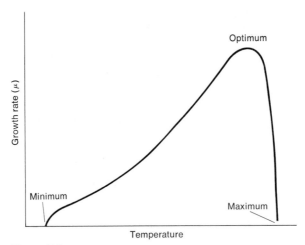

**Figure 5.9**
Generalized relationship of bacterial growth rate to temperature.

From the effect of temperature on the rate of a chemical reaction, one would predict that all bacteria would continue to grow (although at progressively lower rates) as the temperature is reduced, until the system freezes. However, most bacteria stop growing at a temperature (the *minimum temperature of growth*) well above the freezing point of water. Every microorganism has a precise minimum temperature of growth below which growth will not occur however long the period of incubation.

At low temperature, all proteins undergo slight conformational changes, attributable to the weakening of their hydrophobic bonds, which play an important role in determining three-dimensional structure. All other types of bonds in proteins become stronger as the temperature is lowered. The importance of precise conformation to the proper function of allosteric proteins and to the self-assembly of ribosomal proteins makes these two classes of proteins particularly sensitive to cold inactivation. Such inactivation is undoubtedly a major contributor to establishing the minimum temperature for growth.

The numerical values of the *cardinal temperatures* (minimum, optimum, and maximum) and the range of temperature over which growth is possible vary widely among bacteria. Some bacteria isolated from hot springs are capable of growth at temperatures as high as 95°C; others, isolated from cold environments, can grow at temperatures as low as −10°C (if high solute concentrations prevent the medium from freezing). On the basis of the temperature range of growth bacteria are frequently divided into three broad groups: *thermophiles,* which grow at elevated temperature (above 55°C); *mesophiles,* which grow well in the midrange of temperature (20 to 45°C); and *psychrophiles,* which grow well at 0°C.

As is often true of systems of biological classification, this terminology implies a clearer distinction among types than is found in nature. The tripartite classification of temperature response does not take fully into account the variation among bacteria with respect to the extent of the temperature range over which growth is possible.

The data describing the temperature ranges of growth of many different bacteria (Table 5.4) show the somewhat arbitrary nature of the designations *thermophile, mesophile,* and *psychrophile*. The range of temperature over which growth is possible is as variable as are the maxima and minima. The temperature range of some bacteria is less than 10 degrees, while for others it is as much as 50 degrees.

## effect of oxygen on microbial growth

The present atmosphere of the earth contains about 20 percent (v/v) of the highly reactive gas oxygen. With the exception of many bacteria and a few protozoa, all organisms are dependent on the availability of molecular oxygen as a nutrient. The responses to $O_2$ among bacteria are remarkably variable, and this is an important factor in their cultivation (see Chapter 2). The aerobes are dependent on $O_2$; the facultative

**Table 5.4**
Temperature Range of Growth of Certain Procaryotes

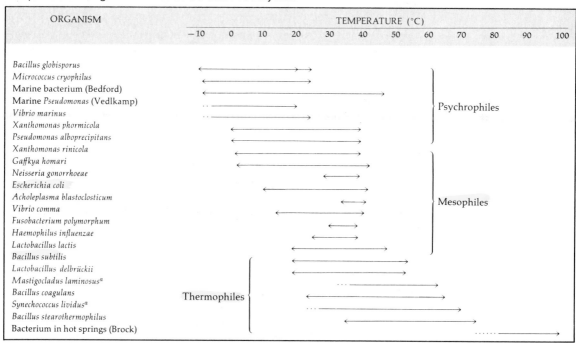

*Note:* Lines terminating in single arrows indicate established temperature limits of growth for at least one strain of the indicated species; variations exist among different strains of some species. Double-headed arrows indicate that the actual temperature limit lies between the arrow points. Solid lines terminating in dotted lines indicate that the minimum growth temperature is not established.
[a] Blue-green bacterium.

anaerobes use $O_2$ if it is available, but they also can grow in its absence; the anaerobes cannot utilize $O_2$. Anaerobes are of two types: the *obligate anaerobes*, for which $O_2$ is toxic, and the *aerotolerant anaerobes*, which are not killed by exposure to $O_2$. Although the toxicity of $O_2$ is most strikingly revealed by its effect in the obligate anaerobes, it is, in fact, toxic at high concentration even for aerobic organisms. Many obligate aerobes cannot grow in $O_2$ concentrations greater than atmospheric (i.e., >20 percent v/v). Indeed, some obligate aerobes require $O_2$ concentrations considerably lower than atmospheric (2 to 10 percent v/v) in order to grow. Aerobic bacteria that show such $O_2$ sensitivity are called *microaerophiles*.

### the toxicity of oxygen: chemical mechanisms

All bacteria contain certain enzymes capable of reacting with $O_2$; the number and variety of these enzymes determine the physiological relations of the organism to oxygen. The oxidations of flavoproteins by $O_2$ invariably result in the formation of a toxic compound, $H_2O_2$, as one major product. In addition, these oxidations (and possibly other enzyme-catalyzed oxidations or oxygenations) produce small quantities of an even more toxic free radical,* superoxide ($O_2^{\cdot-}$).

In aerobes and aerotolerant anaerobes, the potentially lethal accumulation of superoxide ($O_2^{\cdot-}$) is prevented by the enzyme *superoxide dismutase*, which catalyzes its conversion to oxygen and hydrogen peroxide:

$$2O_2^{\cdot-} + 2H^+ \xrightarrow[\text{dismutase}]{\text{superoxide}} O_2 + H_2O_2$$

Nearly all these organisms also contain the enzyme *catalase*, which decomposes hydrogen peroxide to oxygen and water:

$$2H_2O_2 \xrightarrow{\text{catalase}} 2H_2O + O_2$$

One bacterial group able to grow in the presence of air (lactic acid bacteria, see Chapter 12) does not contain catalase. However, most of these organisms do not accumulate significant quantities of $H_2O_2$, since they decompose it by means of *peroxidases*, enzymes that catalyze the oxidation of organic compounds by $H_2O_2$, which is reduced to water.

Superoxide dismutase and catalase or peroxidase therefore play roles in protecting the cell from the toxic consequences of oxygen metabolism. Organisms that can tolerate an exposure to $O_2$ always contain superoxide dismutase, although not all necessarily contain catalase. However, *all strict anaerobes so far examined lack both superoxide dismutase and catalase.* On present evidence, accordingly, *superoxide dismutase is an indispensable enzyme in any organism that comes in contact with air.*

---

* A free radical is a compound with an unpaired electron, indicated by a single dot in the structural formula. Having gained an extra electron, superoxide carries a negative charge.

# THE RELATIONS BETWEEN STRUCTURE AND FUNCTION IN PROCARYOTIC CELLS

# the envelope of procaryotic cells

Very little was known about the composition and functions of the outer layers of bacterial cells until 1952, when M. Salton developed methods for isolating and purifying cell walls (Figure 6.1). Analyses of such preparations revealed their chemical complexity and their diversity. Salton then showed that the hydrolytic enzyme lysozyme, previously known to lyse many Gram-positive bacteria, can completely destroy the isolated cell walls of such organisms as *Bacillus megaterium*.

These observations enabled C. Weibull, in 1953, to perform a simple experiment that clearly revealed the respective functions of the bacterial wall and membrane (Figure 6.2). If cells of *B. megaterium* are suspended in an *isotonic** sucrose solution prior to lysozyme treatment, the enzymatic dissolution of the walls converts the initially rod-shaped cells [Figure 6.2(a)] into spherical *protoplasts* [Figure 6.2(b)] that retain full respiratory activity, have the capacity to synthesize protein and nucleic acid, and are able (to a limited extent) to reproduce by division. When the protoplast suspension is diluted with water, the protoplasts undergo immediate osmotic lysis. The only structural elements that remain after such lysis are the empty cell membranes, termed *ghosts* [Figure 6.2(c)].

Weibull's experiment established a number of important facts about the cell wall: it has the vital mechanical function of protecting the cell from osmotic lysis in a hypotonic environment (the most common condition in the laboratory and in nature); it determines cell shape;

---

* An isotonic solution is one of osmolality equal to that of the cell.

126

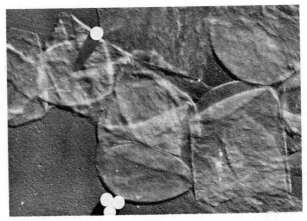

**Figure 6.1**
Isolated and purified cell walls of *Bacillus megaterium*. Electron micrograph.
The white spheres are particles of latex exactly 0.25 $\mu$m in diameter, in-
cluded to show the scale of magnification. From R. Y. Stanier, "Some Sin-
gular Features of Bacteria as Dynamic Systems" in *Cellular Metabolism and
Infection*, E. Racker (editor). New York: Academic Press, 1954.

and it does not (at least in Gram-positive bacteria) contribute vital metabolic
activity nor constitute part of the osmotic boundary.

In addition to the wall and membrane, the sur-
face structures of procaryotic cells may include a loose outer layer known as a
*capsule* or *slime layer* (see p. 137) and two classes of thread-shaped organelles,
the *flagella* and *pili* (see p. 138).

Figure 6.3 summarizes the location and dimen-
sions of the various surface structures associated with the procaryotic cell.

**Figure 6.2**
*Bacillus megaterium* (phase contrast, ×3,000). (a) The intact cells; (b) the spherical protoplasts, formed by
enzymatic dissolution of the cell wall with lysosyme in an isotonic medium; (c) the ghosts (i.e., the empty
cytoplasmic membranes), formed by osmotic rupture of protoplasts. Courtesy of C. Weibull.

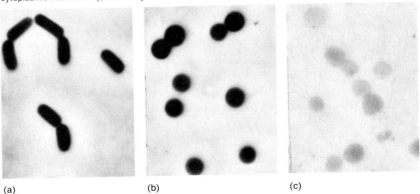

(a)                                    (b)                                    (c)

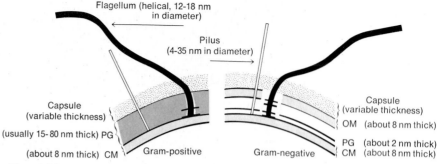

**Figure 6.3**

The location and dimensions of surface structures associated with the procaryotic cell. PG, peptidoglycan; CM, cell membrane; OM, outer membrane.

## the cell membrane

### structure of the membrane

The cell membrane, which bounds the protoplast and is its osmotic barrier, can be visualized in electron micrographs as a double line, about 8 nm wide; this is the fine structure of all so-called *unit membrane* systems. It is made up of a bilayer of phospholipids (see Figure 6.4), the chemical properties of which appear to establish the general molecular architecture of membranes. Phospholipids are linear molecules the two ends of which have markedly different affinities for water: the phosphate "head," being polar, has a high affinity (i.e., is *hydrophilic*), and the aliphatic fatty acid groups that constitute the "tail," being nonpolar, have low affinity (i.e., are *hydrophobic*). The arrangement of phospholipids in the membrane corresponds to the one that might be predicted from energy considerations alone. The hydrophilic "head" regions of the phospholipids are located at the outer surfaces of the bilayer (exposed to water); the hydrophobic "tails" extend into the center of the membrane that is devoid of water (Figure 6.5). The membrane proteins, which account for more than one-half of the dry weight of the membrane, are embedded in this phospholipid bilayer.

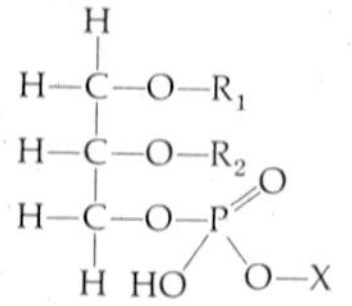

Isolated membranes exhibit remarkable properties of self-assembly. They can be disaggregated by treatment with detergents, and subsequently reassembled to form new membranelike structures. Furthermore, membrane fragments spontaneously reseal at their edge to produce closed vesicles with permeability properties similar to those of the cells from which they were derived.

Although the width of the cell membrane is fixed (being determined by the molecular configuration of the phospholipid bilayer), its area is not. In some bacteria, the membrane appears to have a simple contour, which closely follows that of the enclosing cell wall. In others, it is infolded, at one or more points, into the cytoplasmic region.

**Figure 6.4**

General structure of phospholipids. The 3-carbon polyalcohol, glycerol is esterified with two types of substituent: (1) Fatty acids ($R_1$ and $R_2$) which are usually long-branched or unbranched carbon chains: these constitute the hydrophobic tail. (2) A phosphate group to which any of several types of small molecules (X) may be attached: these constitute the hydrophilic head.

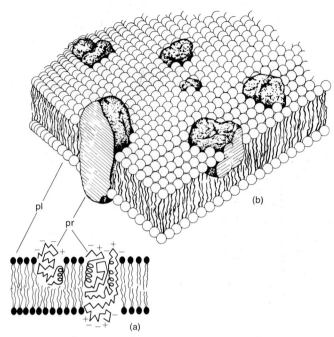

**Figure 6.5**

Schematic drawings of a unit membrane. (a) Phospholipid molecules, pl, are arranged with their "heads" (circles) at the two surfaces and their "tails" (two wavy lines) extended toward the interior. Folded polypeptide molecules (jagged lines, pr) are visualized as being embedded in the phospholipid bilayer with their polar regions (indicated by charges) extending beyond the bilayer on one or both surfaces (b) A three dimensional representation of the cross sectional view shown in (a). From S. J. Singer and A. L. Nicholson, "The Fluid Membrane Model of the Structure of Cell Membranes," *Science* **175,** 720 (1972).

Complex, localized infoldings known as *mesosomes* occur in many bacteria, often at or near the site of cell division (Figure 6.6); they may participate in the formation of the transverse septum. The continuity of the mesosome with the external surface of the membrane, not always evident in thin sections, is revealed in electron micrographs of whole cells, negatively stained with a heavy metal salt that penetrates through the wall but does not enter the cytoplasm (Figure 6.6).

### metabolic activities associated with the cell membrane

The bacterial cell membrane is an important center of metabolic activity; it contains many different kinds of proteins, each of which probably has a specific catalytic function. Most of these proteins are integrated into the hydrophobic region of the membrane, from which they can be separated only by methods (e.g., detergent treatment) that usually destroy their activity. Major classes of proteins known to be localized in the

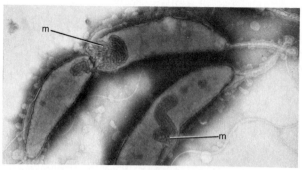

**Figure 6.6**
Mesosomes of *Caulobacter crescentus:* electron micrograph of whole cells negatively stained with phosphotungstate, which has penetrated the mesosomal (m) involutions of the cell membrane, clearly outlining their positions (×22,100). Courtesy of Germaine Cohen-Bazire.

membrane include (1) the permeases (see page 131) responsible for the transport of many organic and inorganic nutrients into the cell, and (2) the biosynthetic enzymes that mediate terminal steps in the synthesis of the membrane lipids and of the various classes of macromolecules that compose the bacterial cell wall. In addition, the bacterial cell membrane often contains the components of the electron transport chain (see Chapter 4). Lastly, circumstantial evidence indicates that the procaryotic cell membrane contains attachment sites for the chromosome and for plasmids, and that it plays an active role both in the replication and in the subsequent segregation of these genetic elements (see Chapter 8). In view of its numerous and varied functions, it is not surprising that the bacterial membrane contains from 10 to 20 percent of the total cell protein.

### transport of nutrients across the membrane

Since the cell membrane contains a lipid bilayer, chemical considerations alone would suggest that this structure should offer an absolute barrier to the passage of polar molecules. However, most nutrients are polar molecules; therefore, the membrane must contain certain chemically modified regions through which those essential polar molecules can flow; these regions are, therefore, operative in *membrane transport*.

PASSIVE DIFFUSION. Membrane transport occurs by a variety of mechanisms, the simplest of which is *passive diffusion*. Net flow of a compound by passive diffusion occurs only in response to a difference in its concentration across the membrane (a concentration gradient), and as a result of such flow, the difference diminishes. The rate of flow is a direct function of the magnitude of the gradient, and it does not approach a limiting value even when the concentration difference is great. Passive diffusion implies that certain compounds can pass freely through the membrane. Water is the principal nutrient that enters and leaves the cell by passive diffusion.

FACILITATED DIFFUSION. Passive diffusion exhibits a minimum of substrate specificity. For example, passive diffusion cannot discriminate between isomers. A related transport mechanism, known as *facilitated diffusion*, is mediated by specific membrane proteins, collectively known as *permeases*, that bind to the substrate molecule on the exterior of the membrane and, by mechanisms still largely unknown, allow its passage through the membrane. Recent evidence suggests that permeases span the membrane. They catalyze the general reaction:

substrate (outside the cell) $\longleftrightarrow$ substrate (inside the cell)

Facilitated diffusion is similar to simple diffusion in the sense that the substrate moves through a concentration gradient in the thermodynamically favorable direction, i.e., from a higher to a lower concentration. Hence, *this process does not require the expenditure of energy*. It differs from passive diffusion in that it is mediated by a specific protein catalyst that has many properties in common with enzymes: it exhibits considerable substrate specificity; the catalysts are often inducible; and the rate of the reaction approaches a limiting value with increasing concentrations of substrate (i.e., it obeys normal enzyme kinetics).

Although facilitated diffusion is a common mechanism of transport in eucaryotes, it appears to be relatively rare in procaryotes. For example, sugars, which characteristically enter eucaryotic cells by facilitated diffusion, enter procaryotic cells by another group of mechanisms, *active transport*, which are described below. One process of transport in a procaryote that is mediated by facilitated diffusion is the entry of glycerol into the cells of bacteria of the enteric group.

ACTIVE TRANSPORT. The mechanisms of transport known collectively as *active transport* permit a solute to enter the cell *against* a thermodynamically unfavorable gradient of concentration. At equilibrium, active transport systems create concentrations of particular solutes within the cell that may be several thousand times as great as those outside the cell. Since active transport concentrates solutes within the cell against a thermodynamically unfavorable gradient, *it requires an expenditure of metabolic energy*. The mechanism by which metabolic energy is coupled to active transport remains one of the most intriguing and important questions of cell biology.

GROUP TRANSLOCATION. The process of active transport catalyzes the movement of chemically unmodified nutrients across the cell membrane against a concentration gradient. Bacteria possess other transport systems that, during the process of transport, convert the nutrient to a chemically modified form to which the membrane is impermeable. Such systems of group translocation do not mediate active transport since the concentration of the unmodified nutrient inside the cell, is always very low. The overall process of group translocation does, however, resemble active transport since the concentration of the chemically modified nutrient inside the cell can greatly exceed the concentration of free nutrient in the medium.

One example of a group translocation system, widely distributed in bacteria, is the *phosphotransferase system*. It mediates the transport of many sugars and sugar derivatives, which are phosphorylated during the process, entering the cell as sugar phosphates. Since the membrane is highly impermeable to most phosphorylated compounds, the sugar phosphates, once formed, are trapped within the cell.

SUMMARY OF MEMBRANE TRANSPORT MECHANISMS. The various transport systems known to operate in bacteria are schematically shown in Figure 6.7. Simple diffusion is the net passage of a nutrient along the thermodynamically favorable concentration gradient through areas of the membrane which offer no barrier to its passage. Facilitated diffusion is mediated by specific carrier proteins or permeases. Active transport resembles facilitated diffusion in that a specific permease mediates passage of a nutrient through the membrane, but differs in that passage can occur against a concentration gradient. Group translocation involves the participation of specific permeases, but is accompanied by a chemical change of the nutrient undergoing transport.

**Figure 6.7**

Comparison of bacterial transport systems. Relative lengths of arrows indicate the direction of equilibria. S and s designate high and low concentrations, respectively, of solutes. C designates a carrier protein (permease). The diagram of group translocation is based on the properties of the phosphotransferase system, where R represents a phosphate carrier.

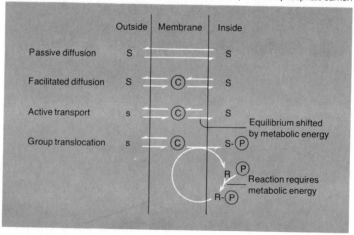

# the bacterial cell wall

## peptidoglycans

Lysozyme, the enzyme capable of dissolving the cell walls of *Bacillus megaterium* (p. 127) and many other bacteria, acts by hydrolyzing a class of macromolecules, termed *peptidoglycans*, which are found only in the walls of procaryotes. Only a few procaryotes are presently known to possess walls devoid of *peptidoglycan*; these include extreme halophiles and methane-producing bacteria.

Knowing the site of action of lysozyme to be peptidoglycan, Weibull's experiment (p. 126) can be further interpreted as establishing that this macromolecule contributes the structural strength characteristic of the procaryotic cell wall. Such strength derives from the fact that a single molecule of peptidoglycan surrounds the protoplast, forming an intact mesh-like sac.

There are two types of strands in the mesh: (i) the long *glycan strand** consisting of alternating residues of two acetylated amino sugars (N-acetylglucosamine and N-acetyl muramic acid), and (ii) the short tetrapeptide strand consisting of four amino acids joined by conventional peptide bonds. Although there is some variation among groups of bacteria, the most common sequence of amino acids in the tetrapeptide strand is shown in Figure 6.8.

The two types of strands are joined to form a mesh by two additional chemical linkages: (i) the amino group of the N-terminal L-alanine residue in the tetrapeptide strand is *always* linked to the carboxyl group of N-acetylmuramic acid; (ii) sometimes the carboxyl group of the C-terminal D-alanine in the tetrapeptide strand is linked by a peptide bond to the free amino groups of *meso*-diaminopimelic acid on an adjacent tetrapeptide strand. The chemical composition of the two types of polymer strands and their mode of linkage to form the intact peptidoglycan sac of *E. coli* are shown in Figure 6.8.

The primary peptidoglycan structure shown in Figure 6.8 is the most widespread one; it occurs in the walls of nearly all Gram-negative bacteria and in many Gram-positive ones. However, dozens of minor variations on this structure occur, particularly among Gram-positive bacteria. Other diamino acids may replace *meso*-diaminopimelic acid, and additional amino acids may be inserted in the linkage between adjacent tetrapeptide chains. These variations make the structure of the cell wall peptidoglycan an important taxonomic character among Gram-positive bacteria. Among Gram-negative bacteria, departures from the structure shown in Figure 6.8 are rare.

---

*Glycan strands typically contain from 20 to 130 monomeric units.

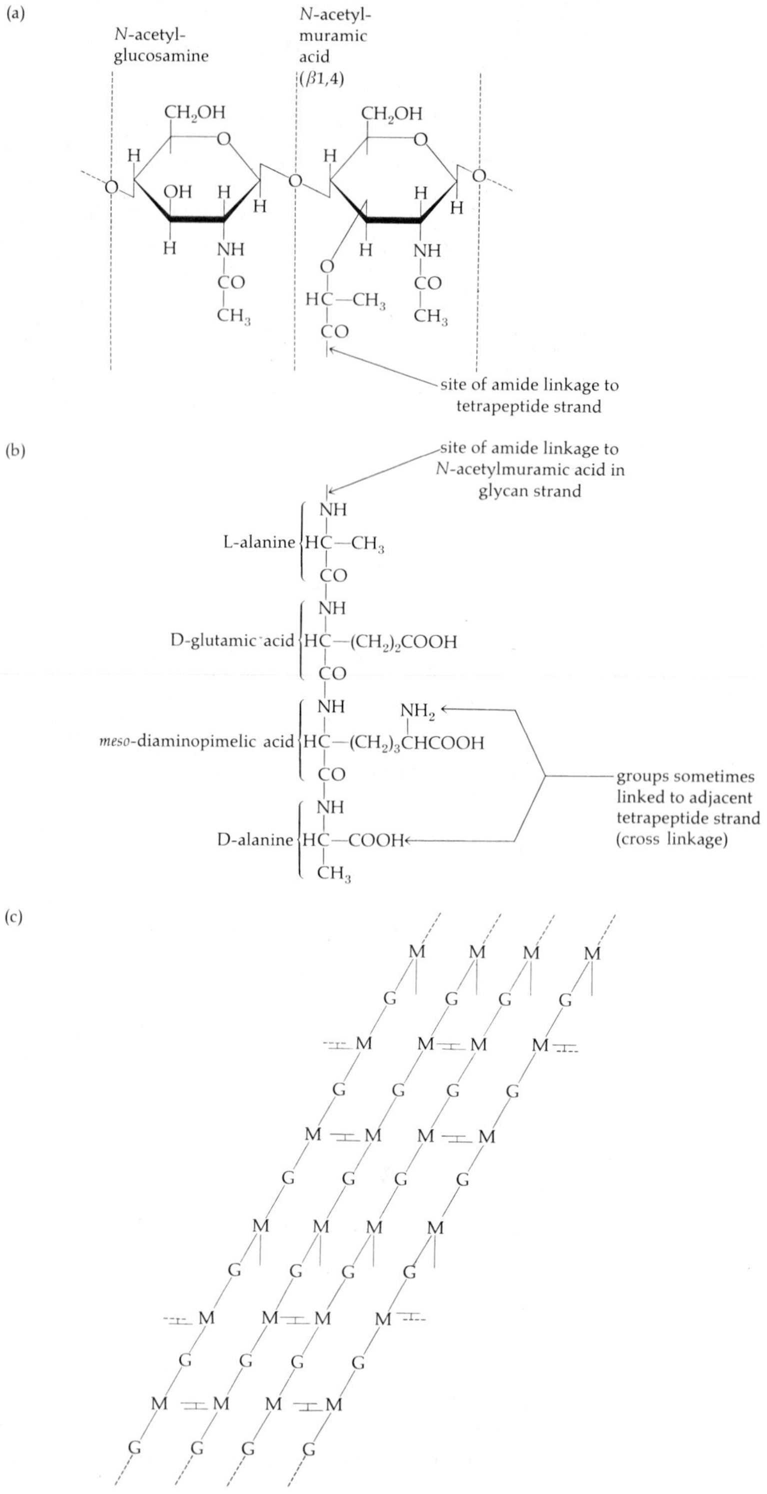

(a)

N-acetyl-glucosamine

N-acetyl-muramic acid
(β1,4)

CH₂OH

CH₂OH

OH H H

H H

H NH

H NH

CO

CO

CH₃

HC—CH₃

CH₃

CO

site of amide linkage to tetrapeptide strand

(b)

site of amide linkage to N-acetylmuramic acid in glycan strand

L-alanine  NH
HC—CH₃
CO

D-glutamic acid  NH
HC—(CH₂)₂COOH
CO

meso-diaminopimelic acid  NH    NH₂
HC—(CH₂)₃CHCOOH
CO

D-alanine  NH
HC—COOH
CH₃

groups sometimes linked to adjacent tetrapeptide strand (cross linkage)

(c)

M M M M
G G G G
M M M M
G G G G
M M M M
G G G G
M M M M
G G G G
M M M M
G G G G
M M M M
G G G G

**Figure 6.8**

General structure of peptidoglycan from *E. coli*. (a) A portion of a glycan strand showing the repeating disaccharide subunit and the site (arrow) of amide linkage to the tetrapeptide strand. (b) A tetrapeptide strand showing site of linkage to glycan strand and sites of cross-linkage to adjacent tetrapeptide strands. (c) Schematic representation of a portion of the intact peptidoglycan sac showing how the two types of strand are joined to form a surface. Diagonal lines represent glycan strand composed of repeating *N*-acetylglucosamine (G) and *N*-acetylmuramic acid (M) residues. The vertical lines represent tetrapeptide strands which are not cross linked. The symbol (—⊤—) represents cross-linked tetrapeptide strands. It should be noted that neither D amino acids nor *meso*-diaminopimelic acid occur in proteins.

### the walls of gram-positive bacteria

The cell walls of Gram-positive bacteria (Figure 6.3) are relatively thick (10 to 100 nm), often ultrastructurally homogeneous structures composed principally of peptidoglycan. They also contain polymers of alternating polyalcohol (either glycerol or ribitol) and phosphate groups known collectively as *teichoic acids*.

The exact location of the teichoic acids in the cell wall is not certain, but a variety of evidence suggests that some teichoic acid is covalently linked to lipids in the membrane, some to peptidoglycan, and some lies on the outer surface of the peptidoglycan layer.

The function of teichoic acids is unknown, but their almost universal distribution among Gram-positive bacteria suggests that they play an important role in the physiology of the group.

### the walls of gram-negative bacteria

The walls of Gram-negative bacteria (Figure 6.3) have comparatively low peptidoglycan content, seldom exceeding 5 to 10 percent of the dry weight of the wall. Calculations suggest that in many Gram-negative organisms, peptidoglycan occurs as a monomolecular (or at most a bimolecular) layer. Even so, as is the case with Gram-positive bacteria, the form of the peptidoglycan layer determines cell shape.

Superimposed on this thin peptidoglycan sac of Gram-negative bacteria is an outer wall layer, termed an *outer membrane* because it has the width and fine structure typical of the unit membrane. This outer membrane is composed of protein, phospholipid, and lipopolysaccharide. The phospholipids are the same as those that occur in the cell membrane, but the proteins are largely, if not entirely, different.

LIPOPOLYSACCHARIDES. The lipopolysaccharides, major components of the outer membrane, are extremely complex molecules with molecular weights over 10,000. They contain two distinct components: (1) a complex lipid (termed *lipid A*) to which is attached; (2) a polysaccharide chain (composed of a variety of unusual sugars), the detailed composition of which varies among bacterial strains. These lipopolysaccharides, although much larger in size than the phospholipids, have certain properties in common with them: one end of the molecule (lipid A) is hydrophobic, and the other (the polysaccharide) is hydrophilic. The hydrophobic moiety is embedded in the membrane, and the hydrophilic one extends into the aqueous environment.

Lipopolysaccharides are the major antigenic (see Chapter 16) determinants of the wall surface of Gram-negative bacteria; they also serve as receptors for the adsorption of many bacteriophages, and constitute the major component of endotoxins, agents that are highly toxic to man and other mammals.

STRUCTURE OF THE OUTER MEMBRANE. The fine structure and width of the outer membrane suggest that it consists, like the cell membrane, of a lipid bilayer composed of a mixture of phospholipids and lipopolysaccharides in which proteins are embedded.

The molecular arrangement of the envelope of Gram-negative bacteria is schematized in Figure 6.9.

THE PERIPLASMIC SPACE. The outer membrane of Gram-negative bacteria interposes a barrier to the passage of some substances, including a number of antibiotics, certain dyes, bile salts, and proteins.

The lipopolysaccharides seem to play a role in this barrier function, since mutants in which the polysaccharide moiety is largely eliminated are much more sensitive than the wild type to penicillin and certain other antibiotics. Similarly, treatment of Gram-negative bacteria with the chelating agent, EDTA, which causes the liberation of much of the lipopolysaccharide from the outer membrane, likewise makes the cells more sensitive to these antibiotics.

Thus the region of Gram-negative bacteria exterior to the cell membrane and bounded by the outer membrane has a special solute composition, differing from both the cytoplasm and the external envi-

**Figure 6.9**
Schematic representation of the envelope of Gram-negative bacteria showing the cell membrane and the two major portions of the wall, the peptidoglycan layer and the outer membrane.

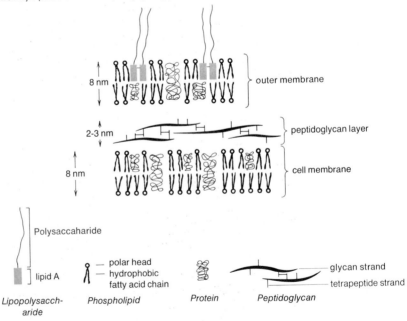

ronment. This region is termed the *periplasmic space*. Many enzymes that are *exoenzymes* (i.e., excreted into the extracellular environment) in Gram-positive bacteria have a periplasmic location in Gram-negative bacteria.

### excretion of enzymes

Eucaryotic cells devoid of walls are able to engulf particulate matter and liquid droplets from the environment by the processes of phagocytosis and pinocytosis. Bacteria as well as eucaryotic microorganisms completely enclosed by walls cannot do so, but they are able to use as nutrients a wide variety of macromolecules, including polysaccharides, proteins, and nucleic acids. Most of these macromolecules cannot cross the cell membrane and are degraded outside the cell by *exoenzymes* synthesized within the cell and excreted into the medium. Most exoenzymes catalyze the hydrolysis of macromolecular substrates to soluble end products, which enter the cell and serve as carbon and energy sources. When microbial colonies develop on an agar medium containing particles of an insoluble macromolecule that is digestible, each colony is surrounded by an expanding clear area in which the insoluble substrate has been hydrolyzed and solubilized by the action of exoenzymes.

PERIPLASMIC ENZYMES. The periplasmic space of Gram-negative bacteria contains enzymes called *periplasmic enzymes*. Since they are external to the cell membrane, they are clearly exoenzymes. However, they cannot traverse the outer layer of the cell wall and are trapped in the periplasmic space. This location is inferred from the fact that they are released by treatments that damage the outer membrane, leaving the cell membrane intact. Most of these enzymes are phosphatases that convert phosphorylated compounds, to which the cell membrane is impermeable, into phosphate-free derivatives.

In addition to periplasmic enzymes, Gram-negative bacteria produce some true extracellular enzymes; it is not known how these traverse the outer membrane while others are trapped in the periplasmic space.

## capsules and slime layers

Many procaryotes synthesize organic polymers (usually polysaccharides), which are deposited outside the cell wall as a loose, more or less amorphous layer called a *capsule*, or *slime layer*. The term *capsule* is usually restricted to a layer that remains attached to the cell wall, as an outer investment of limited extent, clearly revealed by negative staining. However, these *exopolymers* often form much more widely dispersed accumulations, in part detached from the cells that produce them. Such variations in the location and extent of the layer are caused primarily by the abundance with which the exopolymer is formed, and by its degree of water solubility. This layer is clearly not essential to cellular function; many bacteria do not produce it, and those that normally do so can lose the ability as a result of mutation, without any effect on growth. The capsule does, however, contribute to survival in nature by allowing certain saprophytes to attach to areas where the availability of nutrients is favorable and certain pathogens to avoid engulfment by phagocytes in the bloodstream of mammals (Chapter 16).

## flagella and pili

Although the flagella and pili differ both in function and gross form, these two classes of filiform bacterial appendages share many common structural features: both originate from the cell membrane and extend outward through the cell wall for a distance that may be as much as 10 times the diameter of the cell; both are made up of relatively small proteins (termed *flagellins* and *pilins*, respectively); and both have been shown to consist of helical chains of monomers wound to form a hollow core (Figure 6.10).

The gross structure of both flagella and pili is a reflection of the specific types of protein subunits from which they are built. Indeed, isolated flagella or pili can be completely disaggregated to a solution of monomers which, if the ionic environment is made suitable, will spontaneously reassemble to form structures with the same width and shape as the original flagella or pili.

Flagella are agents of locomotion in both Gram-positive and Gram-negative bacteria. Their function can be demonstrated by

**Figure 6.10**
Electron micrograph of a bacterial flagellum. Note the helical structure. From M. P. Starr and R. C. Williams, "Helical Fine Structure of Flagella of a Motile Diphtheroid," *J. Bacteriol.* **63,** 701 (1952).

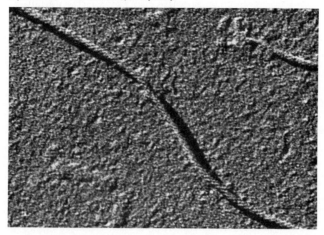

**Figure 6.11**
Electron micrograph of a negatively stained lysate of the purple bacterium, *Rhodospirillum molischianum*, showing the basal structure of an isolated flagellum (×181,000). Note the hook and basal body with attached paired discs. The other objects in the field are fragments of the photosynthetic membrane system. From G. Cohen-Bazire and J. London, "Basal Organelles of Bacterial Flagella," *J. Bacteriol.* **94,** 458 (1967).

a simple experiment. If cells are mechanically deflagellated by shearing, they become immotile. Following such treatment, regrowth of flagella is rapid, the normal number and length being restored in about one generation. As the flagella regrow, the cells at first show rotary motion; translational movement begins only after the flagella have reached a critical length. Although the flagella of different bacteria differ slightly in width (12 to 18 nm) and form, in some cases even being encased by an extension of the outer membrane to form a *sheathed flagellum* (see Figure 11.25, p. 273), they are all composed of the same three distinct regions: the helical flagellar filament (already described), the slightly wider *hook* located near the cell surface, and the *basal body* located entirely within the cell envelope (Figure 6.11).

Pili, which are restricted to Gram-negative bacteria, play no role in locomotion. Many types have been described, differing in width from 3 to as much as 30 nm, but all are ridged structures the role of which is to confer the property of adhesiveness; piliated bacteria tend to stick to one another and thereby form a coherent pellicle on the surface of unagitated liquid cultures. One special class of pili, the *sex pili*, confer on the cell the ability to act as genetic donors in conjugation (see Chapter 8).

## the tactic behavior of motile bacteria

The cells of a suspension of flagellated bacteria are normally in a state of continuous but random active movement. However, if certain chemical gradients are imposed on the population, the cells migrate to and accumulate in that part of the gradient that provides an optimal concentration of the chemical. Some of the substances to which they can respond (for the most part, nutrients) act as *attractants*, in the sense that the cells will

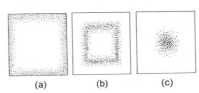

**Figure 6.12**
Aerotactic responses of motile bacteria (after Beijer-inck). Suspensions of various bacteria were placed on slides under cover slips. (a) Aerobic bacteria accumu-late near the edges of the cover slip, where oxygen concentration is greatest. (b) Microaerophilic bacteria accumulate at some distance from the edge. (c) Obli-gate anaerobes accumulate in the central, almost an-aerobic region.

(a)        (b)        (c)

accumulate in the more concentrated region of the gradient. Others (for the most part, toxic substances) act as *repellants,* in the sense that the cells avoid regions of high concentration, and accumulate in that part of the gradient in which the concentration is lowest. This behavior is known as *chemotaxis.* The specific substances that can elicit tactic responses differ for different bacteria; a particular chemical spectrum is characteristic for each species.

*Molecular oxygen* elicits *aerotactic* responses in most motile bacteria. Aerotactic patterns of accumulation can be readily ob-served in wet mounts in which an oxygen gradient is established by diffusion from the edges of the cover slip (see Figure 6.12).

Motile purple bacteria can respond to a *gradient of light intensity,* a phenomenon known as *phototaxis.* This behavior can be readily demonstrated by projecting a narrow spot of bright light onto an otherwise weakly illuminated suspension of motile purple bacteria, in which the cells are evenly dispersed and moving in a random fashion. Within 10 to 30 min-utes most of the population accumulates in the bright spot, which acts as a "light trap." The mechanism of this accumulation is shown in Figure 6.13. Swimming cells enter the light spot by random movement. Once within it, they are prevented from leaving again by an abrupt reversal of the direction of swimming, which occurs every time they penetrate the sharp gradient of light intensity that separates the brightly illuminated spot from the surrounding dim area.

If a wet mount of motile purple bacteria is illu-minated not with white light but with a spectrum produced by focusing light that has passed through a prism on the preparation, the bacteria rapidly accu-

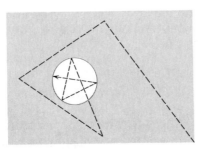

**Figure 6.13**
A diagrammatic illustration of the operation of a "light trap." The dashed line indicates the trajectory of a pur-ple bacterium in a darkened field that contains a single illuminated area (white circle).

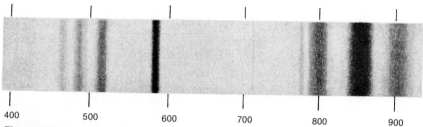

**Figure 6.14**
The pattern of phototactic accumulation of motile purple bacteria in a wet mount, which has been exposed to illumination in a spectrum. The cells accumulate massively at wavelengths corresponding to the absorption bands of their chlorophyll and carotenoids, which are both photosynthetically effective. The relatively weak accumulation around 500 nm corresponds to the positions of carotenoid absorption bands; the accumulations at 590, 800, 850, and 900 nm correspond to the positions of chlorophyll absorption bands. After J. Buder, "Zur Biologie des Bakteriopurpurins und der Purpurbakterien," *Jahrb. wiss. Botan.* **58,** 525 (1919).

mulate in a series of bands corresponding to the principal absorption bands of their photosynthetic pigment system (Figure 6.14). A careful quantitative study of the relative effectiveness of different wavelengths in mediating phototaxis has shown that the action spectrum for phototaxis by purple bacteria corresponds exactly to the action spectrum for the performance of photosynthesis.

## cell inclusions

### gas vacuoles

Since cells have a slightly higher density than water, they sink in an aqueous medium. Some bacteria can swim against the gravitational pull. Other aquatic procaryotes have developed a different device to counteract it: their cells contain gas-filled structures known as *gas vacuoles*. By light microscopy, gas vacuoles appear highly refractile, and have an irregular contour (Figure 6.15). Electron microscopy shows that gas vacuoles are compound organelles made up of a variable number of individual hollow cylinders with conical ends (Figure 6.16).

### the procaryotic cellular reserve materials

A variety of cellular reserve materials may occur in procaryotic organisms; they are frequently detectable as granular cytoplasmic inclusions.

NONNITROGENOUS ORGANIC RESERVE MATERIALS. Two chemically different kinds of nonnitrogenous organic reserve materials, each of which can provide an intracellular store of carbon or energy, are widespread among procaryotic organisms (Table 6.1). They are glucose-containing polysaccharides such as

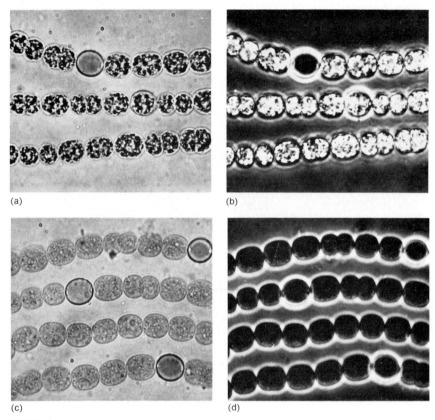

**Figure 6.15**
Filaments of a blue-green bacterium containing gas vacuoles, as visualized by bright field (a) and phase-contrast (b) illumination. Filaments from the same culture, after collapse by pressure of the gas vacuoles, are visualized by bright field (c) and phase contrast (d) illumination. The clear cells are heterocysts, which never contain gas vacuoles. From A. E. Walsby, "Structure and Function of Gas Vacuoles," *Bact. Revs.* **36**, 1 (1972).

starch and glycogen and a polyester of $\beta$-hydroxybutyrate, poly-$\beta$-hydroxybutyrate:

$$-\text{O}-\underset{\underset{\text{CH}_3}{|}}{\text{CH}}-\text{CH}_2-\underset{\underset{\text{O}}{\|}}{\text{C}}-$$

Although starch and glycogen also occur as reserve materials in many eucaryotic organisms, poly-$\beta$-hydroxybutyric acid is uniquely found in procaryotes. Fats, which commonly occur as reserve materials in eucaryotic organisms, never occur in procaryotes.

As a general rule, only one kind of reserve material is formed by a given species. Thus, many bacteria of the enteric group and

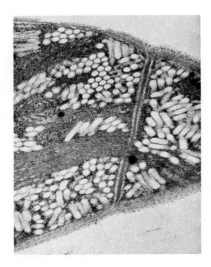

**Figure 6.16**
Electron micrograph of a thin section of *Oscillatoria*, showing the intracellular arrangement of the cylindrical gas vesicles which compose gas vacuoles ($\times$25,800). Courtesy of Germaine Cohen-Bazire.

anaerobic sporeformers (*Clostridium*) synthesize only glycogen or starch as a reserve material, whereas many *Pseudomonas, Azotobacter, Spirillum,* and *Bacillus* species synthesize only poly-$\beta$-hydroxybutyrate. Certain bacteria can, however, synthesize both types of reserve material; this is characteristic of the purple bacteria. Finally, it should be noted that a few bacteria (e.g., the fluorescent species of the genus *Pseudomonas*) do not synthesize any specific nonnitrogenous organic reserve material.

**Table 6.1**
**The Distribution of Nonnitritrogenous Organic Reserve Materials Among Procaryotes**

A. Glycogen
　　Blue-green bacteria (most representatives)
　　Enteric bacteria (most genera)
　　Sporeformers (many species)
B. Poly-$\beta$-hydroxybutyrate
　　Enteric bacteria (two genera)
　　*Pseudomonas* (many species)
　　Azotobacter group
　　*Rhizobium*
　　*Spirillum*
　　*Bacillus* (some species)
C. Both glycogen and Poly-$\beta$-hydroxybutyrate
　　Blue-green bacteria (a few species)
　　Purple bacteria
D. No detectable reserve material
　　Green bacteria
　　*Pseudomonas* (many species)
　　*Acinetobacter*

*Note:* This list is partial.

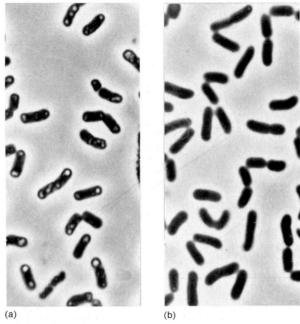

(a)                                                    (b)

**Figure 6.17**
The formation and utilization of poly-β-hydroxybutyric acid in *Bacillus meg-aterium* (phase contrast, ×2,200). (a) Cells grown with a high concentration of glucose and acetate. All cells contain one or more granules of poly-β-hydroxybutyric acid (light areas). (b) Cells from the same culture after incubation for 24 hours with a nitrogen source, in the absence of an external carbon source. Almost all the polymer granules have disappeared. Courtesy of J. F. Wilkinson.

**Figure 6.18**
Volutin("metachromatic granules") in the cells of a *Spirillum*, demonstrated by staining with methylene blue (×850). From George Giesberger, *Beitrage zur Kenntnis der Gattung Spirillum Ehbg* (1936), p. 46.

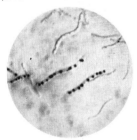

The cellular content of these reserve materials is relatively low in actively growing cells: these reserves accumulate massively when cells are limited in nitrogen but still have a carbon and energy source available. Under such circumstances, nucleic acid and protein synthesis are impeded and much of the assimilated carbon is converted to reserve materials, which may accumulate until they represent as much as 50 percent of the cellular dry weight. If such cells are then deprived of an external carbon source and furnished with an appropriate nitrogen source (e.g., $NH_4Cl$), the reserve materials can be used for the synthesis of nucleic acid and protein (Figure 6.17). Essentially, the synthesis of polysaccharide or poly-β-hydroxy-butyrate represents a device for accumulating a carbon store in a form that is *osmotically inert*.

VOLUTIN GRANULES. Many microorganisms, both procaryotic and eucaryotic, may accumulate *volutin granules*, which are stainable with basic dyes such as methylene blue (Figure 6.18). These bodies are also sometimes termed *meta-*

*chromatic granules*, because they exhibit a *metachromatic effect*, appearing red when stained with a blue dye. In electron micrographs of bacteria they appear as extremely electron-dense bodies. Volutin granules are long linear polymers of orthophosphate.

The conditions for volutin accumulation in bacteria have been studied in some detail. In general, starvation of the cells for almost any nutrient leads to volutin formation. Sulfate starvation is particularly effective and leads to a rapid and massive accumulation of polyphosphate. When cells that have built up a polyphosphate store are again furnished with sulfate, the polyphosphate rapidly disappears, and the phosphate is incorporated into nucleic acids. The volutin granules therefore appear to function primarily as an intracellular phosphate reserve, formed under a variety of conditions when nucleic acid synthesis is impeded. The formation of polyphosphate occurs by the sequential addition of phosphate residues to pyrophosphate, ATP serving as the donor:

$$\text{P—P} + \text{ATP} \longrightarrow \text{P—P—P} + \text{ADP}$$

$$(\text{—P—})_n + \text{ATP} \longrightarrow (\text{—P—})_{n+1} + \text{ADP}$$

The degradation of polyphosphate, if it occurs by the reversal of this reaction, might also provide a source of ATP for the cell, although this function has not so far been firmly established.

SULFUR INCLUSIONS. Inclusions of inorganic sulfur may occur in two physiological groups: the purple sulfur bacteria, which use $H_2S$ as a photosynthetic electron donor, and the filamentous, nonphotosynthetic organisms, such as *Beggiatoa* and *Thiothrix*, which use $H_2S$ as an oxidizable energy source. In both these groups, the accumulation of sulfur is transitory and takes place when the medium contains sulfide; after the sulfide in the medium has been completely utilized, the stored sulfur is further oxidized to sulfate.

# THE VIRUSES

Viruses are a class of acellular organisms that possess a molecular organization and growth pattern distinctly different from that of cellular organisms. They are *obligate intracellular parasites*. Although viruses most probably should be regarded as living, they lack cellular organization, a basic property of living things. Viruses are dependent on their host cells for growth and reproduction.

## the general properties of viruses

A virus alternates in its life cycle between two phases: one extracellular and the other intracellular. In its *extracellular phase,* a virus exists as a metabolically inert infectious particle, or *virion*. The virion consists of *either* DNA or RNA (but not both) contained within a protein coat or *capsid*. The virion may include a few enzymes, but they are far from sufficient to reproduce another virion. No ATP-generating or protein synthetic machinery is present.

In its *intracellular phase,* the virus exists in the form of a replicating molecule of nucleic acid. The host ATP-generating and synthetic abilities are utilized by the virus to achieve the replication and expression of its nucleic acid. The component parts of the virion then assemble and the virions are released, usually by lysis of the host cell.

Viruses thus differ from cellular organisms in two fundamental respects, one structural and the other functional:

1. Cellular organisms contain a genetic message encoded in DNA and the machinery necessary to translate the message into protein in an orderly and regulated fashion.

A virus (in the virion form) contains a genetic message encoded in either DNA or RNA, but few or none of the components necessary to translate that message into protein.

2. Growth of cellular organisms consists of an orderly increase of all cellular components, during which time the integrity of the cell is continuously maintained. New cells are produced directly from preexisting ones by division, which accomplishes the partitioning of all cellular constituents into the daughter cells. Viruses do not undergo division. Instead, their multiplication is accomplished by the separate synthesis of viral constituents, followed by their assembly into virions. Thus in the life cycle of a virus there is a stage during which the virus, as an identifiable entity, ceases to exist.

## virion structure

The genome of a virion consists of either single-stranded or double-stranded DNA or RNA. The most common are double-stranded DNA or single-stranded RNA, but some viruses (principally bacterial viruses) contain single-stranded DNA, and some plant and animal viruses contain double-stranded RNA. The viral genome is always small relative to that of most cellular organisms, ranging from as few as three genes to several hundred (*Escherichia coli* contains about 3,000 genes).

The viral capsid is composed of a single or several kinds of monomers termed *capsomers*. The arrangement of capsomers determines the *geometrical symmetry* of the capsid, normally either *polyhedral* or *filamentous*. In polyhedral viruses, the nucleic acid is contained within a hollow polyhedral capsid. In the filamentous viruses, the nucleic acid and protein are organized into a rod-shaped virion in which the capsomers have a helical arrangement (Figure 7.1). The virions of many bacterial viruses possess both

**Figure 7.1**
(a) A drawing of the structure of tobacco mosaic virus. For clarity, part of the ribonucleic acid chain is shown without its supporting framework of protein. From A. Klug and D. C. D. Caspar, "The Structure of Small Viruses," *Adv. Virus Res.* **1,** 225 (1960). (b) Electron micrograph of tobacco mosaic virus particles. From S. Brenner and R. W. Horne, "A Negative Staining Method for High Resolution Electron Microscopy of Viruses," *Biochim. Biophys. Acta* **34,** 103 (1959). Courtesy of R. W. Horne.

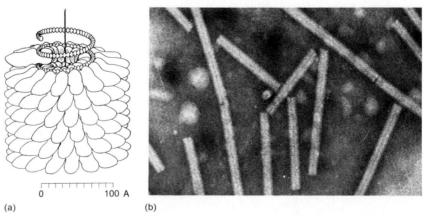

0          100 A

(a)                              (b)

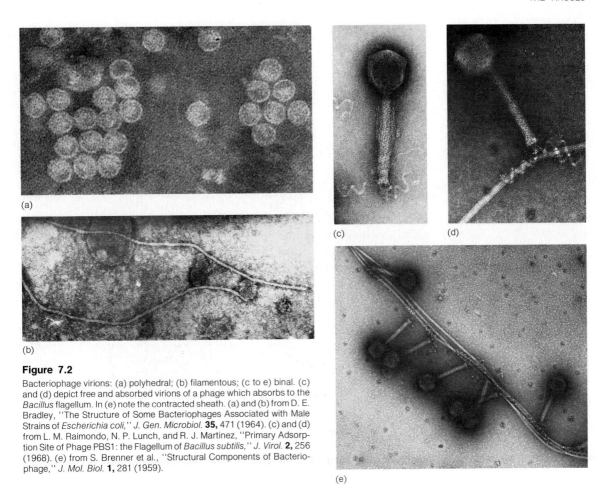

**Figure 7.2**

Bacteriophage virions: (a) polyhedral; (b) filamentous; (c to e) binal. (c) and (d) depict free and absorbed virions of a phage which absorbs to the *Bacillus* flagellum. In (e) note the contracted sheath. (a) and (b) from D. E. Bradley, "The Structure of Some Bacteriophages Associated with Male Strains of *Escherichia coli*," *J. Gen. Microbiol.* **35**, 471 (1964). (c) and (d) from L. M. Raimondo, N. P. Lunch, and R. J. Martinez, "Primary Adsorption Site of Phage PBS1: the Flagellum of *Bacillus subtilis*," *J. Virol.* **2**, 256 (1968). (e) from S. Brenner et al., "Structural Components of Bacteriophage," *J. Mol. Biol.* **1**, 281 (1959).

types of symmetry (*binal* symmetry): a polyhedral head containing the nucleic acid is joined to a filamentous tail. The different types of viral capsids are illustrated in Figure 7.2.

The capsids of many animal and a few bacterial viruses are further enclosed in a membrane or *envelope*. This envelope is derived from the cell membrane of the host by the process of extrusion of virions from the cell. The envelope may also contain virus-specific proteins, but the lipids of the envelope appear identical to those of the host cell membrane. Figure 7.3 depicts the process of virus extrusion from an animal cell, showing the origin of the envelope.

In addition to the envelope, animal viruses may have their surfaces coated with a variety of complex structures (e.g., Figure 7.4); thus while the basic structural plan of a virus is very simple, considerable molecular complexity is sometimes found.

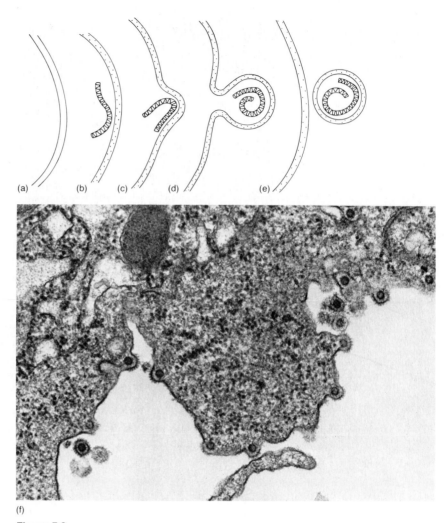

(f)

**Figure 7.3**
(a to e) Schematic representation of the liberation of an enveloped virus. The black dots represent new membrane proteins coded by viral genes. Modified from B. Davis, et. al., *Microbiology*. New York: Harper and Row, 1967. (f) Electron micrograph of the liberation of an enveloped virus (Monkey Foamy Virus) from its animal host cell. Note the "spikes" on the surface of the virions and in the cytoplasmic membrane; these are virus encoded glycoproteins which embed within the host cell membrane only at extrusion sites. Courtesy of JaRue Manning. ×72,630.

## the classification of viruses

The most widely used system for the classification of viruses, introduced by A. Lwoff and his colleagues in 1962, groups viruses according to the properties of their virions: type of nucleic acid, capsid architecture, presence or absence of an envelope, and capsid size. Further subdivisions are based on other features of the virion, such as the number of nucleic acid strands (one or two); on features of viral development, such as the

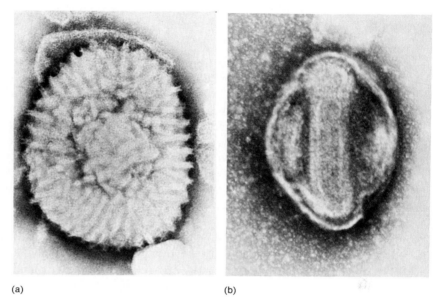

(a)                                                      (b)

**Figure 7.4**

(a) A whole vaccinia particle, negatively stained with phosphotungstic acid. The ridges on the surface may be long rodlets or tubules. (b) A negatively stained vaccinia particle that has been centrifuged in a sucrose gradient. The particle has been partially disrupted, and it has lost its outer membrane. The remaining structure includes a biconcave inner core, containing the nucleic acid, two elliptical bodies, and a surrounding membrane. From S. Dales, "The Uptake and Development of Vaccinia Virus in Strain L Cells Followed with Labeled Viral Deoxyribonucleic Acid," *J. Cell Biol.* **18,** 51 (1963).

site of viral synthesis in the cell; and on host-virus interactions, as exemplified by the range of susceptible hosts.

This system is not intended to represent a natural or phylogenetic classification; that is, it does not express the evolutionary relationships between viruses, relationships that are completely obscure. Instead, it groups viruses according to common sets of chemical and structural features, which are constant properties that can be determined with precision.

It will be noted that host range plays only a minor role in this classification. The different methodologies required for the study of viruses in widely different hosts, however, have led to the general practice of grouping them according to whether their normal hosts are bacteria, animals, or plants.

# the reproduction of viruses

Although all viruses share the same basic pattern of reproduction, there is considerable variation in the molecular details of the process. We shall illustrate the basic pattern using bacterial viruses as examples, then briefly discuss the principal variations exhibited by other groups.

### the detection and enumeration
### of bacterial viruses

When a small number of virulent bacterio-phage* particles is added to a growing culture of susceptible bacteria, in a liquid medium some of the bacterial cells become infected. After a period of time, the infected cells undergo *lysis*. The lysis of the infected cell liberates a large number of new phage particles. These particles can in turn infect other cells in the population, with a repetition of the same cycle. Consequently, even if the number of infectious phage particles originally introduced is small relative to the number of bacterial cells, practically the entire bacterial population may be destroyed in a few hours.

It is very easy to determine the number of phage particles or infected bacterial cells in a suspension by spreading an appropriate dilution of this suspension over the surface of an agar plate that is evenly inoculated with a thin suspension of susceptible bacteria. After appropriate incubation, the surface of the plate shows a confluent layer of bacterial growth, except at those points where a phage particle or an infected cell has been deposited. Around such sites of infection, clear zones of lysis, or *plaques*, are formed as a result of the localized destruction of the film of bacteria by successive cycles of phage growth (Figure 7.5).

### the isolation and purification
### of bacterial viruses

The plaque method may be used to isolate from nature phages that are capable of attacking a particular species or strain of bacterium. A sample of material from the natural habitat of the bacterium is shaken with water; the supernatant is then freed of cellular organisms by membrane filtration or by treatment with chloroform (to which most bacteriophage virions are resistant), and samples are mixed with suspensions of the bacterium and plated on agar. Any plaques that appear represent phage particles that were present in the natural material. The phage from a single plaque can be isolated by stabbing the plaque with a sterile inoculating needle and suspending the adhering material in a small volume of sterile diluent; such a suspension usually contains between $10^4$ and $10^6$ phage particles. Purification of the phage is then achieved by repeated serial isolations from single plaques.

**Figure 7.5**
Bacteriophage plaques. Courtesy of G. S. Stent.

### the one-step growth experiment

The essential features of the viral growth process and its nonhomology with that of cellular organisms are clearly demonstrated by the *one-step growth experiment* [Figure 7.6(a)]. A culture of susceptible bacteria is mixed with a suspension of viruses (at a ratio such that few cells are

*Bacterial viruses are often termed *bacteriophages* or *phages*.

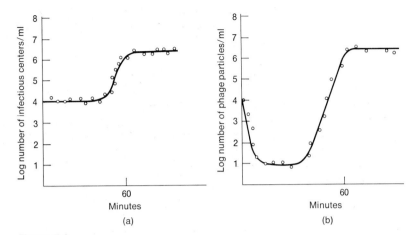

**Figure 7.6**
One-step growth of bacterial viruses. In (a) samples of the infected culture are removed and the number of infectious centers (free phage particles and infected cells) determined by the plaque method. In (b) each sample is treated with chloroform prior to plating, so that only free phage particles are determined.

infected by more than one virus particle). Following a brief period during which the viruses adsorb to the cells, the mixture is diluted making further adsorbtion improbable (since adsorbtion is a second-order reaction, its rate is proportional to the product of concentration of cells and viruses). At intervals, samples are removed from the culture, and the number of virus particles and infected bacteria (collectively termed *infectious centers*)* is determined by plaque counts. For a time (the *latent period*) the number of infectious centers remains constant. This number then suddenly increases (the *burst period*) as the infected cells lyse, each releasing numerous virus particles. When all infected cells have lysed, the number of infectious centers again remains relatively constant, further cycles of growth being largely prevented by the initial dilution of the culture. Assuming that adsorbtion was efficient, the *burst size*, or number of virus particles released from a single infected cell, can be determined by dividing the number of particles present after the burst by the number present before. In the example shown in Figure 7.6(a) the burst size is about 300.)

The one-step growth experiment clearly shows that viral growth follows kinetics not normally encountered in cellular systems, but the experiment as performed gives no information regarding the processes occurring during the latent period. Additional information can be gained by exposing each sample briefly to chloroform before plating. Such treatment allows a distinction to be made between infected cells and phage

---

* An *infectious center* is a particle capable of initiating the formation of a plaque. This can be either a phage particle or an infected cell whose lysis will result in the localized release of phage particles.

particles because bacterial cells (including phage-infected cells) are lysed, whereas virions are unaffected. Plaque counts of chloroform-treated samples reflect only the number of phage particles present in the sample. If a one-step growth experiment is done in which samples are exposed to chloroform prior to performing the plaque count [Figure 7.6(b)], a striking feature of growth emerges: *viruses, as stable infectious entities, disappear immediately after infection, and reappear in large numbers late in the latent period.* The period in which infectious particles are not recoverable from the cells is termed the *eclipse period.* (The small number detected during the eclipse period are virus particles that did not adsorb to cells prior to the initial dilution.)

The discovery of the eclipse period provided striking evidence that growth of cellular organisms and of viruses are fundamentally different processes. We now know that the eclipse is the period during which viral components are being synthesized prior to assembly into virions.

### reproduction of DNA phages

Through a combination of techniques, the broad outlines of viral reproduction have been clarified. Four stages may be discerned: (1) entry into the host cell, (2) synthesis of virion components, (3) assembly of the components to form virions, and (4) release of the mature virions from the cell.

Entry into the bacterial cell requires penetration of both the wall (including the outer membrane in Gram-negative bacteria) and the cell membrane. The details of this process are obscure; however, it appears that the tail of some phages with binal symmetry acts in a syringelike fashion, penetrating through the wall (Figure 7.7). The helical outer portion

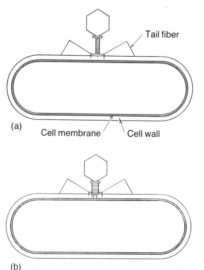

**Figure 7.7**

The syringelike action of the T-even phages. (a) Phage adsorbed to bacterial cell wall; the sheath is extended. (b) The sheath has contracted, driving the tail core through the cell wall.

(a) Cell membrane    Cell wall    Tail fiber

(b)

of the tail (*sheath*) contracts, driving the central *core* through the wall. Figure 7.2(e) shows such a phage with the tail sheath contracted, revealing the inner core. The mechanisms by which the nucleic acid of other phages gain entry to the host cytoplasm are unknown.

Once in the cytoplasm, the viral DNA is transcribed by host enzymes to produce messenger RNA. The mRNA directs the synthesis (on host ribosomes) of polymerases necessary for replication of the viral DNA, of the capsid proteins, and of other proteins necessary for viral growth. These activities soon lead to an accumulation of viral DNA and capsomeres, which then assemble into virions. Ultimately, the host cell is lysed by an enzyme (lysozyme) encoded by the virus, and the progeny virions are released (Figure 7.8).

### filamentous DNA phages

The single-stranded DNA phage with helical capsid symmetry (the *filamentous DNA phages*) provides an exception to the generalization that bacterial viruses are released by lysis of the host cell. In

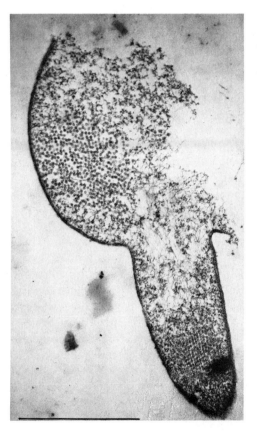

**Figure 7.8**
Bacteriophage particles (small circular objects) are released from their host cell by lysis. Note that the local concentration of phage particles may be sufficiently high for them to form crystalline aggregates. Scale marker = 1 μm (×50,000). From David E. Bradley, "Ultrastructure of Bacteriophages and Bacteriocins," *Bact. Rev.* **31,** 230–314 (1967).

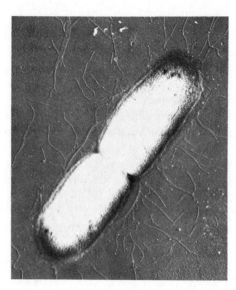

**Figure 7.9**
The liberation of filamentous phage by extrusion from living bacterial cells. The micrograph was made 30 minutes after the cells had been infected and washed free of unadsorbed phage particles. From P. H. Hofschneider and A. Preuss, "M13 Bacteriophage Liberation from Intact Bacteria as Revealed by Electron Microscopy," *J. Mol. Biol.* **7,** 450 (1963).

this case, a pool of capsomers is contained within the host cell membrane. Viral nucleic acid is extruded through the membrane, becoming coated with capsomers in the process (Figure 7.9). The host cell remains viable and continues to grow during the process of virion liberation, although it is not known if indefinite growth is possible. The infected cell is clearly compromised; although viable, its growth rate is lowered to such a degree that plaques may be formed.

### RNA phages

When the nucleic acid of the virion is RNA, it is recognized by the host ribosomes as a messenger RNA. It thus participates in translation immediately upon entry. One of the products of the translation process is a novel enzyme *RNA replicase,* which replicates the viral RNA. A double-stranded RNA molecule is an intermediate in this replication process (Figure 7.10).

### lysogeny

The pattern of viral growth described above is termed *lytic growth,* because it results in the lysis of the host cell, with the concomitant release of progeny virus particles.

Many bacteriophages are capable of an alternative interaction with their host: following penetration, the viral genome may reproduce in synchrony with that of the host, which survives and undergoes normal cell division to produce a clone of infected cells. In most of the progeny cells no viral structural proteins are formed, most viral genes have become *repressed*. In an occasional cell, however, derepression occurs spontaneously and the viral genome initiates a lytic cycle of development; the cell in which this occurs lyses and liberates mature virions.

This virus-host relationship is called *lysogeny*; infected cells that possess the latent capacity to produce mature phage particles are said to be *lysogenic*. Bacteriophages capable of entering this relationship are called *temperate*, and the viral genome present in the cells of a lysogenic culture is called *prophage*.

Thus, when a temperate phage particle infects a susceptible host, the entering phage genome has two alternative fates: it may commence rapid vegetative multiplication, culminating in the formation of mature virions and lysis of the host cell; or it may enter the prophage state, giving rise to a clone of lysogenic cells. When a suspension of phage particles is added to a culture of a susceptible host strain, some of the cells undergo

**Figure 7.10**

The replication of an RNA virus. (a) The infecting parental strand is labeled (+). (b) Viral RNA replicase converts the parental strand to a double-stranded intermediate. (c, d) A second replicase uses the double-stranded intermediate as a template to synthesize a sucession of (+) progeny strands. (e) The displaced parental and progeny strands are either incorporated into virions or used by the replicase to form new double-stranded intermediates.

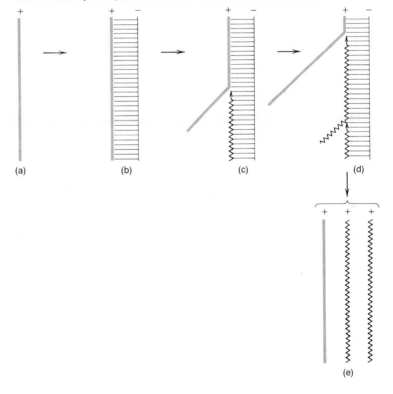

*productive infection* and lyse, while the remainder of the cells are lysogenized. The fraction that encounters each fate is determined by the genetic constitution of the host, by the genetic constitution of the virus, and by the environment. By experimenting with each of these variables, it has been possible to investigate the factors that determine the fate of the individual cell.

From such experiments, the following picture has emerged. In every newly infected cell, three processes take place more or less simultaneously: (1) the viral genome is transcribed and translated to form viral proteins, including one or more types of *repressor* molecules that inhibit both viral replication and viral gene expression; (2) the viral genome commences replication; and (3) one or more of the genomes enters the prophage state. The fate of the cell in which these three processes have begun then depends on the outcome of a "race" between repressor production and viral maturation: if mature virions and lysozyme are produced before repressor action can interfere, the cell will be lysed; if repressor molecules accumulate in time to shut off viral replication before mature virions and phage lysozyme have been produced, the cell will be lysogenized. In the lysogenized cell, phage repressor molecules continue to be produced by the prophage and prevent both vegetative replication and maturation of the phage.

Repressor molecules do not interfere with the replication of prophage in synchrony with the cell. *Prophage replication* and *vegetative replication* are thus different processes.

There are two known modes of prophage replication, which differ in the relationship between the prophage and the host cell chromosome. In one, that characteristic of the phage *lambda* (λ), the prophage is physically integrated into the chromosome at a specific site by recombination between the host and virus genomes. The prophage is thus replicated passively as part of the host chromosome. The establishment of lysogeny by λ is shown in Figure 7.11.

Other phages, such as P1, lysogenize bacteria by a mechanism that does not involve recombination with the host chromosome. In this case, the circular prophage replicates autonomously, but in synchrony with the bacterial chromosome.

Although the process of lysogenization does not result directly in the production of progeny virions, it is nevertheless an effective mechanism of viral multiplication, since the number of prophages (each with the potential to produce virions) increases along with the number of cells.

### growth of plant and animal viruses

The broad outlines of reproduction of plant and animal viruses are similar to those presented for bacterial viruses. Major differences occur at the stages of penetration and, in the case of animal viruses, at the release of progeny virions.

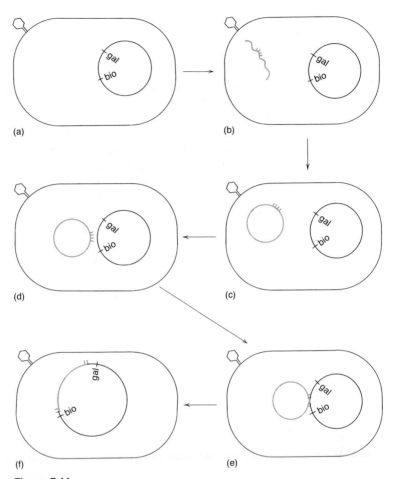

**Figure 7.11**
The formation of λ prophage. (a) Adsorption of the virus. (b) Injection of viral DNA. (c)
Circularization of the viral genome. (d) Pairing of homologous regions on the viral and bacterial
genomes. (e) A crossover event occurs within the region of pairing. (f) The two genomes have
been integrated, forming a single circle. Note that the attachment site for λ is at a specific
location, between the loci *gal* and *bio* (see Figure 8.14). The specific attachment site on λ is
indicated by four short vertical lines on the DNA strand.

Plant viruses, like bacterial viruses, must pene-
trate a host cell wall to initiate infection. Unlike bacterial viruses, they appear
not to have specific machinery for penetration. Instead, they depend on rupture of the host wall by some other agent. In the laboratory, this is most
frequently abrasion; in nature, insects whose bite penetrates the wall are
probably the principal agents responsible for transmission.

Animal viruses adsorb to the host cell membrane
and may then be taken into the cell by phagocytosis. As a result of phagocy-

tosis, the ingested virion is contained within a membrane-bounded food vacuole; it still must penetrate the cell membrane to reach the cytoplasm or nucleus, where it will be reproduced. In the case of a virus that possesses an envelope derived from the cell membrane of its previous host, penetration is facilitated by fusion of the viral envelope with the membrane of the new host cell (approximately the reverse of the process shown in Figure 7.3). In animal viruses that lack an envelope, the mechanism of penetration is unknown.

Thus in plant and animal viruses, both the nucleic acid and the capsid enter the host cytoplasm. Before transcription or replication can commence, the viral nucleic acid must be stripped of the capsid proteins, a process termed *uncoating*. In some cases, uncoating may begin in the phygocytic vesicle.

A common characteristic of the life cycle of animal viruses is their release by extrusion through the host cell membrane (Figure 7.3) in a process that leaves the host cell fully viable. Some animal viruses (those without an envelope) are released by lysis of the host cell. Bacterial and plant viruses are usually released by lysis of their host cells, a consequence, presumably, of the necessity to breach the cell wall to insure dissemination of the virions.

## the tumor viruses

### the discovery of viruses as etiological agents of tumors

The tissues of animals are formed by the *regulated, limited growth* of their component cells. As a rare event, a cell may escape the normal regulatory processes and divide without restraint, forming an abnormal mass of tissue. Such masses are called *tumors* or *neoplasms*; the development of a tumor is called *neoplasia*.

Some tumors, such as most *papillomas* (warts) are *benign:* they remain localized and the animal is unharmed. Other tumors are *malignant:* their growth is invasive, so that the organ in which it occurs is damaged and the animal dies. Often, unregulated cells are released from the tumor and establish new neoplastic foci in other parts of the body. The development of malignant tumors is called *cancer*.

Tumors are usually named by appending the suffix -*oma* to the name of the tissue in which the tumor has arisen. Thus, a lymphoma is a cancer of the lymphoid tissue; a carcinoma is a cancer of epithelial tissue; a sarcoma is a cancer of fleshy, nonepithelial tissue; an adenocarcinoma is a cancer of glandular tissue; and so on. An exception to this system of nomenclature is *leukemia*, or cancer of the white blood cells.

The first evidence of a relationship between viruses and cancer was obtained as early as 1908, when V. Ellerman and O. Bang demonstrated that certain chicken leukemias could be transmitted to healthy birds by cell-free filtrates of diseased blood. Some years later, P. Rous was able similarly to transmit a chicken sarcoma. These discoveries received little attention at the time. Only in 1932, when R. E. Shope demonstrated a viral origin of rabbit papillomas, was importance attached to viruses as causative agents of tumors.

An important step was taken when a *natural route of transmission* was found for a virus-induced cancer. In 1936 J. J. Bittner showed that mammary tumors of mice are caused by a virus that is transmitted from mother to offspring through the milk. Bittner's work led to an understanding of several important aspects of virus-induced cancer. First, an animal that is infected with a tumor virus at the time of birth may not develop a tumor until it has become an adult. Second, the ability of a virus to induce a tumor depends on certain environmental factors, such as the physiology of the host; mammary tumors of mice, for example, occur at high frequency only in animals undergoing hormonal stimulation characteristic of pregnancy. Even male mice will develop mammary tumors if injected with the hormone estradiol over a long period of time, provided that they are infected with the virus.

### the transformation of normal cells into tumor cells

The induction of a tumor by a virus in vivo can be reproduced in tissue cultures. When a suspension of tumor virus particles is used to infect a susceptible tissue culture, some of the infected cells are transformed into a type that exhibits unregulated growth.

The process of *transformation*, as it is called,* causes a number of striking changes in the cell. Most animal cells in tissue culture exhibit the phenomenon of *contact inhibition:* the cells move about randomly by ameboid motion and divide repeatedly until they come into contact with one another; contact between cells inhibits both movement and cell division. The result is that normal cells in tissue culture form a monolayer on the surface of the glass vessel; transformed cells, however, do not exhibit contact inhibition, and they form tumorlike cell masses in tissue culture. Furthermore, cells transformed in tissue culture will initiate tumors when inoculated into host animals.

### types of tumor viruses

Oncogenic (tumor-producing) members are found within several groups of DNA viruses. Most attention has been focused on polyoma virus, because of its small size: it has only enough DNA to code

---

* The term *transformation* is also used in biology to describe a totally different process: the formation of bacterial recombinants by the transfer of DNA extracted from donor cells (Chapter 8).

for five proteins, and five genes have been identified by genetic tests. Two of the genes, which are transcribed late in infection, are involved in capsid formation, and two others, which are transcribed early, are involved in the transformation process. The product of one of these genes is probably involved in bringing about the integration of the viral DNA into the host genome (see below); the product of the other is apparently involved in the changes that free the host cell from its normal controls and confer on it the properties characteristic of the transformed state.

A limited group of RNA viruses are capable of oncogenesis: the leukemia, lymphoma, and sarcoma viruses of mice, cats, and chickens, and the mammary tumor virus of mice. All of these are similar in their physicochemical properties and are classified together as the *leukovirus group*. The ability of these viruses to cause tumors is apparently related to their unique mode of reproduction. The virions of these RNA viruses contain an enzyme termed *reverse transcriptase*. Following uncoating, this enzyme produces a strand of DNA complementary to the single strand of viral RNA. The RNA strand is then replaced with DNA in another round of replication (Figure 7.12). The resulting double-stranded DNA molecule is then transcribed to produce both mRNA encoding viral proteins and single-stranded RNA of the virion.

### the state of the viral genome in transformed cells

Following transformation by DNA viruses, cells of a virus-induced tumor or transformed cells of a tissue culture are completely free of detectable, infectious virus. Nevertheless, it can be demonstrated that part or all of the viral genome persists in the transformed cell and is reproduced at each cell generation.

**Figure 7.12**
The action of leukovirus reverse transcriptase. RNA strands are shown as wavy lines, DNA strands as straight lines. Both individual reactions are mediated by the enzyme. During synthesis of the second DNA strand, the RNA is hydrolyzed.

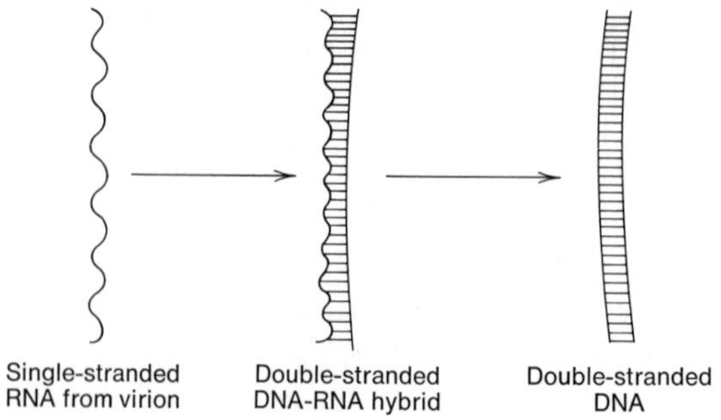

Single-stranded          Double-stranded          Double-stranded
RNA from virion          DNA-RNA hybrid                 DNA

For several of the DNA tumor viruses, it has been shown that the viral DNA is *integrated into the DNA of the host cell*. However, the evidence for the integration of DNA complementary to tumor virus RNA (formed by reverse transcriptase) is less direct. Such viral DNA sequences have been found in the nuclei of leukemic cells, for example, but a direct demonstration of their covalent linkage to host cell DNA has not yet been possible. The integration of viral DNA is suggested, however, by the presence in tumor virions of the enzymes that would be required for inserting a viral genome into host DNA by a series of breakage and rejoining events.

The ability of the tumor viruses to enter the *provirus state* means that they are capable of two modes of transmission: *horizontal transmission*, in which the virus passes from one cell to another by the release and adsorption of virions, and *vertical transmission*, in which the virus is transmitted from one cell generation to the next in form of a provirus. If the provirus enters a cell in the germ line of an animal, then it will also be vertically transmitted from one animal generation to the next. This possibility has formed the basis of a number of theories concerning the etiology of cancer, as discussed below.

### the role of viruses in human cancer:
### DNA viruses

Two groups of DNA viruses have been established as the etiological agents of *naturally occurring* cancers of animals: *papillomaviruses*, which cause malignant warts in rabbits (as well as benign warts in other animals, including man); and *herpesviruses*, which cause lymphomas in chickens (Marek's disease) and adenocarcinomas of the kidney in frogs.

Although a number of viruses found in human tissues can cause tumors when injected into animals, few oncogenic viruses have ever been isolated from human cancers. The major exception to this rule is the herpesvirus called *Epstein-Barr* virus, or EB virus. EB virus was first isolated from a fatal form of human cancer called Burkitt lymphoma, and an apparently identical virus was later isolated from the cells of carcinomas of the human nose and throat. The EB viruses are true tumor viruses, producing malignant lymphomas when injected into monkeys.

Recently, however, it has been found that a large fraction of the population of the United States carries EB viruses in their lymphocytes in the latent state, without producing disease. Liberated EB virions are found in the throats of individuals who are experiencing infectious mononucleosis (a nonmalignant disease), and these individuals exhibit a high titer of antibody against EB virus in their circulation. Thus, EB viruses are present in latent form in many healthy individuals, and they are released as mature virions in individuals suffering from Burkitt lymphoma and from nose or throat carcinoma, as well as from infectious mononucleosis. All of these EB viruses are indistinguishable from one another by existing laboratory methods; they may, however, represent genetically different strains with differing pathogenicities.

It is not possible, despite the association of EB viruses with human diseases, to assign them an etiological role in either infectious mononucleosis or in human cancer. If they do play a role in these diseases, it is not clear why they affect such a small fraction of the individuals in which they occur.

The fact that other oncogenic DNA viruses have not been isolated from human tumors does not rule out a possible role for them in cancer. First of all, the lack of epidemiological evidence for contagion is inconclusive, since experience with tumor viruses in experimental animals shows that a very long time may elapse between infection and the appearance of a tumor. Second, the possibility exists that DNA tumor viruses are transmitted *vertically* from parent to child, causing cancers only when their transforming genes are activated by environmental factors such as chemical carcinogens or radiation.

### the role of viruses in human cancer: RNA viruses

A number of animal cancers have been shown to be caused by RNA tumor viruses: these are the leukemias, sarcomas, and lymphomas of mice, cats, and chickens, and the mammary tumors of mice. Each of these diseases has its analogy in a human cancer. Using extremely sensitive detection techniques, S. Spiegelman and his associates have examined extracts of human cancer tissues for the presence of RNA viruses related to the etiological agents of the corresponding animal diseases. Their procedure was to look for particles with the following properties: (1) a buoyant density falling within the normal range of the RNA tumor viruses, (2) the presence of 70S RNA, typical of RNA tumor viruses, (3) the presence of reverse transcriptase, (4) the ability of the reverse transcriptase to synthesize DNA complementary to the 70S RNA, and (5) sequence homology between the DNA so formed and the RNA of the tumor virus causing the analogous cancer in animals.

In each case they were successful: for example, particles with the above properties were found in human leukemic cells but not in normal cells, and the DNA synthesized endogenously by these particles hybridized to the RNA of mouse leukemia virus but not to other RNA tumor viruses. Similar homologies were found between the RNAs of particles from human mammary cancers, lymphomas, and sarcomas and the RNAs of the agents of the corresponding mouse cancers.

Spiegelman and his associates then went on to ask if similar particles could be isolated from human cancers for which no animal homology exists. Extracts were prepared from human brain, lung, and gastrointestinal tumors, and again particles were found with all of the above properties. (In this case, of course, the final test of homology with a known animal tumor virus RNA could not be made.)

Since the RNA-containing particles found in human cancers have not yet been shown to be infectious or oncogenic, final proof of an etiologic role is still lacking. Nevertheless, the specific homology of their base sequences with those of the RNA viruses causing analogous tumors in animals is fully in accord with such a relationship.

Even if RNA viruses were proved to carry the genes that govern the malignant state, the question of their origin and transmission would remain unanswered. With very few exceptions, believed to be laboratory artifacts, the RNA viruses found in animal tumors are poorly infective; and the infective RNA viruses that are capable of inducing tumors do so with very low efficiency. These facts, coupled with the ability of the tumor viruses to enter the provirus state and undergo vertical transmission, suggest that the epidemiology of human cancer is complex, with horizontal and vertical transmission both playing a role. Regardless of the transmission, long latent periods appear to characterize tumor viruses. Current evidence indicates that environmental insult (specifically, contact with mutagenic chemicals or chronic contact with irritating particulate matter) frequently triggers the transition to the malignant state.

# GENETICS

The study of genetics was brought to the molecular level in one swift stroke when, in 1953, J. D. Watson and F. H. C. Crick published a proposal for the structure of DNA (see Chapter 4).

The great significance of their model is that it provides an explanation in terms of molecular structure for both DNA replication and gene mutation. Since these processes underlie heredity and variation, respectively, it can safely be said that in the entire history of biological science only Darwin's recognition of the existence and fundamental mechanism of evolution has equaled their discovery in importance.

It is now possible to define *gene* and *mutation* in precise molecular terms. *A gene is a segment of DNA in which the sequence of bases determines, by transcription, the sequence of bases in an RNA molecule, and thus—by the process of translation—the sequence of amino acids in a polypeptide chain. A mutation is any permanent alteration in the sequence of bases of DNA,* even if this alteration does not have a detectable phenotypic effect.*

Although polypeptides vary considerably in chain length, a molecular weight of 40,000 is about average. Taking the average molecular weight of an amino acid to be 133, this corresponds to 300 amino acid residues per average peptide chain. Given the triplet genetic code (Chapter 4) the average gene must contain about 1,000 nucleotide pairs.

A given gene can exist in a variety of different forms as a result of mutational changes in its nucleotide sequence; the differ-

---

*The term *gene* is also applied to certain segments of DNA that are transcribed but not translated (such as the genes encoding transfer RNAs and ribosomal RNAs) as well as to certain segments that are neither transcribed nor translated (such as the operator sites described on p. 173). It should be noted that in the case of the RNA viruses, the genes are segments of RNA rather than of DNA.

ent mutational forms of a gene are called *alleles*. The form in which a given gene exists in a microorganism as it is first isolated from nature (the wild-type organism) is defined as the *wild-type allele* of that gene; altered forms resulting from mutations are called *mutant alleles*.

# the arrangement of genes in bacteria

Bacterial genes are regions of a very large circular molecule of DNA, which, by analogy to the complex protein-DNA structures of the eucaryotic nucleus, is termed a *chromosome*. Again, by analogy to the eucaryotic cell, the region of the bacterial cell in which the chromosome is concentrated is termed the *nucleus*. By electron microscopy, it appears (Figure 8.1) as a region closely packed with fine fibrils of DNA. This region is not separated from the cytoplasm by a membrane, nor does it contain any evident structures, apart from the fibrils themselves. Recently, however, the intact nucleoplasm has been isolated, and shown to contain some protein and RNA in addition to DNA (Figure 8.2).

### the bacterial chromosome

By 1960 the cytological information about the structure of the bacterial nucleus had been complemented by genetic studies

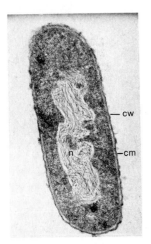

**Figure 8.1**
Thin section of a dividing cell of a unicellular procaryotic organism, *Bacillus subtilis* (×20,000): n, nucleus; cm, cytoplasmic membrane; cw, cell wall. Courtesy of C. F. Robinow.

**Figure 8.2**
Electron micrograph of the isolated folded chromosome of *Escherichia coli*, attached to a fragment of the cell membrane (dark, irregular area in center of figure). The bar represents 2 μm. From H. Delius and A. Worcel, "Electron Microscopic Visualization of the Folded Chromosome of *Escherichia coli*," *J. Mol. Biol.* **82,** 107 (1974).

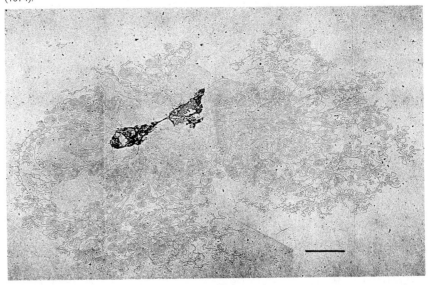

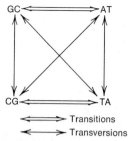

**Figure 8.3**
Base pair substitutions. Those in which a purine is replaced by a different purine or a pyrimidine by a different pyrimidine are called transitions. Those in which a pyrimidine replaced a purine or vice versa are called transversions.

of *E. coli*, which suggested the presence of a single circular array of genes. This implied that each nucleus should contain a single, circular chromosome. If such were indeed the case, the fibrils of DNA revealed in the nuclear region by electron microscopy should represent a section of an extremely long circular molecule of DNA highly folded to form a compact mass.

In 1963 J. Cairns succeeded in extracting DNA from *E. coli* under conditions that minimized its shearing. He was able to visualize the isolated chromosome and thereby directly confirmed its circularity.

# the chemical basis of mutation

### types of mutations

The sequence of nucleotides within a gene can be altered in any of several ways: *base-pair substitutions, frame-shift mutations, large deletions*, and *duplications*.

In base-pair substitution (see Figure 8.3), a base pair at a specific site in the wild-type allele such as GC is replaced in the mutant allele by a different base pair, *i.e.*, CG, AT, or TA. Such substitutions can be subdivided into two classes: those in which a purine is replaced by a different purine or a pyrimidine by a different pyrimidine (*transitions*), and those in which a purine is replaced by a pyrimidine and vice versa (*transversions*) (Figure 8.3).

In frame-shift mutations, so named because they shift the "reading frame" of the translation process (Figure 8.4), one or a few base pairs are inserted into or deleted from the gene.

In large deletions, a long sequence of base pairs representing a major segment of the gene or sometimes spanning several genes is removed. Large deletions are irreversible; i.e., the wild-type allele

**Figure 8.4**
The effect of a frame-shift mutation on transcription and translation. Through mutation a new base pair is inserted at the position of the vertical arrow. Note that codons in the mRNA, and amino acids in the protein, are changed beyond the point of the mutation. Also note that the mutation brought a nonsense codon (UAA) in reading frame which caused premature termination of the peptide (see p. 172).

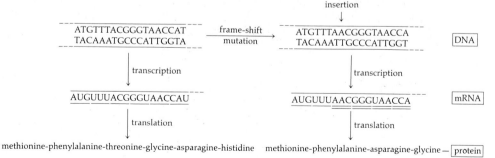

**Table 8.1**
Classification of Chemical Mutagens

| CLASS | EXAMPLES | TYPE OF MUTATION CAUSED |
|---|---|---|
| I. Agents that become incorporated into DNA (Base analogues) | | |
|   A. Purine analogues | 2-Aminopurine | GC $\longrightarrow$ AT and AT $\longrightarrow$ GC transitions |
|   B. Pyrimidine analogues | 5-Bromouracil | GC $\longrightarrow$ AT and AT $\longrightarrow$ GC transitions |
| II. Agents that react chemically with DNA | | |
|   A. Alkylating agents | Mustard gas, EMS, nitroso-guanidine | All transitions and transversions |
|   B. Deaminating agent | Nitrous acid | GC $\longrightarrow$ AT and AT $\longrightarrow$ GC transitions |
|   C. Hydroxylating agent | Hydroxylamine | GC $\longrightarrow$ AT transitions |
| III. Agents that intercalate between pairs of bases in DNA | Proflavine | Insertion and deletion-type frame-shifts |

cannot be reestablished by a subsequent mutation. In contrast, base-pair substitutions and frame-shifts are reversible.

Duplications are the insertion into the chromosome of a second copy of a segment of DNA. Duplications are readily reverted to wild type.

### mutagenesis

A number of chemical and physical treatments dramatically increase the normal rate of mutation of bacteria. Such chemical or physical agents are termed *mutagens*. Although chemical mutagens are quite varied in structure, the mechanisms by which they cause mutation are limited and hence provide a basis for their classification (Table 8.1).

The use of mutagens has made possible the development of physiological genetics from which much of our current knowledge of microbial metabolism and regulation was obtained. Studies on the mode of action of mutagens have provided information concerning the basis of mutation in natural populations, upon which evolution depends.

In their original paper,* Watson and Crick pointed out that mutations of the type now called *transitions* should occur if, at the time they are acting as templates, any one of the bases underwent a *tautomeric shift of electrons* from the more stable forms (shown in Figure 4.27) to the less stable ones, the consequence of which is the reversal of the base-pairing

*J. D. Watson and F. H. C. Crick. "Genetic implications of the structure of deoxyribonucleic acid," *Nature* **171,** 964 (1953).

rules. Thymine then pairs with guanine rather than adenine, and cytosine then pairs with adenine rather than guanine. Such tautomeric shifts seem to account for mutations caused by base analogues, because they undergo tautomeric shifts (and therefore pair incorrectly) more readily than do the natural bases. When added to a culture medium, base analogues become incorporated into newly synthesized DNA. They may pair incorrectly at the time of incorporation or later. Hence, they cause transition mutations of both types: GC $\longrightarrow$ AT or AT $\longrightarrow$ GC.

Mutagenic agents that react chemically with DNA change the structure of previously incorporated bases, thereby altering their pairing properties and causing mutation. Some such agents are relatively nonspecific in their action. The alkylating agents, for example, transfer alkyl groups to nitrogen atoms in the bases of DNA. The mutagenic consequence of alkylating agents is nonspecific, resulting in transitions and transversions.

Frame-shift mutations are caused by *intercalating agents* that distort the structure of the DNA in such a way that replication frequently leads to the insertion or deletion of one or a few base pairs.

Of the physical agents that cause mutations, ultraviolet is the most commonly employed experimentally. It causes a reaction between adjacent thymine residues in one strand of the DNA helix, which chemically links them together. The product of this reaction is termed a *thymine dimer*. Although cells possess the biochemical capacity to excise these dimers and repair the damaged DNA, errors are frequent during such repair. Mutation is the consequence.

### spontaneous mutations

The term *spontaneous mutation* refers to those mutations that occur in the absence of known mutagenic treatment. The rate of spontaneous mutation is not constant; many environmental conditions will affect it. There are many different mechanisms of spontaneous mutation. Many products or intermediates of cell metabolism are demonstrably mutagenic, *e.g.,* peroxides or nitrite. Some spontaneous mutation may thus in reality be induced by endogeous mutagens.

A mutation in one of the genes of E. coli and in a similar gene in *Salmonella typhimurium* has been shown to cause an increase in the spontaneous mutation rate for all loci by a factor of 100 to 1,000. The mutations caused by the new allele of the *mutator* locus are transversions. The product of the bacterial mutator gene has not been identified, but a similar mutator gene has been discovered in phage T4 and the product of this gene has been identified as DNA polymerase, the enzyme responsible for replication of the chromosome. The implications of this discovery are profound, since it demonstrates that the selection of nucleotides for the synthesis of a new strand of DNA depends not only on template-directed hydrogen bonding, but also on the fidelity of action of the polymerase. Since the mutationally

altered polymerase causes a very high rate of "mispairings" leading to transversions, there is every reason to believe that the wild-type polymerase of
phage and of bacteria is also responsible for spontaneous mispairings, although at a much lower rate than that of the altered polymerase.

A significant fraction (about 15%) of spontane-
ous mutations has proved to reflect *deletions* of DNA segments. Their span
varies from a large part of a single gene to several genes.

# the effects of mutation
# on the translation process

### nonsense mutations

A codon is said to make "sense" if it is translated
into the correct amino acid of the polypeptide for which the gene codes. A
mutant codon that is translated into a different (incorrect) amino acid is said to
contribute *missense* to the message, and the mutation that produced that codon
is called a *missense mutation*.

Three mutant codons (UAG, UAA, and UGA)
for which there are no corresponding tRNA species have been found to cause
*premature chain termination* (see Figure 8.4). When the ribosome reaches such a
codon, the process of polypeptide chain elongation is terminated, and the
incomplete polypeptide is released. Such codons are called *nonsense codons,*
and the mutations producing them are called *nonsense mutations*.

The existence of three codons that can bring
about chain termination raises the strong probability that one or more of them
is the *natural chain terminator* in the translation process. As we shall see later,
some groups of genes are transcribed into a single *polygenic messenger* RNA
molecule; when a ribosome proceeds along a polygenic messenger, it must be
given a signal to terminate one polypeptide chain and to start another. The
termination signal is probably one of the nonsense codons.

# the genetic aspects of regulation

### the induction and repression
### of enzyme synthesis

The physiological aspects of the regulation of
enzyme synthesis were discussed in Chapter 5.

Much of our current knowledge about regulation
of enzyme synthesis, at both the physiological and genetic levels, is derived
from the intensive studies on the $\beta$-galactosidase system of E. coli carried out
by J. Monod, F. Jacob, and their collaborators. Their work culminated in the
development of a general hypothesis to explain the control, at the genetic

level, of the rate at which gene expression occurs. The following discussion will, accordingly, be centered on the studies on the *β-galactosidase system* of *E. coli*, the system of enzymes responsible for the catabolism of lactose.

The β-galactosidase system consists of *structural genes* that encode the enzymes catalyzing the first reactions of the pathway of dissimilation of lactose, and *regulatory genes* that control expression of these structural genes. In wild-type cells, one key enzyme of the system, β-galactosidase, is *inducible* by the presence in the medium of the primary substrate of the pathway (lactose).

The first clue to the mechanism by which this regulation is achieved was provided by the discovery of mutants of *E. coli* in which the synthesis of β-galactosidase had become *constitutive*, i.e., producing β-galactosidase in the absence of lactose. Genetic experiments showed that many of the mutations that make β-galactosidase synthesis constitutive, map in a gene termed *lacI* (for inducibility). The wild-type allele of the *lacI* locus (*lacI*$^+$) determines the inducible state; the mutant allele (*lacI*$^-$) determines the constitutive state.

Later it was found that *lacI*$^+$ is dominant over *lacI*$^-$. Evidently the dominant *lacI*$^+$ allele produces a product that actively inhibits β-galactosidase formation; this product has been designated as a *repressor*, and has been shown to be a protein.

The discovery of the *lac* repressor led to the general theory that inducible enzymes are ones for which the cell constantly synthesizes repressors; *the role of the inducer is to combine with and inactivate the repressor*. The theory was extended to explain end-product repression as well, by postulating the existence of genes that form repressors of biosynthetic enzymes. A repressor of this type would be nonfunctional as an inhibitor of enzyme formation, *unless activated by combination with the end product of the biosynthetic pathway*.

### operator genes

The existence of a repressor implies the existence of a target for its action. The target is a segment of DNA that binds the repressor; it is called the *operator gene*. This gene (*lacO*) is adjacent to the structural gene (*lacZ*) that encodes β-galactosidase. The repressor, when bound at this site, prevents transcription of *lacZ*. Induction occurs when the inducer combines with the repressor causing an allosteric change that renders the repressor unable to bind at *lacO*, thus allowing transcription to proceed. The operator locus (*lacO*) can mutate to a form (*lacO*$^c$) that cannot bind the repressor, thus making enzyme synthesis constitutive.

An "operator constitutive" mutant can generally be distinguished from a constitutive mutant of the *lacI*$^-$ type in partial diploids. Whereas a *lacI*$^+$/*lacI*$^-$ diploid is inducible, by virtue of the production of the repressor by the *lacI⁺* allele, a *lacO⁺/lacOᶜ* diploid is constitutive, since the β-galactosidase gene adjacent to the *lacOᶜ* allele continues to function, even though the repressor is present and able to bind to the *lacO⁺* allele on the other chromosome of the diploid.

### operons

Early in the work on the genetic control of lactose metabolism, it was discovered that some mutants that are unable to ferment lactose contain normal amounts of β-galactosidase but are deficient for a specific permease that allows lactose to enter the cell.

The mutations that affect the lactose permease were found to map in one locus that is designated *lacY*. Certain mutations were also observed to affect a third locus, *lacA,* governing an enzyme that chemically modifies lactose but plays no role in its metabolism. Experiments showed that the genes *lacO-lacZ-lacY-lacA* are adjacent, forming a linear array in the order shown. When the experiments described above on the functions of *lacI* and *lacO* were performed, it was observed that the syntheses of β-galactosidase, permease, and transacetylase are affected in an identical manner; all are inducible in *lacI⁺* strains and constitutive in *lacI⁻* strains; all are inducible when their respective genes are adjacent to *lacO⁺*, and constitutive when their respective genes are adjacent to *lacOᶜ*.

The genes *lacZ, lacY,* and *lacA* thus behave as a unit of *coordinated expression*. Such a unit—together with its operator—is called an *operon*. Many other operons have since been discovered that function in both catabolic and biosynthetic pathways.

The operon provides an efficient mechanism for the regulation of metabolic pathways. The enzymes of a pathway function as a unit: if the pathway is not operative, none of its enzymes is required; if the pathway is operative, all of its enzymes must function.

It only remains to ask how an operator, which has bound a molecule of repressor, can inhibit the transcription of a sequence of adjacent genes. This question has been answered by the discovery that all the genes of the operon are transcribed as a unit, forming a very long molecule of *polygenic* messenger RNA. The transcribing enzyme, RNA polymerase, binds to a site on the DNA called the *promoter,* and transcribes sequentially all the genes of the operon. This process is blocked when a repressor molecule is bound to the operator that is immediately adjacent to the promoter. This mechanism is illustrated in Figure 8.5.

### diversity in the mechanism of genetic regulation

We have presented above only one mechanism of genetic regulation—one that is operative for the *lac* genes. Many other operons have been detected, including sets of genes for the enzymes of bio-

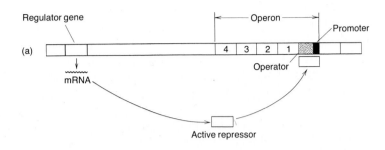

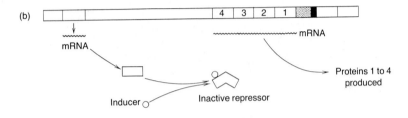

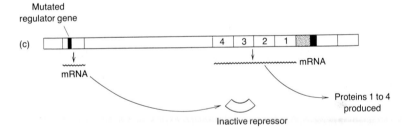

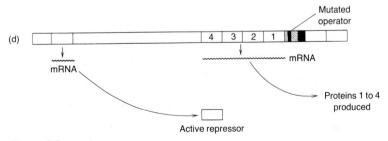

**Figure 8.5**

Regulation of an operon. The segmented horizontal bar represents a section of a chromosome, with each segment representing a gene. (a) Active repressor, produced by a regulator gene, binds to its specific operator and blocks transcription of the adjacent set of four genes. (b) An inducer molecule combines with repressor and inactivates it, preventing its binding to the operator. The operon is then expressed. (c) Inactive repressor is produced as a result of a mutation in the regulator gene; the operon is constitutively expressed. (d) A mutation in the operator prevents the binding of active repressor; the operon is constitutively expressed.

synthetic pathways. In these cases, the repressor is normally inactive and is *activated* by combination with the end product of the pathway.

The genes for the enzymes of metabolic pathways are not always organized in operons. In *E. coli,* for example, many of the pathways are governed by genes that are scattered along the chromosome. In some of these systems, enzyme synthesis is coordinated by the product of a single regulator gene, indicating that one repressor can bind to a number of different operators. In other cases of metabolic pathways governed by scattered genes, a common repressor is not involved; the different genes are separately regulated.

Some regulator gene products *activate,* rather than repress, the formation of enzymes. Several such cases have been discovered, including the regulation of alkaline phosphatase synthesis and of the genes that bring about the catabolism of arabinose in *E. coli.*

# the effects of mutation on phenotype

All the properties of an organism are ultimately determined by its genes, including the genes of the organelles (the mitochondria and chloroplasts of eucaryotic cells) and plasmids, as well as the chromosomal genes in the nucleus. The allelic states of all the genes in a cell collectively constitute its *genotype.*

The structural and physiological properties of a cell constitute its *phenotype.* In the following sections we shall discuss the general ways in which the genotype of the cell determines its phenotype, and the manner in which changes in the genotype (mutations) bring about changes in the phenotype.

### the relation of phenotype to genotype

Each gene in the cell determines the structure of a single protein.* The structural and catalytic properties of the proteins determine in turn the anatomical and metabolic properties of the cell. Since the proteins interact with each other structurally to form the complex organelles of the cell, as well as functionally to form coordinated and regulated metabolic pathways, by altering the structure of a single protein, a mutation can bring about a profound change in cell structure and function.

### the effects of mutation on primary gene products

For the purposes of this discussion, we shall consider the primary gene products of the cell to be proteins, since mutational

---

*With the exception of operator genes, and genes that determine ribosomal RNA and tRNA.

effects on phenotype are usually the result of changes in protein structure. Such changes always consist of alterations in amino acid sequence resulting from mutational alterations in the base-pair sequence of DNA. The amino acid sequence may be altered by the substitution of one amino acid for another or by the deletion or insertion of a set of amino acids.

The alteration in amino acid sequence of a polypeptide may produce any of a number of alterations in the properties of the protein, including the following:

1. The protein may have an altered catalytic site; such alteration may cause the enzyme to be partially or completely inactive.
2. The protein may become unusually sensitive to high or low temperature, salts, a metal ion, and so on.
3. There may be no detectable change in the physiochemical or catalytic properties of the protein.

PHENOTYPIC CHANGES IN DISPENSABLE CELL FUNCTIONS. Depending on the environmental conditions, many cell functions are dispensable; mutations conferring the loss or alteration of these functions may thus be studied under conditions that permit the mutants to grow and divide.

The cell functions that are dispensable under some environmental conditions include the utilization of alternative forms of carbon, nitrogen, sulfur, or phosphate; the synthesis of the precursors of macromolecules and of coenzymes; the regulation of enzyme synthesis and activity; and the synthesis of various components of the cell surface: peptidoglycan, capsule, flagella, and pili.

### phenotypic changes in nutrition: growth factors

When mutation causes the inactivation of an enzyme of a biosynthetic pathway, the cell becomes dependent on an exogenous supply of the end product. This condition is termed *auxotrophy*; the wild-type state is termed *prototrophy*. The most common growth factors that may become required as a result of auxotrophic mutations are amino acids, purines, pyrimidines, and vitamins.

### resistance and sensitivity to antimicrobial agents

Mutational change in protein structure can lead to increased (or decreased) resistance to antimicrobial agents, both chemical and physical. Such resistance may reflect any of several different changes in primary gene products: the cell membrane may become impermeable to the agent; the cell may gain the ability to degrade or inactivate the agent; or the component of the cell that is the primary target of the agent's action may acquire a reduced sensitivity to the agent.

# the conditional expression of the phenotype of a gene mutation

The mutations discussed in the preceding sections are expressed as functionally altered gene products. The phenotypes of many mutations, however, are *conditionally expressed:* under one set of conditions, called *permissive,* the phenotype is essentially *normal;* the mutant cell forms a gene product that is functionally equivalent to that of the wild type; under a different set of conditions, called *nonpermissive,* the mutant phenotype is expressed; a functionally altered product is produced. Some mechanisms of conditional expression are described below.

The phenomenon of conditional expression has made possible the study of mutations that when expressed, deprive the cell of *indispensable functions;* i.e., if expressed, they are *lethal under all conditions* of cultivation. Mutations that inactivate DNA or RNA polymerases are examples of this class of mutations; the products of the reactions catalyzed by these enzymes cannot be supplied from the environment.

The phenotype of *nonlethal mutations* can also be conditionally expressed. Thus, some mutants are auxotrophic under nonpermissive conditions, forming an inactive biosynthetic enzyme; and prototrophic under permissive conditions, forming an active one. The expressed mutation is not lethal because the end product can be supplied from the medium.

The great utility of conditionally expressed lethal mutations is that the mutant can be cultured under permissive conditions. When transferred to the nonpermissive environment, the consequences of expression of the mutation can be studied. Two classes of conditionally expressed mutations are: *temperature-sensitive mutations* (including both heat- and cold-sensitive types) and *salt-remedial mutations.*

*Heat-sensitive mutations* are those that render the protein thermolabile at temperatures approaching the higher end of the physiological range of the organism. For example, many proteins of *E. coli* may be altered by mutation such that they are denatured at 40°C, while remaining intact at 30°C. Conversely, mutant proteins that cannot function at the lower end of the temperature range, but remain functional at higher temperatures, are termed *cold-sensitive mutations.*

A number of mutants have been found that grow normally in media of high osmotic strength, but not in media of low osmotic strength. Called *salt-remedial mutants,* they have mutationally altered enzymes that are denatured unless stabilized by high salt concentrations.

# the time course of phenotypic expression of mutation

When the primary effect of a mutation is the loss of a stable gene product, there may be a delay of several generations before the mutation is phenotypically expressed. This delay is called *phenotypic lag;* it reflects the time required for nuclear segregation and the time required for the dilution of the active gene product that is no longer synthesized.

### phenotypic lag: nuclear segregation

Unicellular bacteria may contain as many as four nuclei per cell. These nuclei are genetically identical since they are derived from a single nucleus that existed in the cell line one or two generations earlier. If a mutation in one nucleus causes the loss of ability to produce a functional gene product, the mutation will be recessive since the other nuclei continue to make the gene product. Thus, one or two cell generations are required for nuclear segregation to permit a *loss mutation* to be phenotypically expressed. This situation is shown in Figure 8.6.

**Figure 8.6**

Phenotypic expression. If a mutation first occurs in a tetranucleate cell, two generations are required before the mutant nucleus has completely segregated from wild-type nuclei.

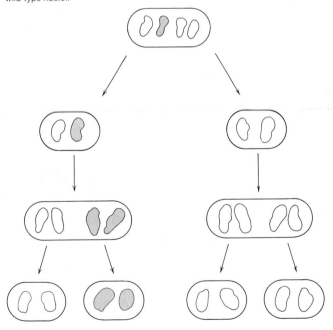

## phenotypic lag: dilution of active gene product

When a suspension of phage-sensitive bacteria is treated with a mutagenic agent and the survivors are plated immediately on phage-coated agar, virtually no induced phage-resistant mutants develop. If, however, the survivors are permitted to undergo several generations of growth in nutrient broth before plating with phage, a large number of resistant mutants are obtained. The results of a typical experiment are shown in Figure 8.7, which shows that some of the induced mutations take as long as 14 generations to be expressed.

The basis of this delay in phenotypic expression became clear when the mechanism of phage resistance was elucidated. A sensitive bacterium adsorbs phage by means of specific receptors in the cell wall; resistant mutants lack these receptors. At the time that the sensitive cell undergoes the genetic change to resistance, its wall still possesses the preexisting receptor sites, and the cell remains phenotypically sensitive to phage. During subsequent growth, however, receptors are no longer synthesized, and the old ones are diluted by the formation of new cell wall material. Phenotypic expression of the mutation to resistance is thus delayed until the mutated cells possess too few receptors to allow phage adsorption.

In general, a phenotypic lag occurs whenever the primary effect of a mutation is the loss of a stable gene product. Phenotypic lags will thus be observed in mutations from prototrophy to auxotrophy and in mutations that deprive the cell of the ability to use a particular form of carbon or nitrogen source. In each of these cases, the primary effect of the mutation is the loss of an enzyme activity; at the time of the mutation, the cell possesses a large number of enzyme molecules that must be diluted out by cell growth and division before the effect of the mutation is observed.

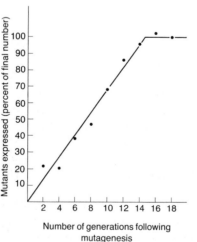

**Figure 8.7**

Delay in phenotypic expression of mutation. A suspension of phage-sensitive bacteria is treated with a mutagen and the survivors are plated on phage-coated agar. Only induced phage-resistant mutants appear. In this experiment the survivors were allowed to reproduce for varying numbers of generations of growth before plating.

When a mutation results in the gain, rather than the loss, of a gene product, phenotypic expression is for all practical purposes immediate. If an auxotroph, for example, reverts to prototrophy, active enzyme molecules begin to be synthesized immediately and to function in the biosynthetic pathway that had previously been blocked.

# the selection and detection of mutants

There are three general methods for selecting microbial mutants: selection based on relative growth; selection based on relative survival; and selection based on visual detection.

### selection based on relative growth

In the first method, the cells are plated on an agar-solidified medium, the composition of which permits only the desired type of mutant to form a visible colony; the wild-type parental cells remain nondividing or are killed. For example, $10^6$ or more $Lac^-$ cells (unable to utilize lactose for growth) may be spread on agar containing lactose as the sole source of carbon; only the rare $Lac^+$ mutants will grow and form colonies, the $Lac^-$ cells being unable to divide. Similarly, $10^6$ or more $His^-$ cells (unable to synthesize histidine) may be spread on agar lacking histidine; only the rare $His^+$ cells will grow and form colonies. The selective medium may, alternatively, be designed to inhibit or kill the parental cell population, rather than merely not to support its growth. For example, when a streptomycin-sensitive population of cells is spread on agar containing bactericidal concentrations of streptomycin, the parental cells are killed and the rare streptomycin-resistant mutants grow into visible colonies.

### selection based on relative survival

In the second method, selection is based on a condition in which growth of the desired mutant type is inhibited, and an agent is added that kills only the growing (parental) cell type. Thus, the desired mutant cells *survive* the lethal treatment, although they do not multiply during the selection process; they are transferred to a medium that supports their growth *after* the selection process has been completed. For example, penicillin kills only growing cells; when penicillin is added to a logarithmic phase culture of bacteria in a liquid minimal medium,* the wild-type bacteria continue to grow until they are killed by the penicillin. If auxotrophic mutants are present, they fail to grow in the minimal medium and thus survive the penicillin treatment. At the end of the incubation period, the survivors are

---

* A minimal medium contains the minimal set of nutrients required for growth of the wild-type organism. For E. coli, this consists of mineral salts plus a carbon source such as glucose.

plated on a nutritionally complete medium, and the resulting colonies are picked and tested to confirm their auxotrophy.

### selection based on visual detection

In the third method, plating conditions are designed so as to make colonies of the desired mutant type visually distinguishable from colonies of the wild type. For example, the colorless compound tetrazolium is reduced intracellularly to the brilliantly red, insoluble product formazan, only within a narrow pH range. Thus, in an otherwise complete medium supplemented with a high concentration of fermentable sugar, cells able to ferment the sugar lower the pH to the point where the dye is not reduced, and form white colonies. Mutant cells unable to ferment the provided sugar, however, reduce the tetrazolium intracellularly to formazan and produce bright red colonies. By this technique it is possible to detect a single fermentation-deficient mutant colony among $10^5$ wild-type colonies on a petri dish.

## isolation of mutant strains

We have emphasized the many contributions to the understanding of fundamental aspects of biology that have come from the study of microorganisms. There are many reasons why these studies have played such a significant role in the development of biology, but important among them is the ease with which particular mutant strains can be isolated. The preceding sections of this chapter constitute a summary of the steps usually employed in the isolation of mutants: *mutagenesis, phenotypic expression, selection,* and *detection.* An example of how these steps might be applied to isolate a particular type of mutant strain, one auxotrophic for the amino acid arginine, is outlined in Table 8.2.

## selection and adaptation

### the genetic variability of pure cultures

As a general rule, any bacterial gene has only one chance in about $10^8$ of mutating at each cell division. At first sight, therefore, mutation might appear too rare to be of much significance. Suppose, however, that we have a "pure culture" of a bacterium in the form of 10 ml of a broth culture that has grown to the stationary phase. Such a culture will contain about $10^{10}$ cells; for any given gene, there may well be several thousand mutant cells present in the culture. Even during the growth a single bacterial colony, which may contain between $10^7$ and $10^8$ cells, a large number of mutants will arise.

**Table 8.2**
Schematized Protocol Suitable for the Isolation of an Arginine Auxotrophic Mutant

| STEP | EXAMPLE OF TREATMENT | CONSEQUENCE |
|---|---|---|
| I. Mutagenesis | Culture is treated with nitrosoguanine. | Increases mutant frequency 100–1000-fold. |
| II. Phenotypic expression | Culture is grown for several generations in minimal medium + arginine. | Mutant cells become phenotypically Arg⁻; mutant frequency unaffected. |
| III. Selection | Culture is treated with penicillin in a minimal medium without arginine. | Prototrophs are killed; desired mutants (and other auxotrophs) do not grow and hence survive. Mutant frequency increased 100–1000-fold. |
| IV. Detection | Survivors of III are plated on plates of minimal medium + arginine, and the developing colonies are tested for their ability to grow on unsupplemented minimal medium. | Clones developing on plates supplemented with arginine but not on unsupplemented plates are desired mutants. |

Thus, a large population of bacteria is endowed with a high degree of potential variability, which is ready to come into play in response to changing environmental conditions. Because of their exceedingly short generation times and the consequent large sizes of their populations, these haploid organisms possess a store of latent variation despite the fact that they cannot accumulate recessive genes as can a population of diploid organisms. In practice, this means that no reasonably dense culture of bacteria is genetically pure; even a slight change in the medium may prove selective and bring about a complete change in the population within a few successive transfers.

### selective pressures in natural environments

So far we have considered only the selective forces that may operate in artificial cultures. In nature, however, selection acts in an even more stringent fashion. A microbe in the soil, for example, must be able to survive its environment and to compete with the numerous other microbial forms that occupy the same niche. Any mutation that decreases, even to the slightest extent, the ability of the organism to compete, will be selected against and quickly eliminated. Nature tolerates little variation within microbial populations; the laws of competition demand that each type retain the array of genes that confers maximum fitness.

As soon as an organism is isolated in pure culture, however, the selective pressures resulting from biological competition are removed. The isolated population becomes free to vary with respect to certain of the characters that are maintained stable in nature by selection. In adapting to existence in laboratory media, organisms may undergo genetic modifications that would lead to their speedy elimination in a competitive environment.

# recombination in bacteria

In evolution, natural selection operates not so much on single gene mutations as on the *new combinations of genes* that arise when, in a single cell, genes from two different cells are brought together. This process is called *genetic recombination*.

In molecular terms, recombination is the process by which a chromosome is formed from DNA derived from two different parental cells. Three processes that lead to the formation of recombinant chromosomes are known to occur in bacteria. In order of their discovery, these processes are *transformation, conjugation,* and *transduction*. They differ from the sexual process of eucaryotes in that a true fusion cell is not formed; instead only a part of the genetic material of a donor cell is transferred to a recipient cell. The recipient cell thus becomes diploid for only part of its genetic complement; such partial zygotes are called *merozygotes*.

The original genome of the recipient is termed the *endogenote,* and the fragment of DNA introduced into the recipient cell is termed the *exogenote*. Both the nature and the size of the exogenote differ in the three processes. In transformation, short pieces of double-stranded DNA, released into the medium from donor cells, are adsorbed to the surface of recipient cells and are taken into the cell by a process that results in the degradation of one strand. In transduction, a small, double-stranded fragment of DNA is brought from the donor cell into the recipient cell by a bacterio-phage particle. In conjugation, a single strand of DNA, which may represent a major fraction of the donor genome, is transferred between cells.

Before we discuss these different processes, some features common to all three will be described.

## the fate of the exogenote

If the exogenote has a sequence of base pairs that is homologous with a segment of the endogenote, pairing occurs rapidly, and a recombinant chromosome is immediately formed by the integration of part or all of the exogenote with the endogenote. If pairing and integration are prevented for some reason, the exogenote will usually undergo one of two fates. (1) If the exogenote carries genetic elements necessary for its own repli-cation, i.e., if the exogenote constitutes a *replicon,* it may persist and replicate as a plasmid (see Chapter 3), so that the merozygote gives rise to a clone that is partially diploid cells. (2) The exogenote may be enzymatically degraded.

# bacterial transformation

Transformation was first discovered in the pneu-mococcus (*Streptococcus pneumoniae*). The pneumococci in the sputum or tissues of a victim of pneumonia are always surrounded by large capsules consisting of polysaccharide; on agar plates the encapsulated cells form smooth (S) colonies. Pneumococci can be separated into a great many types on the basis of chemical differences between their capsular polysaccharides. The different types (designated I, II, III, and so on) can be distinguished immunochemically.

When an S strain (forming smooth colonies) is serially subcultured, R cells (forming rough colonies) appear in the population. R cells have no capsules and are avirulent due to their increased susceptibility to phagocytosis. In 1928 F. Griffith observed that when a very large inoculum of R cells, derived from what had originally been a type I smooth culture, was inoculated under the skin of a mouse together with heat-killed S cells of type II, the mouse died within a few days. Blood from such an animal yielded only type II smooth cells. The dead type II cells thus had liberated something that conferred on the R cells the ability to make a new type of capsular polysaccharide. The transferred property proved to be heritable.

A few years afterward, other groups of investigators succeeded in carrying out such transformations in vitro. It was later found that transformations could be brought about with cell-free extracts. In other words, a chemical substance extracted from one cell could mediate heritable change in another.

In 1944 O. T. Avery, C. M. MacLeod, and M. McCarty succeeded in purifying the active chemical agent and identified it as DNA. Until that time, it was generally believed that the specificity of the gene was determined by the protein moiety of the nucleoprotein found in the eucaryotic chromosome; the chemical characterization of the active agent of transformation provided the first direct evidence that DNA is the carrier of genetic information.

Since 1944 similar transformations have been effected in other genera of bacteria, notably in *Neisseria, Bacillus, Acinetobacter,* and *Escherichia.* The genetic fragment that is transferred is usually very small; although it may carry a number of genes, it often carries only one. This is an artifact that results from the fragmentation of transforming DNA which usually occurs during its purification. When crude preparations of transforming DNA are used, as much as a third of the chromosome can be transferred.

## bacterial conjugation

Conjugation is a mechanism of genetic exchange discovered by J. Lederberg and E. L. Tatum in 1946 that requires direct contact between the parental cells.

It was initially assumed that bacterial conjugation involves cell fusion with the formation of true zygotes. Ultimately, however, it was discovered that there are two mating types in *E. coli* and that during conjugation *one partner acts only as genetic donor, or male, and the other only as genetic recipient, or female.*

When the recombinants issuing from a variety of crosses were analyzed for mating type, it was discovered that *maleness in bacteria is determined by a transmissible genetic element:* when male and female bacteria conjugate, *every female cell is converted to a male.* The genetic element governing the inherited property of maleness is called the F *factor* (for "fertility"); it is transmitted only by direct cell-to-cell contact.

Although every conjugating cell transfers the F factor, the transfer of chromosomal markers is relatively rare. F is an autonomous element, separate from the bacterial chromosome. In 1952 Lederberg coined the term *plasmid* as a generic name for all extrachromosomal hereditary determinants, of which F is an example.

The process of conjugation occurs in two steps, both governed by plasmid genes: (1) the formation of a conjugation bridge between the donor and recipient cell, (2) passage of the plasmid across the bridge.

### the role of the cell surface in bacterial conjugation

The reactions occurring between donor and recipient are the consequence of specific properties of the surface of the donor cell, determined by plasmid genes. The most striking of these properties, absolutely essential for conjugation, is the presence of special pili (see p. 139) called *sex pili.*

Sex pili are present in small numbers on the surface of cells that carry transmissible plasmids. Their presence was first revealed when electron microscopists looked for the sites of adsorption of certain male-specific phages (phages which infect only bacteria harboring

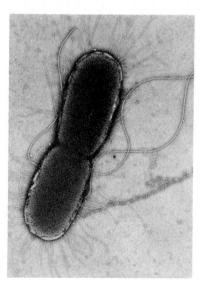

**Figure 8.8**
A dividing cell of an Hfr strain of *E. coli,* showing three types of appendages. The long, curved, thick appendages are flagella. The short, thin, straight appendages are ordinary pili. The long, thin appendage (lower right) coated with phage particles is a sex pilus. Courtesy of Judith Carnahan and Charles Brinton.

transmissible plasmids). When male (donor) cells were treated with a large excess of such phage particles and examined in the electron microscope, the phages were found to be adsorbed exclusively to sex pili (Figure 8.8). A specific receptor site for conjugation also appears to be present on recipient.

The first step in the conjugation process is the attachment of the recipient cell to the tip of a sex pilus [Figure 8.9(a)]. Within a few minutes the two cells move into a position of direct contact [Figure 8.9(b)], probably by retraction of the pilus back into the male cell.

### the transfer of plasmid DNA

Prior to conjugation, the plasmid exists in the donor cell as a double-stranded, circular DNA molecule. Contact between sex pilus and recipient cell wall triggers the process diagrammed in Figure 8.10(a): one strand of the plasmid DNA is broken at the replication origin and the duplex unwinds, the broken strand entering the recipient cell beginning with its 5' end. Complementary strands are synthesized by DNA polymerase in both donor and recipient; the process is thus analogous to normal replication,

**Figure 8.9**
(a) A male and a female cell joined by an F pilus. One sex pilus has been ''stained'' with male-specific RNA phage particles. The male cell also possesses ordinary pili, which do not absorb male-specific phages and which are not involved in conjugation. Courtesy of Judith Carnahan and Charles Brinton. (b) Mating cells of *E. coli* (×25,600). This electron micrograph was taken shortly after mixing together donor (Hfr) cells and recipient (F⁻) cells. Before mixing, the Hfr cells were ''marked'' by causing them to absorb inactive particles of bacteriophage; the F⁻ cells are easily recognized by the fact that in this strain they are covered heavily with pili. The micrograph clearly shows the conjugation bridge that has formed between the Hfr cell and one of the F⁻ cells. Note the bacteriophage particles adsorbed by their tails onto the Hfr cell. From T. F. Anderson, E. L. Wollman, and F. Jacob, ''Sur les Processus de Conjugaison et de Recombination chez *E. coli*. III. Aspects Morphologiques en Microscopie Électronique,'' *Ann. Inst. Pasteur* **93,** 450 (1957).

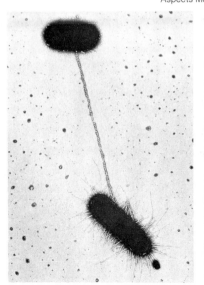

(a)

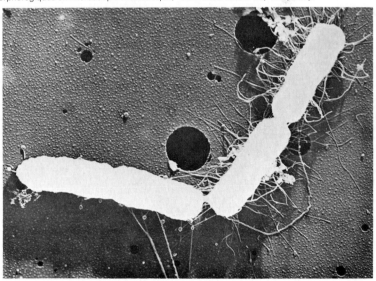

(b)

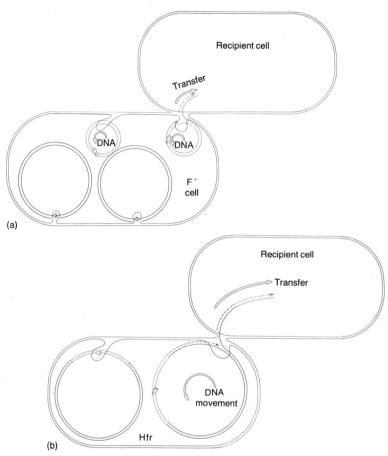

**Figure 8.10**

The hypothetical mechanism of DNA transfer as a consequence of F replication. (a) An F+ cell, containing two autonomous F replicons and two chromosomes, is shown conjugating with a recipient cell. Replication of F is proceeding according to the mechanism outlined in Figure 4.28. The F at the right is being driven into the recipient by replication. (b) The same process takes place when an Hfr cell conjugates. Now, however, chromosomal DNA is also transferred, as a result of its integration with F.

with the difference that, at completion, one copy is located in the recipient while the other remains in the donor cell. Circularization of the plasmid in the recipient then occurs.

## F-mediated chromosome transfer in *E. coli*

### the F+ and Hfr states

Cells that harbor an F factor are called F+; cells that do not are called F−. In a population of F+ cells, the integration of F and chromosome by crossing over occurs about once per $10^5$ cells at each generation; the cells in which this occurs, and the clones that arise from them, are called Hfr, for "high frequency of recombination."

The integration process is reversible. In a population of Hfr cells, detachment by a second crossover occurs at roughly the same rate as integration in an F⁺ population. Thus, every F⁺ population contains a few Hfr cells, and every Hfr population contains a few F⁺ cells.

### DNA transfer by Hfr donor cells

When a suspension of Hfr cells is mixed with an excess of F⁻ cells, every Hfr cell will attach to an F⁻ cell and initiate DNA transfer. Since F and chromosome have integrated, chromosomal DNA, as well as F DNA, passes into the recipient [Figure 8.10(b)].

The order in which chromosomal genes move into the recipient depends on the chromosomal site at which F has integrated, as well as on the polarity of the integration event. In Figure 8.11, for example, the orientation of F at the time of integration is such that, in the resulting Hfr cell, chromosomal markers will be transferred in the order *thr, leu, pro, lac, . . . , met*. Integration may occur at other sites, however, and with the opposite polarity, producing different orders of marker transfer as shown in Figure 8.12. The polarity of integration is presumably determined by the homologous base sequences in the attachment sites on F and chromosome.

The linear transfer of chromosomal genes by Hfr donors can be followed by withdrawing samples from a mating mixture of Hfr and F⁻ cells and shearing apart the mating pairs. Each sample is plated on several different selective media to determine the numbers of different types of recombinants formed by the time of sampling.

The results of such an experiment are shown in Figure 8.13; the genes are designated A, B, C, and D. As this figure shows, the longer the mating couples are allowed to conjugate before being separated, the more genetic material is transferred from the Hfr cells to the F⁻ cells. For example, if the mating pairs are separated after 10 minutes, only gene A will have been transferred. If conjugation is allowed to proceed for 15 minutes, genes A and B are transferred; after 20 minutes, genes A, B, and C are transferred; and so on.

These experiments show that the donor cell slowly injects a chromosomal strand into the receptor cell, so that the chromosomal genes enter the receptor cell in sequence. If conjugation is not interrupted, the entire chromosome may be injected, a process that in *E. coli* requires about 90 minutes at 37°C.

The slopes of the lines in Figure 8.13 show that the donor cells are not synchronized with respect to the initiation of transfer. Some cells begin to transfer almost immediately, but in others there is a variable delay. Within 25 minutes, however, all donor cells begin to transfer. The intercept of each curve with the abscissa indicates the time at which a

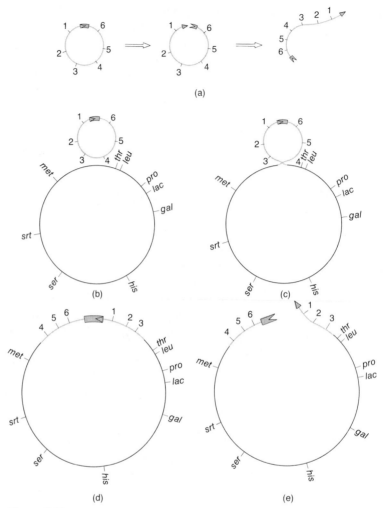

**Figure 8.11**

The breakage and transfer of the Hfr chromosome as a consequence of its integration with F. (a) F is shown as a circle, which has a special site of breakage between markers 1 and 6. Breakage is followed by transfer (during conjugation), such that F markers penetrate the recipient in the order 1-2-3-4-5-6. (b) F has paired with a chromosomal site between *met* and *thr*. (c) A crossover in the region of pairing integrates the two circles. (d) Same as (c), but redrawn as a single circle. (e) Breakage at the special F site leads to transfer causing F markers 1-2-3 to enter the recipient first, followed by chromosomal markers *thr, leu, pro, lac,* and so on; F markers 4-5-6 enter last. Recombinants will be males if they receive all six F markers; thus the terminal end of the chromosome must be transferred to produce an Hfr recombinant.

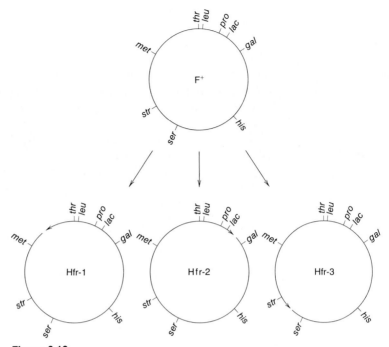

**Figure 8.12**
Diagram showing the origin of three different Hfr males from a common F+ parent. Arrowheads on chromosomes indicate the leading point and direction of chromosome transfer in each case.

**Figure 8.13**
The kinetics of recombinant formation. See text for explanation. After E. Wollman and F. Jacob.

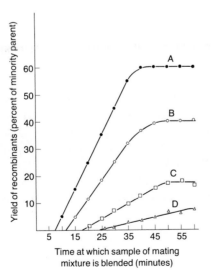

given gene is transferred in the first mating couples to initiate conjugation.

If a mating is allowed to proceed without artificial interruption, considerable spontaneous chromosome breakage occurs. The time at which any given break occurs is random; very few transfers continue as long as 90 minutes. Thus, the farther away a given marker is from the leading point in transfer, the less probability it has of being transferred before breakage occurs. Most couples will achieve the transfer of the earliest marker, A; fewer will achieve the transfer of B; still fewer the transfer of C; and so on. This phenomenon is reflected in the relative heights of the plateaus of the different curves in Figure 8.13.

The order of marker transfer can thus be recognized in two ways: by the *time of entry* of each marker in interrupted mating experiments and by the *frequency of recombination* for each marker in uninterrupted mating experiments.

The speed at which the chromosome is transferred has been found to be relatively constant throughout the mating process, so that the times of entry of markers provide a reliable measure of the distance between them. A genetic map of the chromosome of E. coli K12 is shown in Figure 8.14. Although, of cellular organisms, E. coli is the best understood at the genetic level, fewer than one fourth of the genes that could be accommodated on its chromosome have been identified and mapped.

### the origin and behavior of F′ strains

The change from the $F^+$ to the Hfr state depends on the integration of F and chromosome by a recombinational event (crossover). As pointed out earlier, this process is reversible: in any culture of an Hfr strain, F may detach from the chromosome in an occasional cell, giving rise to a clone of $F^+$ cells.

For a true reversion to the $F^+$ state to occur, recombination must take place within a region of pairing identical to that which existed when integration occurred [Figure 8.15(a)]. Much more rarely, however, exceptional pairing takes place. A crossover in the region of exceptional pairing does not regenerate an ordinary F factor but rather an *F factor that contains within its circular structure a segment of chromosomal DNA* [Figure 8.15(b)]. As shown in Figure 8.15, the *F-genote,* as it is called, may or may not lack a segment of F DNA.

The cell in which this event takes place and the clone derived from it are called *primary F′* (F-prime) *cells.* These cells carry a chromosomal deletion corresponding to the segment that is now an integral part of the F factor.

When primary F′ cells are mated with cells of an $F^-$ strain, F is transferred (together with the integrated chromosomal segment) with high efficiency. Chromosomal transfer, however, occurs with an efficiency of $10^{-5}$ per cell or less, the primary F′ cells having the same low probability of F integration with chromosome as do ordinary $F^+$ cells.

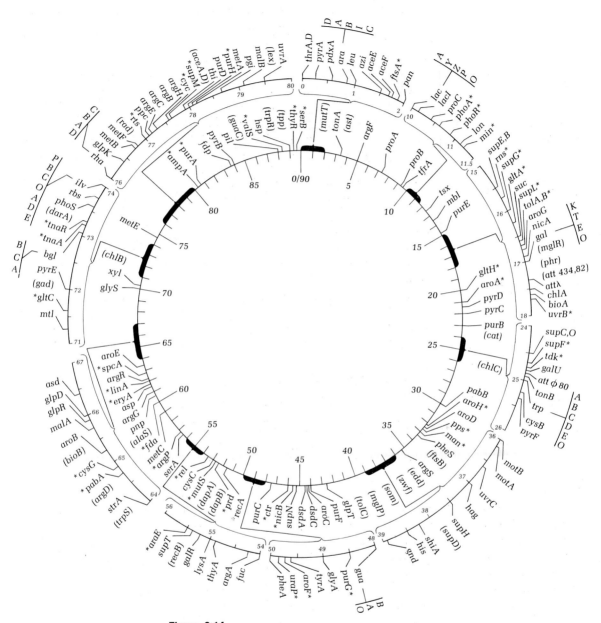

**Figure 8.14**

The genetic map of *E. coli* K12. The units of distance along the inner circle represent the relative times of entry of the markers during conjugal transfer at 37°C. The zero point has arbitrarily been chosen as the *thrA* locus. Certain parts of the map (e.g., the region from 9 to 10 minutes) are displayed on arcs of the outer circle to provide an expanded time scale for crowded regions. Markers in parentheses are only approximately mapped; those marked with an asterisk are more precisely mapped than markers in parentheses, but their positions relative to adjacent markers are not exactly known. Arrows show the direction of messenger RNA transcription for the loci concerned. From A. L. Taylor and C. D. Trotter, "Linkage Map of *Escherichia coli* Strain K12," *Bacteriol. Rev.* **36,** 504 (1972).

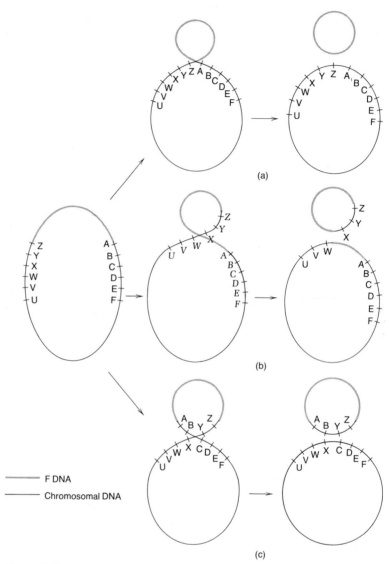

**Figure 8.15**

The generation of F-genotes, in primary F′ cells. At the left is an Hfr chromosome, with the integrated F DNA at the top. Letters A to F and U to Z represent chromosomal markers. (a) Crossing over within the original region of pairing between F and chromosome regenerates a normal F. (b) Pairing in an exceptional region, followed by crossing over, generates an F-genote carrying the chromosomal markers XYZ. The chromosome of the primary F[ contains a segment of F DNA and has a deletion of the XYZ segment. (c) Exceptional pairing in a different region has generated an F-genote containing a full complement of F DNA, plus chromosomal genes from both sides of the former attachment site.

An F-genote that is transferred to an F⁻ cell reestablishes its circular form and becomes an autonomous replicon, producing a *secondary F' cell*. The secondary F' cell differs from the primary F' cell in that the chromosomal segment of the F-genote is also present in the host chromosome; the secondary F' cell is thus a *partial diploid* (Figure 8.16), and the F-genote is a dispensable element.

## occurrence of conjugation in other bacteria

### gram-negative bacteria

In general, a Gram-negative bacterium carrying a conjugational plasmid can mate with a wide variety of other Gram-negative bacteria and transfer plasmid DNA. The efficiency of interspecific and intergenic mating varies widely, however: F⁺ and F' strains have been created in many bacteria of the enteric group by the transfer of suitable plasmids from *E. coli* K12. In some cases, these plasmids have undergone integration with the chromosome of the recipient, forming Hfr donor cells; such Hfrs have been produced in *Salmonella, Yersinia pseudotuberculosis,* and *Erwinia amylovora.*

When Hfr cells are used as donors in intergeneric matings, recombinants are not found, unless there is sufficient base-pair homology between donor and recipient chromosomes to permit pairing and crossing over. Thus, chromosomal recombinants are formed in matings among *Escherichia, Salmonella,* and *Shigella,* among which there is some homology, but never in matings between *Escherichia* and *Proteus,* in which no homology is detectable.

Plasmid-mediated conjugation also occurs in the pseudomonads. Some degradative plasmids mentioned in Chapter 3 mediate conjugation between different species of *Pseudomonas,* and F-like sex factors

**Figure 8.16**
A secondary F' cell. The cell shown is a heterozygous diploid for genes X, Y, and Z.

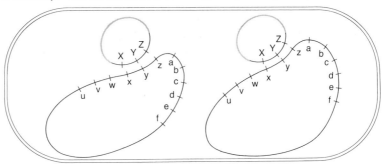

----- F DNA

—— Chromosomal DNA

have been discovered that promote chromosomal transfer at low frequency between strains of *Pseudomonas aeruginosa*.

### gram-positive bacteria

The genetics of the actinomycete *Streptomyces coelicolor* have been explored in detail by D. Hopwood, G. Sermonti, and their associates. Recombination requires direct contact between hyphae, and is mediated by sex factors. Although there are certain fundamental differences between mating systems of *S. coelicolor* and *E. coli*, sex-factor states analogous to those of $F^+$, Hfr, and $F'$ cells are found.

## transduction

In *transduction*, small pieces of bacterial chromosome are incorporated into some maturing phage particles. When these particles infect a new host cell, they inject this small piece of genetic material from the former host.

Certain phages, *e.g.*, P22, transduces all genes with roughly equal efficiency. Other phages, which occupy fixed positions on the bacterial chromosome when in the prophage state, transfer only certain genes. For example, $\lambda$ always attaches between the *bio* and *gal* genes as shown in Figure 8.17, and can transduce only *gal* or *bio*.

Transduction of the type mediated by P22 is called *generalized transduction*. Transduction of the type mediated by $\lambda$ is called *specialized transduction*. Specialized and generalized transduction differ in the mechanism by which the genetic material of the transducing phage particles is formed.

### the mechanism of specialized transduction

In a culture lysogenic for $\lambda$, every cell carries a prophage inserted into the chromosome at a site between the *gal* and *bio*. When such a culture is induced, the prophage detaches by a recombinational event. As a rare event, a recombinational event generates a circle of DNA consisting of a major fraction of $\lambda$, together with a segment of chromosomal DNA from one or the other side of the attachment site (Figure 8.17).

The total amount of DNA that can be packaged in a phage particle is fixed; thus, if *gal* or *bio* genes are included, a corresponding amount of $\lambda$ DNA must be missing. The transducing particle is always defective for some $\lambda$ functions; *gal* and *bio* genes are never found in the same transducing particle.

The induction of a $\lambda$-lysogenic *gal⁺* culture produces a lysate containing about one transducing particle for every $10^5$ normal phage particles. When this lysate is used to infect a culture of nonlysogenic *gal* bacteria, some of the cells receive the contents of the transducing particles,

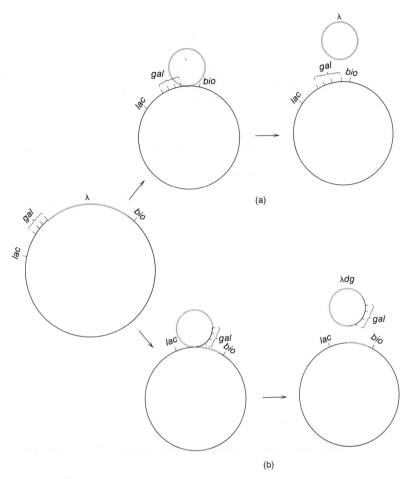

**Figure 8.17**
The formation of λ*dg*. (a) Normal pairing followed by a single crossover event regenerates the wild-type λ genome. (b) Abnormal pairing followed by a single crossover event generates the λ*dg* genetic element.

which become integrated into the chromosome as prophage. Each of these cells, however, still retains its *gal⁻* chromosomal genes and thus becomes a *gal⁺/gal⁻* partial diploid. *Specialized transduction thus involves the addition to the chromosome of a defective prophage carrying a segment of DNA taken from its former attachment site.*

### the mechanism of generalized transduction

In generalized transduction, the transducing phage particles contain a random fragment of bacterial DNA packaged in a phage coat. Thus, the lysate contains two types of particles: a majority type, containing only phage DNA, and a minority type (the transducing particles), containing host DNA.

The difference between the DNA in a specialized transducing particle and in a generalized transducing particle is consistent with the manner in which the transducing lysate is formed. If λ is produced in its host by a lytic cycle of growth, no specialized transducing particles are formed. To obtain such particles, the lysate must be prepared by inducing a lysogenic culture (i.e., λ must be present in the prophage state). Generalized transducing phages, however, can yield transducing lysates by a cycle of lytic growth as well as by induction of a lysogen.

Some phages, including P22, are capable of both generalized and specialized transduction. The former occurs following a lytic cycle of phage growth; the latter follows the induction of lysogenic cells carrying P22 as a chromosomally attached prophage.

## genetic recombination in vitro

Recently a technology (sometimes termed *recombinant DNA technology*) has been developed that makes possible the introduction of genes from any source (viral, procaryotic, or eucaryotic) into bacteria in a form that permits their replication along with the resident genome. This technology derives from the discovery of two types of enzymes that together bring about recombination in vitro: *restriction endonucleases* and *DNA ligases*.

Each restriction endonuclease* is highly specific in its action, recognizing a specific series of base pairs up to six in length and cutting the DNA at specific points within this region. Different bacteria produce different endonucleases with respect to the sequence of base pairs that they recognize and the locations at which they cut the DNA strands. For example, *E. coli* strain K produces a totally different restriction endonuclease from that produced by *E. coli* strain C. Certain restriction endonucleases cleave double-stranded DNA assymetrically in regions termed *palindromes*, i.e., regions that read the same from left to right on one strand as from right to left on the other (Figure 8.18). The consequence of cleavage at such sites is the formation of fragments of DNA each with a single-stranded end that can pair with the end produced by similar cleavage of any DNA. When mixed in the proper ionic environment at the proper temperature, these ends can be caused to pair (*anneal*) at high frequency. The action of a DNA ligase can then reseal the cuts made by the restriction endonuclease. Thus by the combined actions

---

*The biological function of restriction endonucleases is to recognize and cleave DNA from other organisms (foreign DNA), thereby rendering it susceptible to complete destruction by other DNA-degrading enzymes. The DNA of each organism is enzymatically *modified* (usually by attachment of methyl groups on the bases at specific sites) in such a way that it is resistant to cleavage by the particular restriction endonuclease that the organism produces.

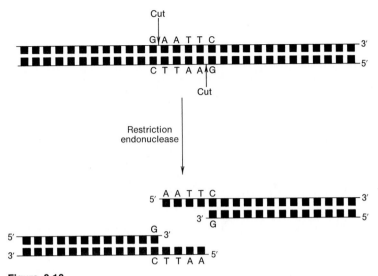

**Figure 8.18**

Cleavage of double stranded DNA by restriction endonuclease EcoR1. A segment of double stranded DNA is shown schematically with the complementary base pairs (G-C and A-T) represented by the vertical bars. The restriction endonuclease, EcoR1, recognizes the sequence G-A-A-T-T-C and cleaves between G and A (vertical arrows). The resulting cuts are, therefore, staggered leaving single-stranded ends which have complementary base-pair sequences, and can be reannealled with each other. After E. A. Adelberg, *Connecticut Medicine* **41**, 397–404 (1977).

of a restriction endonuclease and a DNA ligase, DNA from any source can be joined (recombined) in vitro with DNA from another source (Figure 8.19).

For such in vitro recombined DNA to be replicated in quantity, it must be introduced into a cell (this can be accomplished in bacteria by transformation if the fragment of DNA is relatively small), and the DNA must be able to encode its own replication (if one of the components of the DNA is a virus or a plasmid, i.e. is a replicon [see p. 58] this requirement is also satisfied).

Thus, it is possible to recombine plasmid or viral DNA in vitro and introduce it into a bacterial cell in a form that will be replicated and distributed to progeny cells. The foreign DNA is then said to be *cloned*. In vitro recombination should, in principle, allow the construction of bacterial strains that contain any given gene and, since the genetic code is universal, produce any given protein. Using this technique, certain eucaryotic proteins have already been produced in bacteria. Eucaryotic genes encoding other proteins have been successfully introduced into bacteria, but their protein products have not been made—a consequence, most probably, of the different regulatory controls on expression of eucaryotic and procaryotic genes. It seems reasonable to expect, however, that these barriers can be overcome by genetic manipulation.

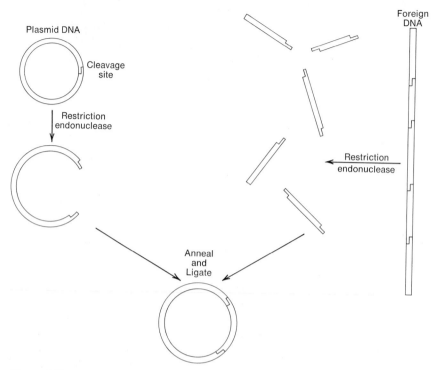

**Figure 8.19**

Schematic representation of the techniques of *in vitro* recombination to clone foreign DNA. Plasmid DNA containing one cleavage site, and foreign DNA containing several randomly located cleavage sites, are treated with the same restriction endonuclease producing fragments which have complementary cohesive ends. When mixed and subjected to annealing conditions some circles will form consisting of a segment of foreign DNA inserted in the plasmid; *in vitro* recombination is completed by sealing the original cuts with DNA ligase. The resulting enlarged plasmid can be transformed into bacteria and replicated by them. After E. A. Adelberg, *Connecticut Medicine* **41**, 397–404 (1977).

The possible benefits to mankind of in vitro recombination techniques are enormous. For example, many eucaryotic proteins, such as insulin and growth hormone, are vital to medical practice and are in short supply. If the genes that encode such proteins were introduced into bacteria, unlimited quantities might be produced inexpensively. Important food components might also be produced in this manner. It has been pointed out, however, that there could be hazards associated with the introduction of foreign DNA into bacteria; i.e., potentially dangerous strains of bacteria might be constructed by design or even by accident.

# TAXONOMY OF CELLULAR MICROORGANISMS

The art of biological classification is known as *taxonomy*. It has two functions: the first is to identify and describe as completely as possible the basic taxonomic units, or *species;* the second, to devise an appropriate way of arranging and cataloguing these units.

## species: the units of classification

The notion of a species is complex. Speaking broadly, a species consists of an assemblage of individuals (or, in microorganisms, of clonal populations) that share a high degree of phenotypic similarity, coupled with an appreciable dissimilarity from other assemblages of the same general kind. The recognition of species would not be possible if natural variation were continuous so that an intergrading series spanned the gap between two assemblages of markedly different phenotype. However, it became evident early in the development of biology that, among most groups of plants and animals, reasonably sharp discontinuities do separate the members of a group into distinguishable assemblages. Hence, the notion of the species as the base of taxonomic operation has proved workable.

Every assemblage of individuals shows some degree of internal phenotypic diversity because genetic variation is always at work. Hence, it becomes a matter of scientific tact to decide what *degree* of phenotypic dissimilarity justifies the breaking up of an assemblage into two or more species; or, to put the matter another way, How much internal diversity is permissible in a species? Opinions on this question vary. Taxonomists themselves can be broadly divided into two groups: "lumpers," who set wide limits to a species, and "splitters," who differentiate species on more slender grounds.

202

For plants and animals that reproduce sexually, a species can be defined in genetic and evolutionary terms. As long as a sexually reproducing population is free to interbreed at random, its total gene pool undergoes continuous redistribution; and new mutations, the source of phenotypic variation, are dispersed throughout the population. Such an interbreeding population may evolve in response to environmental changes, but it will evolve with reasonable uniformity. *Divergent evolution*, eventually leading to the emergence of new species, can occur only if a segment of the population becomes reproductively isolated in an environment that is different from that occupied by the rest of the population. Reproductive isolation is probably usually geographic in the first instance; a physical barrier of some sort (for example, a mountain range or a body of water) is interposed between two parts of the initially continuous population. Within each of these subpopulations, a common gene pool is maintained by interbreeding, but through chance mutation and selection, the two subpopulations are now free to evolve along different lines. They will continue to diverge as long as the geographical barrier persists. Eventually, the cumulative differences become so great that *physiological* isolation is superimposed on geographic isolation, and members of the two populations are no longer capable of interbreeding if they are brought together. Hence, even if the two populations subsequently commingle once more, their gene pools remain permanently separated; a point of no return has been reached. These evolutionary considerations lead to a dynamic definition of the species as *a stage in evolution at which actually or potentially interbreeding arrays have become separated into two or more discrete breeding populations*. This definition is, in fact, an *explanation* of the origin of specific discontinuities in nature. At the same time, it provides an experimental criterion for the recognition of species' differences: the failure to interbreed.

Since most microorganisms are haploid, and reproduce predominantly by asexual means, the concept of the species that has emerged from work with plants and animals is evidently inapplicable to them. A microbial species cannot be considered an interbreeding population: the two offspring produced by the division of a bacterial cell are reproductively isolated from one another, and, in principle, they are free to evolve in a divergent manner. Genetic isolation is to some degree reduced by sexual or parasexual recombination in eucaryotic microorganisms and by the special mechanisms of recombination distinctive of procaryotes. However, it is very difficult to assess the evolutionary effect of these recombinational processes since the frequencies with which they occur in nature are unknown. In procaryotes the problem is further complicated by plasmid transfer, which is relatively nonspecific and permits exchanges of genetic material among bacteria of markedly different genetic constitution.

Since the dynamics of microbial evolution are so unlike the dynamics of evolution of plants and animals, there is no theoretical basis for the assumption that microbial evolution has led to phenotypic discontinuities that would justify the recognition of species. However, the experience of microbial taxonomists has shown that when many strains of a given microbial group are thoroughly analyzed, they can usually be divided into a series of discontinuous clusters: it is such *clusters of strains* that the microbial taxonomist recognizes empirically as species. Further insights into the dynamics of microbial evolution may eventually permit a formal definition of the microbial species; if so, this will most likely be different from the species definition applicable to plants and animals.

In bacterial populations, genetic change can occur so rapidly by mutation that it would be unwise to distinguish species on the basis of differences in a small number of characters, governed by single genes. Accordingly, the best working definition of a microbial species is the following: *a group of strains that shows a high degree of overall phenotypic similarity and that differs from related strain groups with respect to many independent characters.*

### the characterization of species

Ideally, species should be characterized by complete descriptions of their phenotypes or—even better—of their genotypes. Taxonomic practice falls far short of these ideals; in most biological groups, even the phenotypes are only fragmentarily described, and genotypic characterizations are extremely rare.

As a general rule, the phenotypic characters that can be most easily determined are structural or anatomical ones that can be directly observed. For this reason, biological classification is still based, at most levels, almost entirely on structural properties. Virtually the only exception is the classification of bacteria. The extreme structural simplicity of bacteria offers the taxonomist too small a range of characters upon which to base adequate characterizations. Hence, the bacterial taxonomist has always been forced to seek other kinds of characters—biochemical, physiological, ecological—with which to supplement structural data. The classification of bacteria is based, to a far greater extent than that of any other biological groups, on functional attributes. Most bacteria can be identified only by finding out what they can do, not simply how they look.

This confronts the bacterial taxonomist with an additional problem. To find out what a bacterium can do, he has to perform experiments with it. The number of possible experiments that can be performed is extremely large, and although all will reveal facts, the facts so revealed will not necessarily be taxonomically significant ones. Consequently, the bacterial taxonomist can never be sure that he has performed the right experiments for taxonomic purposes; he may well have failed to perform certain experiments that would have shown him significant clustering in a collection of strains, and therefore erroneously conclude that he is dealing with a continuous series. There is no obvious way to get around this difficulty except to make phenotypic characterizations as exhaustive as possible.

## the naming of species

According to a convention known as the *binomial system of nomenclature,* every biological species bears a latinized name that consists of two words. The first word indicates the taxonomic group of immediately higher order, or *genus* (plural, *genera*) to which the species belongs, and the second word identifies it as a particular species of that genus. The first letter of the generic (but not of the specific) name is capitalized, and the whole phrase is italicized: for example, *Escherichia* (generic name) *coli* (specific name). In contexts in which no confusion is possible, the generic name is often abbreviated to its initial letter: *E. coli.*

A rigid and complex set of rules governs biological nomenclature; the rules are designed to keep nomenclature as stable as possible. The specific name given to a newly recognized species cannot be changed unless it can be shown that the organism has previously been described under another specific name, in which case the older name has priority. Of necessity, the same stability does not govern the generic half of the name, since the arrangement of related species into genera is an operation that can be carried out in different ways and often changes in the course of time as new information becomes available. For example, *E. coli* has in the past been placed in the genus *Bacterium,* as *Bacterium coli,* and in the genus *Bacillus,* as *Bacillus coli.* These three names are synonyms since they all refer to one and the same species. This consequence of the binomial system can be very confusing, and taxonomic descriptions usually list all such synonyms in order to minimize the confusion. Binomial nomenclature is used for all biological groups except viruses. The virologists are currently divided over the best way to designate members of this group; some wish to extend the binomial system to the viruses, whereas others would prefer another system that gives in coded form information about the properties of the organism.

In bacterial taxonomy, when a new species is named, a particular strain is designated as the *type strain.* Type strains are preserved in culture collections; if one is lost, a *neotype strain,* which resembles as closely as possible the description of the type strain, is chosen. The type strain is important for nomenclatural purposes since the specific name is attached to it. If other strains, originally included in the same species, prove on subsequent study to deserve recognition as separate species, they must receive new names, the old specific name resting with the type strain and closely related strains.

In the taxonomic treatment of a biological group, the individual species are usually grouped in a series of categories of successively higher order: genus, family, order, class, and division (or phylum). Such an arrangement is known as a *hierarchical* one, because each category in the ascending series unites a progressively larger number of taxonomic units in terms of a progressively smaller number of shared properties. It should be noted that the genus has a position of special importance since according to the rules of nomenclature, a species cannot be named unless it is assigned to a genus. The allocation of a species to a taxonomic category higher than the genus does not carry any essential nomenclatural information; it is merely indicative of the position of an organism, relative to other organisms, in the system of arrangement adopted.

# the problems of taxonomic arrangement

In dealing with a large number of different objects, some system of orderly arrangement is essential for purposes of data storage and retrieval. It does not matter what criteria for making the arrangement are adopted, provided that they are unambiguous and convenient. Books can be arranged in different ways: for example, by subject, by author, or by title. Different individuals tend to adopt different systems, depending on their particular needs and tastes. Such a system of classification, based on arbitrarily chosen criteria, is termed an *artificial* one.

The earliest systems of biological classification were largely artificial in design. However, as knowledge about the anatomy of plants and animals increased, it became evident that these organisms conform to a number of *major patterns* or *types,* each of which shares many common properties, including ones that are not necessarily obvious upon superficial examination. Examples of such types are the mammalian, avian, and reptilian types among vertebrate animals. The first system of biological classification that attempted to group organisms in terms of such typological resemblances and differences was developed in the middle of the eighteenth century by Linnaeus. The Linnaean arrangement was more useful than previous artificial arrangements since the taxonomic position of an organism furnished a large body of information about its properties: to say that an animal belongs to the vertebrate class Mammalia immediately tells one that it possesses all those properties that distinguish mammals collectively from other vertebrates. Because Linnaean classification expressed the *biological nature* of the objects that it classified, it became known as a *natural system* of classification, in contrast to preceding artificial systems.

### the phylogenetic approach to taxonomy

When the fact of biological evolution was recognized, another dimension was immediately added to the concept of a natural classification. For biologists of the eighteenth century, the typological groupings merely expressed *resemblances;* but for post-Darwinian biologists, the groupings revealed *relationships.* In the nineteenth century the concept of a "natural" system accordingly changed: it became one that grouped organisms in terms of their *evolutionary affinities.* The taxonomic hierarchy became in a certain sense the reflection of a family tree. The analogy between a family tree and a hierarchy cannot be pushed very far, however, since the dimension of *time*, implicit in a family tree, is completely absent from a taxonomic hierarchy of extant organisms. This point was not very clearly grasped by many evolutionary biologists for whom taxonomy suddenly acquired a new goal: the restructuring of hierarchies to mirror evolutionary relationships. Such a taxonomic system is known as a *phylogenetic system*.

Reflection and experience have shown, however, that the goal of a phylogenetic classification can seldom be realized. The course that evolution has actually followed can be ascertained only from direct historical evidence, contained in the fossil record. This record is at best fragmentary and becomes almost completely illegible in Precambrian rocks more than 400 million years old. By the beginning of the Cambrian period, most of the major biological groups that exist had already made their appearance; vertebrates and vascular plants are the principal evolutionary newcomers in Postcambrian time. For these two groups, the fossil record is, accordingly, reasonably complete, and the main lines of vascular plant and vertebrate evolution can be retraced with some assurance. For all other major biological groups, the general course of evolution will probably never be known, and there is not enough objective evidence to base their classification on phylogenetic grounds.

### numerical taxonomy

For these and other reasons, many modern taxonomists have explicitly abandoned the phylogenetic approach in favor of a more empirical one: the attempt to base taxonomic arrangement upon quantification of the similarities and differences among organisms. This was first suggested by Michel Adanson, a contemporary of Linnaeus, and is known as *Adansonian* (or *numerical*) *taxonomy*. The underlying assumption is that, provided each phenotypic character is given equal weighting, it should be possible to express numerically the taxonomic distances between organisms, in terms of the number of characters they share, relative to the total number of characters examined. The significance of the numerical relationships so determined is greatly influenced by the number of characters examined; these should be as numerous and as varied as possible, to obtain a representative sampling of phenotype.

Until recently, the Adansonian approach appeared impractical because of the magnitude of the numerical operations involved. This difficulty has been obviated by the advent of computers, which can be programmed to compare data for a large number of characters and organisms and to compute the degrees of similarity.

Numerical taxonomy does not have the evolutionary connotations of phylogenetic taxonomy, but it provides a more objective and stable basis for the construction of taxonomic groupings. Perhaps its greatest advantage is that it cannot be applied at all until a relatively large number of characters have been determined, so that its use encourages a thorough examination of phenotypes. Furthermore, the analyses are open to continuous revision and refinement as more characters in a given group are determined.

## new approaches to bacterial taxonomy

The growth of molecular biology has opened up a number of new approaches to the characterization of organisms, which has had a profound impact on the taxonomy of bacteria. Of particular value are certain techniques that give insights into genotypic properties and thus complement the hitherto exclusively phenotypic characterizations of these organisms. Two kinds of analysis performed upon isolated nucleic acids furnish information about genotype: the analysis of the base composition of DNA and the study of chemical hybridization between nucleic acids isolated from different organisms.

### the base composition of DNA

DNA contains four bases: adenine (A), thymine (T), guanine (G), and cytosine (C). For double-stranded DNA, the base-pairing rules require that $A = T$ and $G = C$. However, there is no chemical restriction on the molar ratio $(G + C):(A + T)$. Early in the chemical study of DNA, analyses showed that this ratio in fact varies over a rather wide range in DNA preparations from different organisms, and subsequent work has revealed that the base composition of DNA is a character of profound taxonomic importance, particularly among microorganisms.

Although DNA base composition may be determined chemically, it can be determined more easily by physical methods, and these are now the ones principally used. The "melting temperature" of DNA (i.e., the temperature at which it becomes denatured, by breakage of the hydrogen bonds that hold together the two strands) is directly related to $G + C$ content, because hydrogen bonding between GC pairs is stronger than that between AT pairs. Strand separation is accompanied by a marked increase in absorbance at 260 nm, the absorption maximum of DNA, and this can be easily measured in a spectrophotometer. The $G + C$ content of DNA may also be determined by subjecting a DNA sample to centrifugation in a CsCl gradient, and determining the position of the DNA bands in the gradient, which affords a precise measure of the density of the DNA. This method can be used because the density of DNA is also a function of the $(G + C):(A + T)$ ratio.

Even when DNA has been considerably fragmented by shearing, preparations from most organisms remain relatively homogeneous when examined by these techniques, which indicates that the mean $G + C$ content varies little in different parts of the genome. However, in many eukaryotic organisms, DNA of mitochondrial or chloroplast origin may differ appreciably in G + C content from the nuclear DNA; and there is sometimes a marked molecular difference between the DNA of the chromosome of a bacterium and that of a plasmid it harbors.

The mean DNA base compositions characteristic of the chromosomal DNA in major groups of organisms are shown in Figure 9.1. In both plants and animals the ranges are relatively narrow and quite similar, centering about a value of 35 to 40 mole percent G + C. Among the protists the ranges are much wider. The widest range of all occurs among the procaryotes, in which the range extends from about 30 to 75 mole percent G + C. If, however, one examines the G + C content of many different strains that belong to a *single microbial species,* the values are closely similar or identical. Each bacterial species, accordingly, has DNA with a characteristic G + C content; this can be considered one of its important specific characters.

Although the selective pressures that have led to divergence are unknown, one point is clear: *a substantial difference between two organisms with respect to mean DNA base composition reflects a large number of individual differences between the specific base sequences of their respective DNAs.* This is prima facie evidence for a major genetic divergence and hence for a wide evolutionary separation. The very broad span of values characteristic of the procaryotes, accordingly, reveals the great evolutionary diversity of this particular biological group, and also suggests its evolutionary antiquity.

**Figure 9.1**

The ranges of DNA base composition characteristic of major biological groups.

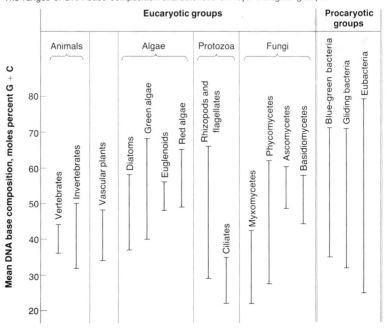

However, two organisms with identical DNA base compositions may differ greatly in genetic constitution. This is evident from the very similar base ratio values for DNA from all plants and animals. Hence, major evolutionary divergence is not *necessarily* expressed by a divergence of mean base composition. When two organisms are closely similar in their DNA base composition, this fact can be construed as indicative of genetic and evolutionary relatedness only if the organisms *also* share a large number of phenotypic properties in common or are known to resemble one another in genetic constitution (e.g., different strains that belong to a single bacterial species). In such a case, near-identity of DNA base composition provides supporting evidence for their genetic and evolutionary relatedness.

### nucleic acid hybridization

Upon rapid cooling of a solution of thermally denatured DNA, the single strands remain separated. However, if the solution is held at a temperature from 10 to 30°C below the $T_m$ (temperature causing denaturation), specific reassociation (annealing) of complementary strands to form double-stranded molecules occurs. There is always some random pairing, but since it is the complementary regions that form the most stable duplexes, their reassociation is favored.

Shortly after this phenomenon had been discovered, it was shown that when DNA preparations from two related strains of bacteria are mixed and treated in this manner, *hybrid DNA molecules are formed*. When similar experiments were conducted with DNA preparations from two unrelated bacteria, no hybridization could be detected; upon annealing, duplexes were formed only by specific pairing between single strands originally derived from the same DNA.

The discovery of the reassociation of single-stranded DNA molecules from different biological sources to form hybrid duplexes laid the foundations of an entirely new approach to the study of genetic relatedness in bacteria. In vitro experiments on DNA-DNA reassociation permit an assessment of the overall degree of genetic homology between two bacteria. Furthermore, since duplexes can also be formed between single-stranded DNA and complementary RNA strands, analogous DNA-RNA reassociations can be performed. If the RNA preparations consist of either tRNAs or rRNAs, such experiments permit an assessment of the genetic homology between two bacteria *with respect to specific, relatively small segments of the chromosome:* those that code the base sequences either of the transfer RNAs or of the ribosomal RNAs (rRNAs).

A variety of relatively simple methods for measuring nucleic acid reassociation have been developed. All are based on the same general principle: formation of duplexes between two denatured DNA samples; separation of the duplexes from residual single-stranded nucleic acid; and measurement of the amount of heteroduplex formed. The amount of reassociation of a reference DNA with itself is determined and assigned an

**Table 9.1**
Nucleic Acid Homologies Between *Pseudomonas acidovorans* and Four Other *Pseudomonas* Species, as Revealed by Parallel DNA-DNA and DNA-rRNA Hybridization Experiments

| | RELATIVE DUPLEX FORMATION | |
| --- | --- | --- |
| | DNA-DNA | DNA-rRNA |
| *P. acidovorans/P. acidovorans* | 100 | 100 |
| *P. acidovorans/P. testosteroni* | 33 | 92 |
| *P. acidovorans/P. delafieldii* | 0 | 89 |
| *P. acidovorans/P. facilis* | 0 | 87 |
| *P. acidovorans/P. saccharophila* | 0 | 79 |

*Note:* Data from N. J. Patleroni, R. Kunisawa, R. Contopoulou, and M. Doudoroff, *Int. J. Syst. Bacteriol.* **23**, 333 (1973).

arbitrary value of 100. The amount of reassociation between the reference DNA and DNAs from heterologous strains can then be measured and expressed as a percentage of the value for the homologous DNA-DNA reassociation. The same principles apply to DNA-RNA reassociation experiments.

Carefully controlled DNA-DNA reassociation experiments can provide much semiquantitative information about the degree of genetic homology between related strains or species of bacteria. However, if evolutionary divergence has led to numerous differences of base sequence in the two genomes, specific DNA-DNA reassociation becomes too weak to be measured. The range of organisms among which genetic homology is detectable can be greatly extended by parallel studies on DNA-rRNA reassociation, because the relatively small portion of the bacterial genome that codes for ribosomal RNAs has a much more highly conserved base sequence than the bulk of the chromosomal DNA. As a result, it is frequently possible to detect by DNA-rRNA reassociation relatively high homology between the genomes of two bacteria that show no significant homology by DNA-DNA reassociation (Table 9.1).

## the study of relatedness at the level of gene translation

### protein sequence
Although the redundancy of the genetic code*
theoretically permits proteins of identical amino acid composition to be specified by radically different base sequences on the DNA, usually changes in gene sequences cause changes in the primary structures of the proteins encoded. Moreover, in view of the near-infinite variety of possible amino acid

---

*Redundancy of the code refers to the fact that in certain cases several different codons specify the same amino acid (see Table 4.6, p. 95).

sequences in proteins, the evolution of similar amino acid sequences from unrelated ancestral proteins is unlikely. Thus similarity of the primary structure of proteins is an indication of evolutionary relatedness. This principle provides a powerful method for determining relatedness of bacterial strains at a nongenetic level. In cases in which it has been possible to compare this approach to genetic approaches and, in the case of higher organisms, to the analysis of the fossil record, the results have been essentially concordant.

Obviously, the most accurate assessment of relatedness of proteins requires the detailed comparison of their primary sequences. However, the determination of primary sequence is tedious, and the complete amino acid sequences of homologous proteins from a variety of microorganisms are available in only a relatively few cases. More widely employed are methods that detect similarities in primary sequence by examing the properties of the proteins. One of the most powerful of such methods involves the use of antibodies. If a purified protein is injected into a rabbit or other warmblooded vertebrate, antibodies are formed (see Chapter 16, p. 400). These antibodies are capable of specifically combining with small areas on the surface of the protein termed, *antigenic determinants*. Since each protein has numerous different determinants on its surface, the animal will produce a population of antibodies, each directed against a different antigenic determinant. The serum from the animal can then be used as a probe to detect the presence of these determinants on other protein molecules. When such cross reaction is detected, it is presumptive evidence for relationship, since the similar determinants reflect similar amino acid sequence.

### regulatory mechanisms

Recent comparative studies on the patterns of regulation of biosynthetic and dissimilatory pathways common to many bacterial groups suggest that the specific mode of regulation of a given pathway is a conservative and group-specific property. A striking illustration is provided by the regulation of two branched biosynthetic pathways: (1) one that leads to the synthesis of aromatic amino acids and (2) one that leads to synthesis of amino acids of the aspartate family. In each, both end-product repression and feedback inhibition are employed by many groups of bacteria to regulate enzyme synthesis and carbon flow through the pathway. However, the details of these regulatory systems (e.g., the identity of allosteric effectors) are different in each group.

# the main outlines of microbial classification

### taxonomy of the bacteria

Until very recently, bacterial systematics was based strictly on phenotypic characterization. Some traits considered to have taxonomic importance were the Gram stain, cell shape and size, mode of flagellar insertion,* and several biochemical tests. Of these, the Gram reaction (or more accurately the structure of the cell wall of which the staining properties are usually a reflection) is still of primary importance, and the dependence on biochemical criteria has been extended.

The recently developed molecular approaches to the analysis of bacterial relationships have provided a very valuable supplement to earlier purely phenotypic characterizations. It is now possible to recognize among the bacteria a considerable number of subgroups that appear to be natural ones, in the sense that their members are all of common evolutionary origin. Many of these groups (e.g., the enteric bacteria and some homology groups of aerobic pseudomonads) contain a large number of species, readily distinguishable by both phenotypic and genotypic criteria. It is often desirable to place such species in two or more genera, in turn united into a multigeneric assemblage, for example, a family. Thus, the family *Enterobacteriaceae* unites the various genera of enteric bacteria.

Up to a certain point, accordingly, it is feasible to organize the classification of bacteria in a hierarchical system of the traditional type. However, apart from the fundamental shared property of procaryotic cellular organization, the bacteria are an assemblage of very great diversity; and one in which, furthermore, major structural and functional attributes are often poorly correlated with one another. Hence, none of the numerous attempts to develop a complete hierarchical classification for these organisms has proved satisfactory. Modern bacterial taxonomists thus have adopted the approach of dividing the bacteria into groups, some of which are natural. Many other groups are provisional, and will undoubtedly require revision as we learn more about their constituent subgroups. The current, generally accepted scheme of bacterial groups can be found in the 8th edition of *Bergey's Manual of Determinative Bacteriology*,† a compendium of bacterial taxonomy revised approximately every decade. The groupings adopted here, and presented in the next three chapters, are broadly similar to those presented in *Bergey's Manual*.

### taxonomy of the protists

Since the protists are predominantly unicellular and frequently lack a sexual cycle, the methodology discussed above is in theory as useful for their taxonomy as it is for that of the bacteria. In practice, however, the much greater structural complexity of these eucaryotic organisms has reduced the dependence of the taxonomist on biochemical or molecular approaches. Description of cell and organelle structure and the elucidation of the details of life cycles have allowed the determination of at least

---

*Flagella may be inserted at any point on the bacterial surface. Although many specific patterns can be recognized, the most useful distinction is between *polar* flagellation (flagella inserted at the end of the cell) and *peritrichous* flagellation (flagella inserted over the entire surface of the cell).

†R. E. Buchanan and N. E. Gibbons, eds., *Bergey's Manual of Determinative Bacteriology*. 8th ed. The Williams and Wilkins Co., Baltimore, 1974.

the broad outlines of an appropriate classification scheme for the protists. There are, nevertheless, numerous areas in which modern molecular techniques could profitably be applied, and their gradual adoption by systematists should greatly clarify our understanding of relationships among the various groups of protists. Chapter 13 outlines the major groups of protists as we now perceive them.

# 10
# THE AUTOTROPHIC PROCARYOTES

The autotrophs are united by their possession of the biochemical pathways necessary to utilize $CO_2$ as the major source of cell carbon. In all cases studied, the Calvin cycle (see Chapter 4, p. 84) is used for $CO_2$ fixation, although there is still some controversy surrounding claims that the green bacteria utilize this route. The Calvin cycle is the mechanism of $CO_2$ fixation in eucaryotic phototrophs, and thus appears to be universal in autotrophs.

Among the procaryotes, two major groups of autotrophs may be discerned: the *photoautotrophs,* which derive energy from light, and the *chemoautotrophs,* which use chemical oxidations as a source of energy. Chemoautotrophic metabolism is found only among the procaryotes. Although not autotrophs, the methylotrophic bacteria (which oxidize reduced one-carbon compounds) share a number of biochemical similarities with the chemoautotrophs and will be discussed in this chapter.

## the phototrophic procaryotes

The procaryotes include three distinct and well-defined photosynthetic groups. The *blue-green bacteria* perform oxygenic photosynthesis and possess a pigment system similar in basic respects to that of photosynthetic eucaryotes. This group has long been treated by botanists as one of the major classes or divisions of the algae. However, the typically procaryotic cell structure of these organisms, which was clearly established about 1960, identifies them unambiguously as bacteria. The group is large and structurally diverse, including many different types of filamentous and unicellular organisms. Movement, when it occurs, is by gliding.

216

*Purple bacteria* and *green bacteria* perform anoxygenic photosynthesis and possess unique pigment systems that confer on them spectral properties unlike those of all other phototrophs. With one exception, they are all unicellular. Most green bacteria are small, immotile, rod-shaped organisms; the purple bacteria are rods, cocci, or spirilla, frequently motile by means of flagella. Because of the evident structural resemblances of these organisms to nonphotosynthetic bacteria, their taxonomic position has never been questioned; they have been included among the bacteria since their discovery in the mid-nineteenth century. However, the nature of their metabolism remained controversial until 1930, when C. B. van Niel first recognized and defined the various metabolic versions of anoxygenic photosynthesis and demonstrated that it is the characteristic mode of energy-yielding metabolism in both purple and green bacteria.

Table 10.1 summarizes and compares some of the major properties associated with photosynthetic function in purple bacteria, green bacteria, and blue-green bacteria. Comparative data for the chloroplasts of two major algal groups are also included.

This tabulation makes clear how closely the photosynthetic machinery of the blue-green bacteria resembles that of eucaryotes, and more particularly that of one algal group, the red algae. Blue-

**Table 10.1**
The Structure of the Photosynthetic Apparatus and the Mechanisms of Photosynthesis in Procaryotes and in Chloroplasts

| | PROCARYOTES | | | CHLOROPLASTS OF | |
| --- | --- | --- | --- | --- | --- |
| | PURPLE BACTERIA | GREEN BACTERIA | BLUE-GREEN BACTERIA | RED ALGAE | GREEN ALGAE AND PLANTS |
| Cell structure that contains photosynthetic apparatus | Cell membrane | Chlorobium vesicles and cell membrane | Thylakoids and phycobilisomes | Thylakoids and phycobilisomes | Thylakoids |
| Photosystems | | | | | |
| I | + | + | + | + | + |
| II | − | − | + | + | + |
| Reductants used for $CO_2$ assimilation | $H_2S$, $H_2$, or organic compounds | $H_2S$, $H_2$ | $H_2O$ | $H_2O$ | $H_2O$ |
| Principal photosynthetic carbon source | $CO_2$ or organic compounds | $CO_2$ or organic compounds | $CO_2$ | $CO_2$ | $CO_2$ |
| Pigments (chlorophylls) | Bacteriochlorophyll a or Bacteriochlorophyll b | Minor Bacteriochlorophyll a Major Bacteriochlorophyll c, d, or e | Chlorophyll a | Chlorophyll a | Chlorophylls a and b |
| Phycobiliproteins | − | − | + | + | − |

green bacteria and red algae share the following unique functional properties: chlorophyll a is their only chlorophyllous pigment; the chromoproteins known as *phycobiliproteins* are their major light-harvesting pigments; and these chromoproteins are not integrated into the thylacoids, as are both chlorophylls and carotenoids, but are instead localized in special structures known as *phycobilisomes*, which are attached to the outer surfaces of the thylacoids.

The purple and green bacteria differ from other phototrophs with respect to nearly all the characters listed. Lacking photosystem II, these organisms cannot use $H_2O$ as a reductant for $CO_2$ assimilation; this function is assumed by other reduced inorganic compounds. The pigments and lipids of the photosynthetic apparatus are largely group-specific in both purple and green bacteria.

The photosynthetic apparatus has a unique intracellular location in each group. In the purple bacteria, it is located within topologically complex infoldings of the cytoplasmic membrane (Figure 10.1). In the green bacteria the reaction centers are also located within the cytoplasmic membrane, but the light-gathering bacteriochlorophyll is housed in special organelles (*chlorobium vesicles*). These vesicles (Figure 10.2) are cigar-shaped structures, about 50 nm wide by 100 to 150 nm long, which are tightly appressed to the inner surface of the cytoplasmic membrane. They are bounded by a single-layered membrane 3 to 5 nm thick, which, although composed of lipid and protein, is chemically distinct from the unit membrane that bounds the cell.

### the chlorophylls of procaryotes

The chlorophylls of purple and green bacteria have the same basic structure and are synthesized through the same biosynthetic pathway as plant chlorophylls. However, because they are confined to these two bacterial groups, they are termed *bacteriochlorophylls*. The bacterio-

**Figure 10.1**
Electron micrographs of thin sections of two purple bacteria, illustrating variations in the structure of the internal membranes. (a) *Rhodopseudomonas sphaeroides*, in which the membranes occur as hollow vesicles (arrows) ($\times$5,640). (b) *Rhodopseudomonas palustris*, in which the membranes occur in regular parallel layers in the cortical region of the cell (arrows) ($\times$5,640). Micrographs courtesy of Germaine Cohen-Bazire.

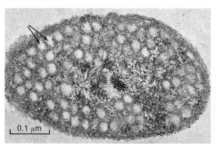
0.1 μm

(a)

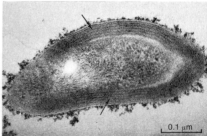
0.1 μm

(b)

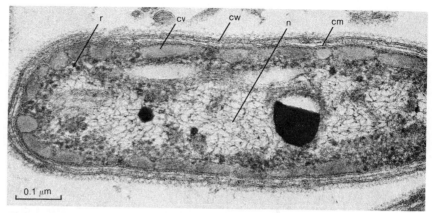

**Figure 10.2**
Electron micrograph of a thin section of the green bacterium *Pelodictyon*, showing the relationship of the chlorobium vesicles (cv) to other parts of the cell (×81,800): cw, cell wall; cm, cell membrane; r, ribosomes; n, nucleoplasm. Photo courtesy of Germaine Cohen-Bazire.

chlorophylls fall into two chemical subclasses: bacteriochlorophylls a and b; and bacteriochlorophylls c, d, and e. This is correlated with their biological distributions. The cells of purple bacteria contain only one form of bacteriochlorophyll: either bacteriochlorophyll a or bacteriochlorophyll b, depending on the species. The pigment is in part responsible for light harvesting and in part associated with the photochemical reaction centers. The cells of green bacteria always contain a major and minor form of bacteriochlorophyll. Depending on the species, the major form is either bacteriochlorophyll c, bacteriochlorophyll d, or bacteriochlorophyll e; it has a light-harvesting role. The minor pigment in all green bacteria is bacteriochlorophyll a; and it is the form of chlorophyll in the photochemical reaction centers.

### the carotenoids of photosynthetic procaryotes

Carotenoids are always associated with the photosynthetic apparatus. They serve as light-harvesting pigments, absorbing light in the blue-green region of the spectrum, between 400 and 550 nm; their relative contribution to this function is major in some photosynthetic organisms, minor in others. Among phototrophs, carotenoid composition tends to be both complex and group specific, although the major carotenoids of all eucaryotic phototrophs are similar. Blue-green bacteria resemble the eucaryotes in terms of carotenoid compositions.

### phycobiliproteins

The phycobiliproteins are the major light-harvesting pigments of both blue-green bacteria and red algae. They are contained in granules termed *phycobilisomes*, which occur in regular array on the outer faces of the thylacoids (Figure 10.3). The light energy absorbed by these

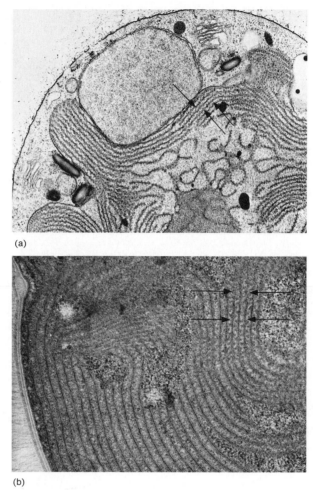

(a)

(b)

**Figure 10.3**
Electron micrographs of cellular thin sections, showing the arrangement of phycobilisomes (arrows) on the surface of the thylacoids in a chloroplast of the red alga *Porphyridium* (a) and in the cell of a blue-green bacterium *Aphanocapsa* (b). (a) Courtesy of Dr. E. Gantt; (b) courtesy of Dr. G. Cohen-Bazire.

pigments is transferred with very high efficiency to the chlorophyll-containing reaction centers in the thylacoids.

### the cellular absorption spectra of photosynthetic procaryotes

Although not all pigments in the photosynthetic apparatus are equally effective in harvesting light (chlorophyll a is far less effective than phycobiliproteins in blue-green bacteria), the cellular absorption spectra of photosynthetic organisms provide a rough indication of the spectral regions that are utilized for the performance of photosynthesis. Figure 10.4 compares the cellular absorption spectra of several different photosynthetic procaryotes; in each case, the specific contributions to light absorption made by chlorophylls, carotenoids, and phycobiliproteins are indicated. It is evident that the cellular absorption spectra characteristic of each group of photosynthetic procaryotes is distinctive, and to a considerable extent complementary to the spectra of other groups.

Broadly speaking, the cellular absorption spectra of photosynthetic eucaryotes resemble those of blue-green bacteria, though only red algae show major peaks in the region between 550 and 630 nm, where phycobiliproteins absorb. The differences in the light-absorbing properties of the various groups of photosynthetic organisms are of profound ecological significance, as will be discussed later (p. 232).

### the colors of photosynthetic procaryotes

The common names of the three groups of photosynthetic procaryotes are not always well correlated with the color of their cells, as judged visually. Since the major chlorophyll absorption bands of purple bacteria lie in the infrared, to which the eye is blind, the visible color of these organisms is determined largely by their carotenoid complement. The green bacteria appear green, yellow-green, orange, or brown, depending on their carotenoid composition. The phycobiliproteins of blue-green bacteria contribute largely to light absorption in the visible region, and the visible color of the cells is therefore much influenced by the phycobiliprotein complement; although usually green or blue-green, masses of these cells may also be brown, orange, yellow, or yellow-green.

### the blue-green bacteria

Although uniform in nutritional and metabolic respects, blue-green bacteria are structurally diverse. Some are immotile unicellular rods or cocci that reproduce by binary fission (Figure 10.5). Certain of these forms develop as colonial aggregates, held together by thin multilayered sheaths formed by detachment of the outer layer of the cell wall (Figure 10.6).

A mode of reproduction characteristic of some blue-green bacteria that is rare in other procaryotes is *multiple fission*. It is the sole mode of reproduction in the unicellular organism *Dermocarpa* (Figure 10.7). The cell increases many times in size, and then it rapidly undergoes many cell divisions, after which the daughter cells are released by breakage of the wall of the mother cell. Each daughter cell is motile for a short period, motility being lost as the cell begins to grow. Other blue-green bacteria can undergo both binary and multiple fission.

Many blue-green bacteria are filamentous organisms, which reproduce by fragmentation of the filament. Some develop

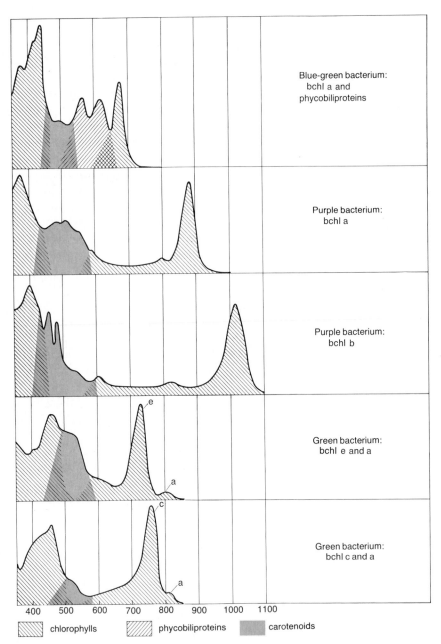

**Figure 10.4**

Cellular absorption spectra of five representative photosynthetic procaryotes, to show the characteristic differences in the positions of the major absorption bands. The approximate contributions to cellular light absorption by the major classes of photosynthetic pigments, and the types of chlorophyll present in each organism are indicated on the figure. The double peak of phycobiliprotein light absorption in the spectrum of the blue-green bacterium illustrated reflects the presence of both phycoerythrin (maximum: 675 nm) and phycocyanin (maximum: 625 nm). The two peaks of bacteriochlorophyll a and bacteriochlorophyll b absorbance at long wavelengths reflect the different physical-chemical states of reaction center bacterio-chlorophyll (small peak) and antenna bacteriochlorophyll (large peak).

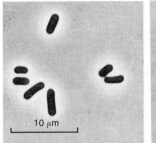
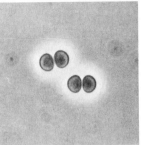

10 μm

**Figure 10.5**
Some unicellular blue-green bacteria. Courtesy of Rosmarie Rippka and
Riyo Kunisawa.

as simple filaments; others can form branched filamentous structures [Figures 10.8(a) and (b)]. Among certain of the filamentous blue-green bacteria, there is some degree of cellular differentiation. Within the filament, vegetative cells may develop into larger, heavy-walled resting cells or *akinetes;* or into cells with a specialized metabolic function, known as *heterocysts* [Figure 10.8(c)].

The structural diversity of blue-green bacteria is paralleled by a considerable genetic diversity, as revealed by their mean DNA base composition (Figure 10.9). The total span is almost as wide as that for procaryotes as a whole, and several distinct compositional groups occur among the unicellular representatives.

With respect to nutritional and metabolic properties, however, the group is relatively uniform. All are photoautotrophs; and growth factors are rarely required.

Many blue-green bacteria are *obligate phototrophs,* being wholly incapable of dark growth at the expense of organic sources of carbon and energy. In members of the group that can grow in the dark, the

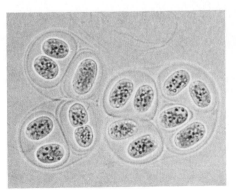

**Figure 10.6**
A large unicellular blue-green bacterium, the cells of which are held together in groups by a multilayered sheath (phase contrast ×935). Courtesy of Rosmarie Rippka and Riyo Kunisawa.

(a)

(b)

(c)

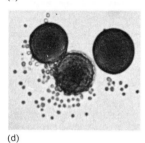

(d)

**Figure 10.7**
*Dermocarpa*, a unicellular blue-green bacterium which reproduces exclusively by multiple fission. Successive photomicrographs, taken over a period of 240 hours, of the development of three cells. (a–c) Cell enlargement; (d) liberation of small daughter cells produced by multiple fission of a vegetative cell, (×400). Courtesy of John Waterbury.

growth rate is very low relative to that in the light; such growth occurs only at the expense of glucose and a few other sugars that are dissimilated by aerobic respiration. The obligate photoautotrophy characteristic of many blue-green bacteria appears to be caused by the absence of the specific permeases necessary for the uptake of exogenous sugars by the cell, since the enzymatic machinery of sugar dissimilation is present in all blue-green bacteria. This pathway permits the generation of ATP in the dark, through endogenous respiration of the cellular glycogen store.

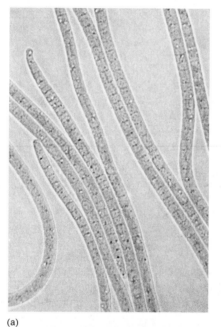

(a)

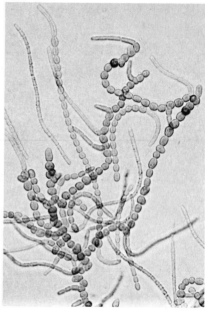

(b)

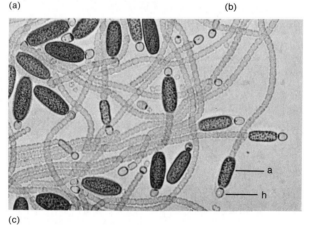

(c)

**Figure 10.8**
Filamentous blue-green bacteria. (a) *Oscillatoria*, a blue-green bacterium which grows as unbranched filaments (×748). (b) *Fischerella*, blue-green bacterium which grows as branched filaments (×467). (c) *Cylindrospermum*, a filamentous, unbranched blue-green bacterium which forms two types of differentiated cells, akinetes (a) and heterocysts (h) (×339). Courtesy of R. Rippka.

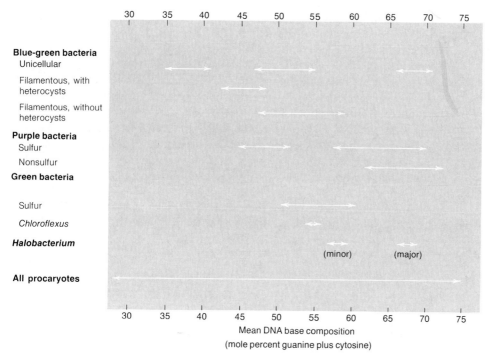

**Figure 10.9**
The ranges of mean DNA base composition among the photosynthetic procaryotes and *Halobacterium*. The range among unicellular blue-green bacteria is particularly wide; they include several subgroups of markedly different DNA base composition. Among purple sulfur bacteria two different base compositional subgroups are also evident. The DNA from *Halobacterium* is heterogeneous, the cells containing a major and a minor DNA component of markedly different compositions.

Blue-green bacteria are the only organisms able to perform oxygenic photosynthesis that can also fix nitrogen; many (though not all) are vigorous nitrogen fixers. Such organisms have the simplest known nutritional requirements since they can grow in the light in a mineral medium, using $CO_2$ as a carbon source and $N_2$ as a nitrogen source.

The coexistence in a single organism of the processes of oxygenic photosynthesis and nitrogen fixation presents an obvious paradox, since nitrogen fixation is an intrinsically anaerobic process; the key enzyme, nitrogenase, is rapidly and irreversibly inactivated by exposure even to low partial pressures of oxygen. With few exceptions, nitrogen-fixing blue-green bacteria are filamentous organisms that produce a specialized type of cell, the *heterocyst*.

Heterocystous blue-green bacteria form few, if any, heterocysts when grown photosynthetically with a combined nitrogen source (nitrate or ammonia); and nitrogenase synthesis is also repressed under these conditions. Both nitrogenase synthesis and heterocyst formation are initiated when cultures are deprived of combined nitrogen. Normally, about 5 to 10 percent of the cells in the filaments develop into heterocysts, following removal of combined nitrogen from the medium (Figure 10.10).

The differentiation of a heterocyst from a vegetative cell is accompanied by the synthesis of a new, thick outer wall layer and the loss of photosystem II; heterocysts can therefore neither fix $CO_2$ nor produce $O_2$ in the light.

The conditions that favor heterocyst formation, coupled with the special properties of these cells, suggest that heterocysts are the specific cellular sites of nitrogen fixation under aerobic conditions in the light. The retention of photosystem I enables the heterocyst to generate ATP

**Figure 10.10**

The effect of the nitrogen source on heterocyst formation in a blue-green bacterium, *Anabaena* sp. (a) Filaments from a culture grown with ammonia as a nitrogen source; (b) filaments from a culture grown with $N_2$ as a nitrogen source; h, heterocyst ($\times 616$). Courtesy of R. Rippka.

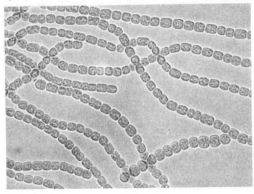

(a)

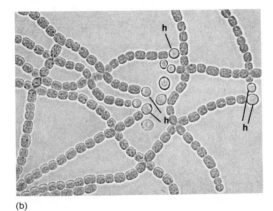

(b)

by cyclic photophosphorylation. The loss of photosystem II provides an intracellular environment in which $O_2$ is not produced, and which is thus favorable for the maintenance of nitrogenase function. However, this loss also renders the heterocyst dependent on adjacent vegetative cells for a supply of intermediary metabolites that can furnish the reducing power necessary to reduce $N_2$ to ammonia. The nongrowing heterocysts in turn furnish the vegetative cells of the filament with the fixed nitrogen compounds necessary for their growth. In most heterocystous blue-green bacteria, heterocysts are spaced at regular intervals along the filament. As the filament elongates, by growth and division of the component vegetative cells, new heterocysts are differentiated midway between the preexisting ones.

The blue-green bacteria occupy a far wider range of habitats than do other photosynthetic procaryotes, occurring in all environments that support the growth of algae: the sea, fresh water, and soil. Furthermore, they develop in certain habitats from which photosynthetic eucaryotes are largely or completely excluded. Nitrogen-fixing representatives are conspicuous in environments where combined nitrogen is a limiting nutrient, notably in tropical soils, which are often nitrogen-poor. Certain thermophilic blue-green bacteria grow abundantly in neutral or alkaline hot springs, where they are the predominant members of the photosynthetic population. The temperature ranges of thermophilic blue-green bacteria vary, but some unicellular forms can grow at temperatures in excess of 70°C. They are excluded by their relatively high pH range from acid hot springs, of which the characteristic photosynthetic inhabitant is a red alga, *Cyanidium caldarum*, which has a low pH optimum and is the only truly thermophilic photosynthetic eucaryote. However, its temperature maximum (approximately 56°C) is considerably below that of many thermophilic blue-green bacteria.

Some blue-green bacteria are capable of switching to nonoxygenic photosynthesis in the presence of high concentrations of sulfide, which replaces water as electron donor. This ability allows them to inhabit illuminated spots from which algae are excluded by the toxic sulfide concentrations and the anaerobic nature of the habitat.

Deserts are an extreme environment in which the microbial photosynthetic population consists almost entirely of unicellular blue-green bacteria. These organisms grow in microfissures just below the surface of rocks, where small amounts of moisture are trapped and where sufficient light penetrates to permit photosynthesis. The ability to tolerate extreme fluctuations of temperature, which often become very high during the day, is important to their survival in the desert habitat.

In lakes that have undergone eutrophication (artificial enrichment with mineral nutrients, notably phosphate and nitrate), a massive development of unicellular and filamentous blue-green bacteria characteristically occurs during the warmer months of the year. These are largely gas vacuolate forms; in calm weather, the population floats to the surface, accumulating there to produce a *bloom*. Subsequent death and decomposition of the bloom promote a massive development of chemoheterotrophic microorganisms, which may have catastrophic effects on the animal population of the lake because they deplete the dissolved oxygen supply.

### the purple bacteria

Taxonomically speaking, the purple bacteria are a small group, consisting of only about 30 species. They are unicellular and reproduce by binary fission or, in a few species, by budding (see p. 105). Most are motile by flagella; a few are immotile. Gas vacuoles are formed by some. Despite the small size of this group, it must be genetically diverse because the mean DNA base composition ranges from 46 to 73 moles percent G + C (Figure 10.9).

All purple bacteria are, at least potentially, photoautotrophs, capable of growing anaerobically in the light with $CO_2$ as the principal carbon source and reduced inorganic compounds as the electron donor. However, some purple bacteria can also develop photoheterotrophically under anaerobic conditions in the light at the expense of organic compounds. Under these circumstances, cell material is derived largely from the organic substrate, although $CO_2$ may also be assimilated.

It is customary to recognize two subgroups among the purple bacteria (Table 10.2); distinctions between them are both physiological and ecological. *Purple sulfur bacteria* have a predominantly photoautotrophic mode of metabolism, based on the use of $H_2S$ as an electron donor, and are strict anaerobes. *Purple nonsulfur bacteria* have a predominantly photoheterotrophic mode of metabolism. They are sensitive to $H_2S$, their growth being inhibited by low concentrations of sulfide, even though some are capable of oxidizing sulfide anaerobically in the light if the concentration is kept very low.

Whereas the purple sulfur bacteria are obligate phototrophs, many purple nonsulfur bacteria can grow well aerobically in the

**Table 10.2**
Characters that Distinguish the Two Subgroups of Purple Bacteria

| | PURPLE SULFUR BACTERIA | PURPLE NONSULFUR BACTERIA |
|---|---|---|
| Principal mode of photosynthesis | Photoautotrophic | Photoheterotrophic |
| Range of photoassimilable organic substrates | Narrow | Broad |
| Aerobic dark growth | − | + or − |
| Ability to oxidize $H_2S$ | + | + or − |
| Accumulation of $S^0$ as intermediate in $H_2S$ oxidation $SO_4^{2-}$ | + | − |
| $H_2S$ toxicity | Usually low | Usually high |

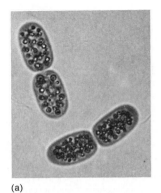

(a)

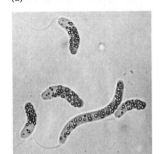

(b)

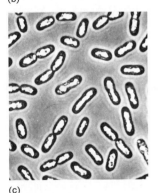

(c)

**Figure 10.11**
Photomicrographs of some representative purple sulfur bacteria. (a) *Chromatium*, ×1400; (b) *Thiospirillum*, ×1190; (c) *Thiodictyon*, ×1400. (a–b) Ordinary illumination; (c) phase contrast. Note the intracellular sulfur granules in (a–b). The phase-bright intracellular areas in (c) are gas vacuoles. Courtesy of Dr. N. Pfennig.

dark. Such strains possess an aerobic electron transport chain, and are thus endowed with respiratory capacity. A few of them can also grow (though very slowly) anaerobically in the dark, by fermenting pyruvate or sugars.

The purple nonsulfur bacteria typically occur in freshwater lakes or ponds, where organic matter is present but sulfide is either absent or present at low concentrations. The typical habitats of the purple sulfur bacteria are sulfide-rich waters, where sulfide is generated by the activity of sulfate-reducing bacteria. Both groups fix $N_2$ under anaerobic conditions; no $N_2$-fixation is performed aerobically.

PURPLE SULFUR BACTERIA. Some typical genera of purple sulfur bacteria are illustrated in Figure 10.11. The characteristic photometabolism of these organisms involves assimilation of $CO_2$, ATP being provided by cyclic photophosphorylation; reducing power is provided by $H_2S$ that is oxidized anaerobically, via elemental sulfur, to sulfate. The overall reaction can be represented schematically as:

$$2CO_2 + H_2S + 2H_2O \longrightarrow 2(CH_2O) + H_2SO_4.$$

Some (but not all) purple sulfur bacteria can use other reduced inorganic sulfur compounds or $H_2$ in place of $H_2S$ as exogenous reductants. The biochemistry of the oxidation of these reduced sulfur compounds by purple sulfur bacteria is complex and not well established. It is probably similar to the oxidation of reduced inorganic sulfur compounds by aerobic chemoautotrophs.

The oxidation of $H_2S$ by the purple sulfur bacteria always leads to a massive but transient accumulation of elemental sulfur ($S^0$) since this first step is much more rapid than the ensuing oxidation of $S^0$ to $SO_4^{2-}$. In most of these organisms the elemental sulfur is deposited within the cell, as refractile globules. However, *Ectothiorhodospira* spp. excrete sulfur into the medium, and subsequently reabsorb it prior to further oxidation.

PURPLE NONSULFUR BACTERIA. Photomicrographs of some typical representatives of purple nonsulfur bacteria are shown in Figure 10.12. The only purple bacteria that reproduce by budding rather than by binary fission are members of this subgroup; they include *Rhodomicrobium* and some *Rhodopseudomonas* spp.

The range of organic compounds that can be photoassimilated by purple nonsulfur bacteria is quite wide. Species capable of respiratory metabolism can grow aerobically in the dark, by oxidation of essentially the same range of organic substrates that they photoassimilate anaerobically in the light. This does not necessarily involve the operation of the same metabolic pathways, however.

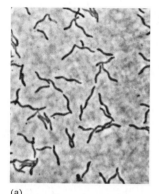

(a)

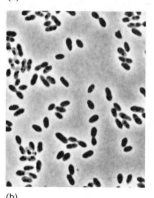

(b)

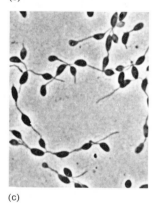

(c)

**Figure 10.12**
Photomicrographs of some representative purple nonsulfur bacteria, all ×1400. (a) *Rhodospirillum;* (b) *Rhodopseudomonas;* (c) *Rhodomicrobium.* (a–b) ordinary illumination; (c) phase contrast. Courtesy of Dr. N. Pfennig.

Most of the purple nonsulfur bacteria require vitamins, and their growth rate is frequently improved by amino acids.

None of the purple nonsulfur bacteria are killed by exposure to air; however, some of these organisms cannot use $O_2$ as a terminal electron acceptor, and therefore cannot grow aerobically in the dark. Others grow at least as rapidly under aerobic conditions in the dark as they do under anaerobic conditions in the light. However, aerobic growth leads rather rapidly to an almost complete loss of the photosynthetic pigment system. This is a consequence of the fact that, even at relatively low partial pressures, $O_2$ is a potent repressor of pigment synthesis by purple bacteria, exerting this effect even in the presence of light. Light itself is not required for pigment synthesis, as shown by the fact that species possessing a fermentative metabolism maintain a high pigment content through many generations of anaerobic growth in the dark.

### the green bacteria

The green bacteria comprise an even smaller taxonomic group than the purple bacteria; there are only nine recognized species placed in five genera. Photomicrographs of typical representatives are shown in Figure 10.13. The span of DNA base composition is comparatively narrow: 48 to 58 moles percent G + C (Figure 10.13). With respect to its physiological and nutritional properties, the group shows interesting parallels to the purple bacteria. Most members of the green sulfur bacteria are counterparts of the purple sulfur bacteria. However, the recently discovered thermophilic green bacterium, *Chloroflexus,* closely resembles the purple nonsulfur bacteria in its metabolic and nutritional properties.

The green sulfur bacteria are small, permanently immotile rod-shaped bacteria; four genera are recognized on the basis of structural characters. The members of this subgroup are strictly anaerobic photoautotrophs, which use $H_2S$, other reduced inorganic sulfur compounds, or $H_2$ as electron donors. Elemental sulfur arising from $H_2S$ oxidation is deposited extracellularly (as in *Ectothiorhodospira*), prior to oxidation to sulfate. Since green sulfur bacteria cannot use sulfate as a sulfur source, they require sulfide to meet biosynthetic needs when growing with $H_2$ as an electron donor. Nitrogen fixation is of common occurrence. In all these respects, the analogies to purple sulfur bacteria are evident. Indeed, purple and green sulfur bacteria commonly coexist in illuminated, sulfide-rich anaerobic aquatic environments and have essentially overlapping natural distributions.

There is a marked difference, however, with respect to carbon nutrition in the two groups. None of the green sulfur bacteria can grow photoheterotrophically, using organic compounds as their sole or principal carbon source in the absence of an inorganic reductant. They can photoassimilate acetate, but only if $H_2S$ and $CO_2$ are simultaneously provided.

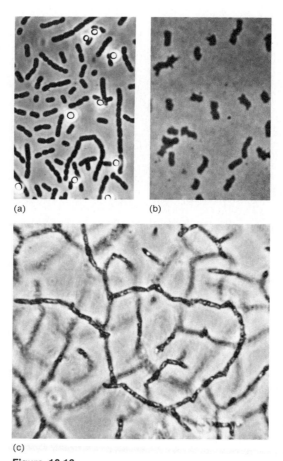

(a)  (b)

(c)

**Figure 10.13**

Photomicrographs (phase contrast) of green sulfur bacteria. (a) *Chlorobium*, ×1500; note extracellular sulfur granules. (b) *Prosthecochloris*, ×2300; the prosthecae are just detectable by light microscopy, conferring an irregular outline on the profile of the cells. (c) *Pelodictyon*, ×1500, showing the characteristic net formation; the phase-bright areas in some of the cells are gas vacuoles. Courtesy of Dr. N. Pfennig.

In 1971 B. Pierson and K. Castenholz discovered a new category of green bacteria, the *Chloroflexus* group. Although these organisms differ from green sulfur bacteria in their structure, nutrition, metabolism, and ecology, they possess two properties that clearly identify them as green bacteria: the presence in the cells of chlorobium vesicles and of bacteriochlorophylls c and a as the major and minor chlorophyllous pigments.

The *Chloroflexus* group consists of filamentous, gliding organisms (Figure 10.14), the filaments sometimes attaining a length of 300 $\mu$m. They are thermophiles and develop abundantly in neutral or alka-

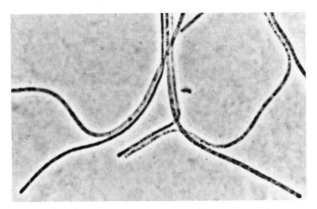

**Figure 10.14**
*Chloroflexus aurantiacus*, a filamentous, gliding green bacterium (phase contrast ×1,040). Courtesy of B. K. Pierson and R. W. Castenholz.

line hot springs at temperatures in the range of 45°C to 70°C. The masses of intertwined filaments form orange to dull green mats several millimeters thick, often closely associated with unicellular blue-green bacteria.

*Chloroflexus* is capable of both anaerobic photo-heterotrophic growth and aerobic chemoheterotrophic growth. Like the purple nonsulfur bacteria, bacteriochlorophyll synthesis is repressed under aerobic conditions. The nutritional requirements of *Chloroflexus* are complex and not yet precisely determined, thus it has not been possible to demonstrate autotrophic growth. However, $H_2S$ can be used as electron donor for the phototrophic reduction of $CO_2$ as principal carbon source.

In its mode of movement and filamentous structure, *Chloroflexus* resembles some filamentous, nonheterocyst-forming blue-green bacteria; and its carotenoids are similar to those of blue-green bacteria. However, its other chemical properties, its cellular fine structure, and its anoxygenic mode of photosynthesis show that its affinities lie rather with the green sulfur bacteria. Since the latter organisms are all immotile, their possible relations to motile groups of procaryotes were previously obscure. However, the gliding motility characteristic of *Chloroflexus* suggests that the green bacteria as a whole should probably be placed among the gliding procaryotes. In this respect, they differ from the purple bacteria, which, if motile, bear flagella.

### ecological restrictions imposed by anoxygenic photosynthesis

For the performance of photosynthesis, anoxygenic phototrophs require anaerobic conditions and either organic compounds or reduced inorganic compounds other than water. These limitations do not apply to blue-green bacteria and photosynthetic eucaryotes. The purple and green bacteria are confined to a limited range of special habitats, and their quantitative contribution to photosynthetic productivity in the biosphere is negligible. They are exclusively aquatic and grow in bodies of water that provide the indispensable combination of anaerobiosis, light, and the nutrients specific for these organisms. These conditions occur principally in two types of aquatic environments, similar in chemical respects but differing markedly in the quality of light available. The first consists of shallow ponds, bays, or estuaries relatively rich in organic matter, $CO_2$, $H_2$, and often $H_2S$, produced by anaerobic bacteria in the underlying sediment. Except near the air-water interface, occupied by blue-green bacteria and algae, the water is essentially oxygen-free. Hence, purple and green bacteria can grow close to the water surface, where light intensity is high, but are usually covered by a growth of oxygenic phototrophs. It is in this environment that the ability of purple and green bacteria to absorb light of very long wavelengths, transmitted by overlying oxygenic phototrophs, becomes of critical importance for their survival. The light used for photosynthesis is almost entirely absorbed by bacteriochlorophylls, in the far red and infrared regions.

The second environment in which purple and green bacteria abound occurs at a considerable depth in lakes, particularly so-called *meromictic* lakes that are characterized by a permanent stratification of the water. The warmer, aerobic upper layer is underlain at depths of 10 to 30 meters by a stagnant layer that is cold and oxygen-free. The anoxygenic phototrophs occur in a narrow horizontal band, situated just within the anaerobic layer (Figure 10.15). Water samples from this depth are often brightly colored as a result of their content of purple and green bacteria, the population density being far greater than that of oxygenic phototrophs in the upper, aerobic layers. However, at the depth where the purple and green bacteria find the anaerobic conditions necessary for development, *the overlying water*

**Figure 10.15**

Diagram of the structure of a meromictic lake, showing (a) the vertical distributions of oxygenetic phototrophs and anoxygenetic phototrophs and (b) the relative concentrations in the water profile of dissolved oxygen and $H_2S$.

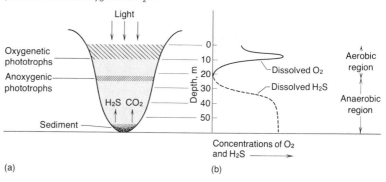

*column itself becomes an effective light filter,* transmitting only green and blue-green light, of wavelengths between 450 and 550 nm. The role of light-harvesting pigments is largely assumed by carotenoids, not by bacteriochlorophylls; and anoxygenic phototrophs from this environment typically have a very high carotenoid content.

### the evolution of photosynthesis

During much of its early history, the surface of the earth remained oxygen-free; the earliest forms of life of necessity obtained energy by anaerobic metabolic processes. It is now believed that the emergence of living organisms occurred at the expense of accumulated organic matter, synthesized in a prebiotic phase of planetary evolution; this organic matter also furnished an energy supply for the first forms of life. However, biological evolution would have come to an early end, had not some members of the community acquired the ability to use light as an energy source by a process analogous to—but no doubt much simpler than—contemporary anoxygenic photosynthesis. Until about 3 billion years ago, anoxygenic photosynthesis remained the dominant form of phototrophic metabolism. About that time, however, molecular oxygen appeared in the biosphere, almost certainly as a consequence of the development, in some members of the primitive photosynthetic community, of oxygenic photosynthesis. This crucial event required a profound modification of the photosynthetic apparatus: the evolution of a second type of photosynthetic reaction center, capable of oxidizing water, which permitted the use of this universally available inorganic compound as a source of reducing power. As the oxygen concentration of the biosphere increased, phototrophs that possessed the older, anoxygenic machinery of photosynthesis were progressively displaced, surviving in anaerobic niches largely as a result of their special light-harvesting pigment systems, which enabled them to avoid direct competition with oxygenic phototrophs for light. Two such isolated groups, the purple and green bacteria, have managed to persist until the present time.

The properties of the contemporary blue-green bacteria provide conclusive evidence that oxygenic photosynthesis emerged within the framework of the procaryotic cell. Indeed, fossil microorganisms resembling blue-green bacteria have been discovered in the oldest known fossil-bearing sedimentary formations (flints about 3 billion years old), which date from before the change (about 2 billion years) from an anaerobic to an aerobic biosphere.

At a later stage of biological evolution, the capacity for the performance of oxygenic photosynthesis became implanted in certain lines of eucaryotes. This probably occurred initially by the introduction into the host cell of photosynthetic procaryotic endosymbionts, which subsequently lost their genetic autonomy and became reduced to organelles (chloroplasts). In the context of this hypothesis, the close structural and functional resemblances between the photosynthetic machinery of blue-green bacteria and the chloroplasts of red algae suggest the likelihood of a direct origin of the red algal chloroplast from blue-green bacterial endosymbionts. The possible procaryotic origins of other types of chloroplasts are more obscure.

### the extreme halophiles

Highly saline environments (salt lakes, brines) harbor large populations of a small and distinctive group of Gram-negative bacteria: immotile cocci (*Halococcus*) and polarly flagellated rods (*Halobacterium*). Despite the differences of cell form, these extreme halophiles share a number of common properties, many of which are clearly adaptations to the high salinity and high light intensity of their natural habitat. Notably, peptidoglycans are absent from the cell walls of both groups.

A characteristic feature of the extreme halophiles is their content of red carotenoids, which are incorporated into the cell membrane. The carotenoids of *Halobacterium* have been shown to protect the cells from photochemical damage by the high light intensities characteristic of their natural environment.

Extreme halophiles are aerobic organisms with complex nutritional requirements and are commonly cultivated in peptone-containing media; amino acids are the preferred carbon and energy sources. These organisms frequently develop as colored patches on salted dried fish and hides as a result of treatment with salt that contains these bacteria.

Until very recently, it was assumed that extreme halophiles were aerobic chemoheterotrophs, with an exclusively respiratory metabolism. However, this interpretation had to be modified in 1971 with respect to *Halobacterium,* as a result of the discoveries of W. Stoeckenius and his collaborators. These discoveries developed from an unexpected observation. When *Halobacterium* is grown in liquid cultures subject to $O_2$ limitation, the cells synthesize a chemically modified cell membrane. In aerobically grown cells, the cell membrane is red, as a result of its high carotenoid content. Oxygen limitation induces synthesis of a new *purple membrane* component. This component is laid down as a series of discrete patches, embedded in the red membrane, and can account for as much as half its total area. The purple areas are readily distinguishable in electron micrographs of freeze-fractured cells (Figure 10.16).

In addition to lipid, the purple membrane contains only one species of protein, a chromoprotein called *bacteriorhodopsin* because of its similarities to the visual pigment of the vertebrate retina, rhodopsin. When intact cells of *Halobacterium* containing purple membrane are

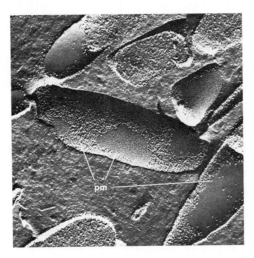

**Figure 10.16**
Electron micrograph of a freeze-fractured preparation of *Halobacterium* cells in 4.3 M NaCl, showing surface structure of the cell membrane. The areas composed of purple membrane (pm) are recognizable by their smooth surfaces. ×35,000. Courtesy of Dr. W. Stoeckenius.

exposed to light, the bacteriorhodopsin catalyzes the release of protons from the cells. In effect, a *proton gradient across the cell membrane is established,* and is maintained as long as illumination is continued. This proton gradient, like that formed during electron transport (see p. 69), results in ATP formation. *Halobacterium* can thus synthesize ATP by a completely new mechanism of photophosphorylation, which does not involve the participation of chlorophyll-containing photochemical reaction centers.

The adaptive value of this device in the strongly illuminated natural environment of *Halobacterium* is obvious. The solubility of $O_2$ in a saturated salt solution is much lower than in pure water; hence, *Halobacterium*, which has a strictly aerobic dark metabolism, is often exposed to low concentrations of dissolved oxygen. This stimulates purple membrane synthesis, just as oxygen deprivation stimulates bacteriochlorophyll synthesis in purple nonsulfur bacteria. *Halobacterium* thereby acquires the ability to make ATP by a light-mediated mechanism in the absence of oxygen.

The extreme halophiles also include one purple sulfur bacterium, *Ectothiorhodospira halophila.* This organism is a strict anaerobe and develops in salt-rich habitats that contain sulfide, which makes these habitats unfavorable for *Halobacterium.* The ecological ranges of the two organisms do not therefore overlap.

## the chemoautotrophs

By definition, a chemoautotroph can grow in a strictly mineral medium in the dark, deriving its carbon from $CO_2$ and its ATP and reducing power from the respiration of an inorganic substrate. This mode of life, which exists only among procaryotes, was discovered between 1880 and 1890 by S. Winogradsky: his pioneering studies on several of the principal subgroups provided a solid foundation for all later work on chemoautotrophy. Winogradsky showed that two other remarkable properties are characteristic of the chemoautotrophs:

1. High specificity with respect to the inorganic energy source.
2. Frequent inability to use organic compounds as energy and carbon sources; indeed, their growth is sometimes adversely affected by organic compounds.

The inorganic materials capable of supporting chemoautotrophic growth include $H_2S$ and other reduced forms of sulfur; ammonia and nitrite; molecular hydrogen; and ferrous iron ($Fe^{2+}$). In the biosphere these materials are in part produced through the metabolic activities of other organisms and are in part of geochemical origin.

The substrate specificities of the chemoautotrophs permit the recognition of four major subgroups (Table 10.3). (1) *Nitrifying bacteria* use reduced inorganic nitrogen compounds as energy sources. The substrate specificity within this subgroup is very high; its members either oxidize ammonia to nitrite, or nitrite to nitrate; none can oxidize both these reduced nitrogen compounds. (2) *Sulfur-oxidizing bacteria* use $H_2S$, elemental sulfur, or its partially reduced oxides, as energy sources; all these substances are converted to sulfate. One member of this group can in addition use ferrous iron as an energy source. (3) *Iron bacteria* can oxidize reduced iron and manganese, but not reduced sulfur compounds; however, their status as true chemoautotrophs remains in some doubt. (4) The *hydrogen bacteria* use molecular hydrogen as an energy source.

**Table 10.3**
Physiological Groups of Chemoautotrophs

| GROUP | | OXIDIZABLE SUBSTRATE | OXIDIZED PRODUCT | TERMINAL ELECTRON ACCEPTOR | TAXONOMIC STRUCTURE |
|---|---|---|---|---|---|
| Nitrifying bacteria | Ammonia oxidizers | $NH_3$ | $NO_2^-$ | $O_2$ | 4 genera; 5 species |
| | Nitrite oxidizers | $NO_2^-$ | $NO_3^-$ | $O_2$ | 3 genera; 3 species |
| Sulfur oxidizers[a] | | $H_2S$, S, $S_2O_3^{2-}$ | $SO_4^{2-}$ | $O_2$; sometimes $NO_3^-$ | 3 genera; 10 species[b] |
| Iron bacteria[c] | | $Fe^{2+}$ | $Fe^{3+}$ | $O_2$ | Several genera |
| Hydrogen bacteria | | $H_2$ | $H_2O$ | $O_2$; sometimes $NO_3^-$ | Classified with chemoheterotrophs in several genera; about 10 species |

[a] One species can also use $Fe^{2+}$ as an energy source.
[b] Includes only those members of the group that have been isolated in pure culture; many other sulfur oxidizers have been described from nature, but never isolated.
[c] There is some doubt whether these organisms are true autotrophs.

### nitrifying bacteria

In the middle of the nineteenth century, circumstantial evidence indicated that the oxidation of ammonia to nitrate in natural environments is a microbial process. However, attempts to isolate the causal agents by the use of conventional culture media failed. This problem was solved in 1890 by S. Winogradsky, who succeeded in isolating pure cultures of nitrifying bacteria by the use of strictly inorganic media. The causal agents proved to be small, Gram-negative, rod-shaped bacteria: the ammonia oxidizer, *Nitrosomonas,* and the nitrite oxidizer, *Nitrobacter.* Both these genera develop best under neutral or alkaline conditions; since the oxidation of ammonia to nitrite results in considerable acid formation, the growth medium for *Nitrosomonas* must be well buffered (for example, by the addition of insoluble carbonates). Growth of both these organisms is slow, minimal generation times approximating 24 hours; and the growth yields, expressed in terms of the quantity of the inorganic substrate oxidized, are low. Winogradsky showed that these bacteria are obligate autotrophs, incapable of development in the absence of the specific inorganic energy source.

In recent years a few other ammonia and nitrite oxidizers have been discovered. They resemble the two prototypes physiologically, but are remarkably diverse in structural respects. The group is a small one, consisting of eight species, assigned to seven genera, distinguished by structural properties and by the specific oxidizable substrate. Both the ammonia oxidizers and the nitrite oxidizers have narrow ranges of DNA base composition, the values for the latter (58 to 62 percent $G + C$) being significantly higher than for the former (50 to 55 percent $G + C$). Most strains possess flagella, although the mode of insertion varies. Division, normally by binary fission, is by budding in some *Nitrobacter* strains.

The diversity of the nitrifying bacteria in gross cell structure is paralleled by a curious diversity in fine structure. In some genera the cell membrane is devoid of intrusions; in others there are extensive intrusions, which may be vesicular, lamellar, or tubular (Figure 10.17).

Obligate autotrophy is the rule in the nitrifying bacteria, with the exception of some *Nitrobacter* strains, which have been shown to use acetate as a carbon and energy source; however, these strains grow much more slowly with acetate than with nitrite.

### sulfur oxidizers

In the course of his pioneering studies on chemoautotrophy, Winogradsky examined the properties of the chemotrophic filamentous gliding bacteria of the Beggiatoa group, which occur characteristically in certain sulfide-rich environments and often contain massive inclusions of elemental sulfur. He showed that these organisms can oxidize $H_2S$, initially to elemental sulfur, which accumulates in the cells; the stored sulfur is subsequently further oxidized to sulfate. A variety of other

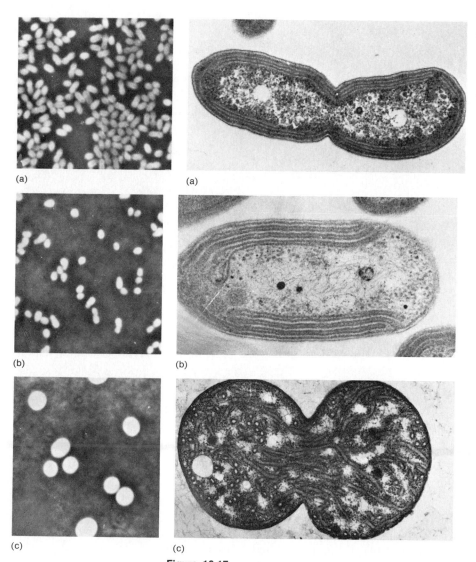

**Figure 10.17**

Phase contrast photomicrographs (at left) and electron micrographs of thin sections (at right) of nitrifying bacteria. (a) *Nitrosomonas*, ×2,200 and ×32,500. (b) *Nitrobacter*, ×2,200 and ×63,200. (c) *Nitrococcus*, ×2,200 and ×21,000. From S. W. Watson and M. Mandel. "Comparison of the Morphology and Deoxyribonucleic Acid Composition of 27 Strains of Nitrifying Bacteria," *J. Bact.* **107,** 563 (1971).

aerobic bacteria (Figure 10.18) have been subsequently shown to oxidize $H_2S$ in a similar manner, with transient intracellular sulfur deposition. However, none of these organisms has been isolated in pure culture because of the technical difficulty of growing them aerobically at the expense of $H_2S$. Current knowledge of this physiological group of chemoautotrophs is based al-

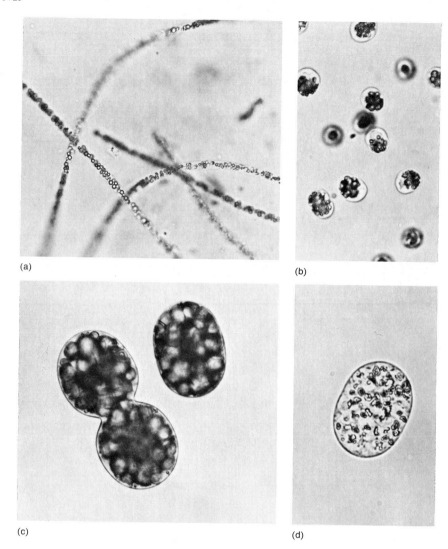

(a)

(b)

(c)

(d)

**Figure 10.18**
Some large, colorless, sulfur-oxidizing bacteria which accumulate sulfur internally. (a) Filaments of *Beggiatoa,* ×900. (b) *Thiovulum,* ×701. (c) and (d) *Achromatium,* ×700. The cells of this very large bacterium contain numerous calcium carbonate inclusions, shown in (c); the cell in (d) has been treated with dilute acetic acid, which has dissolved the inclusions of calcium carbonate, revealing the sulfur granules which are also present. Courtesy of Dr. J. W. M. La Rivière. (a) From J. W. M. La Rivière, ''The Microbial Sulfur Cycle and Some of Its Implications for the Geochemistry of Sulfur Isotopes,'' *Geologischer Rundschau* **55,** 568 (1966). (c) and (d) from W. E. de Boer, J. W. M. La Rivière, and K. Schmidt, ''Some Properties of *Achromatium oxaliferum,''* *Antonie van Leeuwenhoek* **37,** 553 (1971).

most entirely on work with unicellular sulfur oxidizers that have small cells and do not accumulate sulfur within the cell (Figure 10.19). These members of

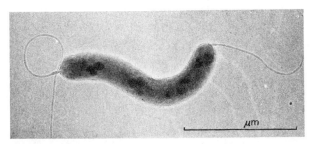

**Figure 10.19**

The colorless sulfur-oxidizing bacterium *Thiomicrospira*. Electron micrograph, showing the polar flagella. From J. G. Kuenen, and H. Veldkamp, "*Thiomicrospira pelophila*, gen. n., sp. n., a New Obligately Chemolithotrophic Colorless Sulfur Bacterium," *Antonie van Leeuwenhoek* **38,** 241 (1972).

the group can be purified and grown without difficulty (Figure 10.20), as a result of their ability to oxidize chemically stable reduced forms of sulfur, notably thiosulfate and elemental sulfur. Most of them are small, polarly flagellated rods, placed in the genus *Thiobacillus* and occur widely in both marine and terrestrial environments. The spiral organism *Thiomicrospira* occurs in marine mud; *Sulfolobus*, an organism of irregular cell form, is confined to thermal habitats (Table 10.4). It may be noted that the range of DNA base composition in the group is rather wide, particularly among the thiobacilli. None of the small-celled sulfur oxidizers contain the extensive membranous intrusions that are characteristic of many nitrifying bacteria.

The chemoautotrophic growth of these organisms is rapid, some having generation times as short as two hours when

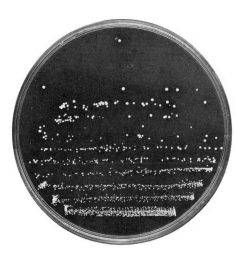

**Figure 10.20**

*Thiobacillus thioparus* colonies growing on a mineral agar plate with thiosulfate as the energy source. The colonies are pale yellow and very refractile, as a consequence of the deposition of elemental sulfur among the cells.

**Table 10.4**
Sulfur-Oxidizing Chemoautotrophs That Have Been Isolated in Pure Culture

|  | THIOBACILLUS (8 spp.) | THIOMICROSPIRA (1 sp.) | SULFOLOBUS (1 sp.) |
|---|---|---|---|
| Cell form | Rods | Spirals | Lobed spheres |
| Flagella | Polar | Polar | Immotile |
| DNA base composition (mole % G + C) | 34–70 | 48 | 60–68 |
| pH range | Variable | 5.0–8.5 | 0.5–6.0 |
| Autotrophy | Variable | Obligate | Facultative |

growing at the expense of thiosulfate. A common and striking feature of the group is their extreme acid tolerance; some species can grow at a pH below 1 and fail to grow at pH values above 6. These organisms are often found in special environments in which the pH is maintained at a low level by their metabolic activities, since the oxidation of reduced sulfur compounds to sulfate results in considerable acid formation. The thermophile *Sulfolobus* is an inhabitant of acid, sulfur-rich hot springs, where sulfide of geochemical origin is immediately converted to elemental sulfur at the high ambient temperature of the water (70 to 85°C) in contact with air. *Sulfolobus* grows attached to the sulfur particles (Figure 10.21), which serve as its oxidizable substrate.

A specialized, man-made environment in which thiobacilli are abundant is the acid drainage water discharge from mines that contain metal sulfide minerals, notably iron pyrite ($FeS_2$). The predominant species in this habitat are the strongly acidophilic species *Thiobacillus thiooxidans,* which rapidly oxidizes elemental sulfur, and *T. ferrooxidans,* which can derive energy from the oxidation both of reduced sulfur compounds and $Fe^{2+}$. Since this organism can grow at pH values of 2 to 4, where ferrous iron is chemically stable, its ability to grow chemoautotrophically at the expense of the reaction:

$$4Fe^{2+} + 4H^+ + O_2 \longrightarrow 4Fe^{3+} + 2H_2O$$

can be rigorously established, but so far this has not been achieved for classical "iron bacteria," which have higher pH ranges (see next section).

Obligate chemoautotrophy is not the rule among sulfur oxidizers as it is among nitrifying bacteria. *Sulfolobus* and several thiobacilli can grow with organic carbon and energy sources; however, the range of utilizable substrates appears to be relatively narrow.

### the iron bacteria

Certain freshwater ponds and springs have a high content of reduced iron salts. It has long been known that a distinctive bacterial flora is associated with such habitats. These *iron bacteria* form natural

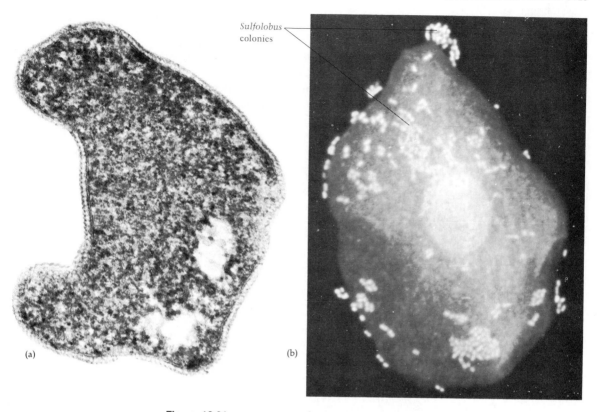

*Sulfolobus* colonies

**Figure 10.21**

*Sulfolobus acidocaldarius.* (a) Electron micrograph of a thin section, ×83,000. From T. D. Brock, K. M. Brock, R. T. Belly, and R. L. Weiss, *Arch. Mikrobiol.* **84,** 64, 1972. (b) Fluorescent photomicrograph of cells stained with acridine orange attached to elemental sulfur crystal, ×750.

colonies that are heavily encrusted with ferric oxide. However, since most iron springs are neutral or alkaline, ferrous iron undergoes rapid spontaneous oxidation in contact with the atmosphere, so it has proved very difficult to ascertain the role that it plays in the metabolism of these bacteria. The most conspicuous iron bacteria are filamentous ensheathed bacteria of the *Sphaerotilus* group (see Chapter 11, p. 259) in many of which the sheaths are encrusted with iron oxide. They can be readily grown as chemoheterotrophs and so isolated in pure culture. Although it is possible to show that such pure cultures will accumulate iron oxide on the sheaths, there is no evidence that such deposition is a physiologically significant process or that these bacteria can develop as chemoautotrophs. Chemoautotrophic growth seems more probable in the case of another structurally distinctive iron bacterium, *Gallionella*. All attempts to obtain cultures of *Gallionella* in organic media have failed, but it has been grown artificially (although not in the pure state) in a mineral medium containing a deposit of ferrous sulfide as a source of reduced iron. The use of this virtually insoluble ferrous salt minimizes the problem of spontaneous oxidation of iron under neutral conditions. In these cultures, *Gallionella* will form cottony colonies attached to the wall of the vessel. Much of the "colony" is inorganic: the small, bean-shaped bacterial cells are located at the branched tips of the excreted stalks, which are heavily impregnated with ferric hydroxide (Figure 10.22).

## hydrogen bacteria

Many species of aerobic bacteria possess the ability to grow chemoautotrophically with molecular hydrogen. In contrast to other groups of chemoautotrophs, the hydrogen bacteria are generally nutri-

**Figure 10.22**
The iron bacterium, *Gallionella*. (a) Flocculent colonies (consisting largely of ferric hydroxide) growing attached to the glass in a liquid culture. (b) Light micrograph of the edge of a colony, showing cells attached to the tips of a branched stalk, ×2,430. (c) Electron micrograph of a single cell, attached to the tip of the stalk, which is impregnated with ferric hydroxide. Courtesy of R. S. Wolfe. Photographs (a) and (b) reproduced from S. Kucera and R. S. Wolfe, "A Selective Enrichment Method for *Gallionella Ferruginea*," *J. Bacteriol.* **74,** 347 (1957). Micrograph (c) from *Principles and Applications in Aquatic Microbiology*, Chap. 5. New York: Wiley, 1964.

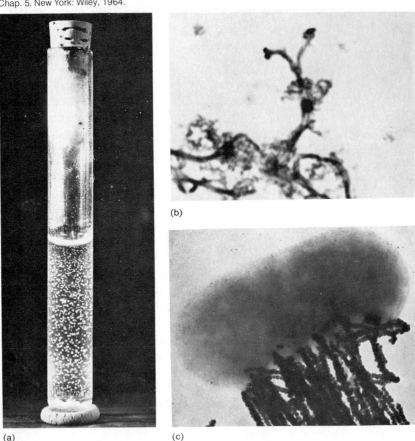

(a)

(b)

(c)

tionally versatile organisms that can use a wide range of organic compounds as carbon and energy sources. Most of these bacteria were formerly placed in a special genus, *Hydrogenomonas*. However, their facultative chemoautotrophy does not appear to justify a generic separation from similar organisms that are chemoheterotrophs, and they are now classified in a series of genera that contain phenotypically similar nonautotrophic bacteria, such as *Pseudomonas*, *Alcaligenes*, and *Nocardia*.

## the methylotrophs

The distinguishing property of the methylotrophs is the ability to derive both carbon and energy from the metabolism of reduced one-carbon compounds or of compounds containing two or more methyl groups that are not directly linked to one another (e.g., dimethyl ether, $CH_3$—O—$CH_3$). By far the most abundant compound of this class in nature is the gas methane ($CH_4$), which occurs in coal and oil deposits and is continuously produced on a large scale in anaerobic environments by the methanogenic bacteria (see Chapter 12, p. 315). Methanol is formed as a breakdown product of pectins and other naturally occurring substances that contain methyl esters or ethers. Methylated amines and their oxides occur naturally in plant and animal tissues.

The dissimilation of these highly reduced methyl compounds is almost always respiratory, being mediated by strict aerobes. Only procaryotes are known to oxidize methane; however, methanol can serve as a growth substrate for certain yeasts.

The procaryotic methylotrophs fall into two primary physiological subgroups: obligate and facultative methylotrophs. The obligate methylotrophs are all able to grow with methane; the only other substrates that support growth are dimethyl ether and methanol. The facultative methylotrophs can grow with methanol and/or methylamines, but rarely with methane. Other growth substrates often include formate, as well as a few simple $C_2$ and $C_4$ compounds; however, the range is always small.

### the obligate methylotrophs

The growth of knowledge about the obligate methylotrophs has been slow. Its first representative, *Methylomonas methanica*, a Gram-negative, polarly flagellated rod, was described almost 70 years ago, but for many decades it remained the only known methane oxidizer. The development of improved methods for the enrichment and purification of methane oxidizers has recently led to the discovery of many new types, all basically similar in nutritional respects, but remarkably varied in cell structure. Electron microscopy of thin sections shows that all possess complex intruded membrane systems, which in their topology and complexity, resemble those found in certain of the nitrifying bacteria. Some form resting stages, resistant to desiccation. These structures are of two types: cysts, resembling the cysts of *Azotobacter*, (see p. 258), and *exospores*, which are small, spherical cells budded at one pole of the mother cell (Figure 10.23).

Methane is the best growth substrate; methanol is toxic for many strains, and hence must be provided at very low concentrations. The growth rates under optimal conditions are low (doubling times of 4 to 6 hours).

### facultative methylotrophs

Although the obligate methylotrophs can grow at the expense of methanol, enrichments with this substrate invariably yield other types of organisms, which are called *facultative methylotrophs*. Aerobic enrichments with either methanol or methylamines lead to the isolation of Gram-negative, polarly flagellated rods, none of which has been well characterized. The best-known facultative methylotroph is *Hyphomicrobium*. It is a powerful denitrifier and can be specifically enriched by the use of a medium containing methanol and nitrate, incubated anaerobically. *Hyphomicrobium*, like the purple bacterium *Rhodomicrobium*, forms buds at the end of the filiform extensions of the cell, termed *prosthecae* (Figure 10.24).

**Figure 10.23**

The production of exospores from one cell pole in *Methylosinus*. Electron micrograph of a negatively stained preparation of whole cells, showing the wrinkled appearance of the surface of the exospore, and the fine fibers which extend out from it, ×12,000. Insert: a similar group of cells forming exospores, negatively stained with India ink and observed by phase-contrast microscopy, ×1,400. From R. Whittenbury, S. L. Davies, and S. L. Davey, "Exospores and Cysts Formed by Methane-Utilizing Bacteria," *J. Gen. Microbiol.* **61,** 219 (1970).

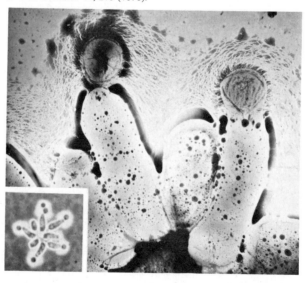

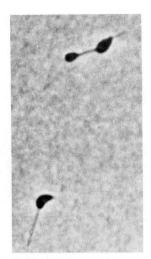

**Figure 10.24**
*Hyphomicrobium vulgare.* Phase contrast micrograph of two budding cells (×2,450). Micrograph courtesy of Peter Hirsch.

## origins of chemoautotrophs and methylotrophs

In view of the simplicity of their nutritional requirements, chemoautotrophs and methylotrophs were once regarded as "primitive" organisms, possibly representative of the earliest forms of life on earth. This notion of their place in evolution is now untenable, for two reasons. First, their biochemical machinery and cellular fine structure are at least as complex as that of most chemoheterotrophic bacteria. Second, there is now good evidence to support the view that the earliest living organisms arose on an anaerobic earth, where the oceans contained an abundance of preformed organic matter. The shift to an oxygen-rich biosphere occurred much later (probably about 2 billion years ago); this major geochemical change can be plausibly explained only as a consequence of the evolution of oxygenic photosynthesis. In this evolutionary scenario the appearance of aerobic chemoautotrophs and methylotrophs on earth was dependent on the development of oxygenic photosynthesis. It is therefore conceivable that the chemoautotrophs and methylotrophs arose from procaryotic ancestors that performed either oxygenic or anoxygenic photosynthesis, by loss of the photosynthetic apparatus and adaptation to a respiratory function of the photosynthetic electron transport chain.

# GRAM-NEGATIVE CHEMO-HETEROTROPHS

Although important in the cycles of matter on earth, and often constituting the predominant flora in extreme environments, the autotrophic bacteria are not nearly as numerous as the heterotrophic procaryotes. We will divide the Gram-negative heterotrophs into three main groups: those whose metabolism is strictly respiratory, those whose metabolism is strictly fermentative, and those that can employ either mode of generating ATP. Several other groups, not readily classified into the above categories, will then be discussed. Finally, the mycoplasmas, an unusual group of frequently pleomorphic bacteria that lack peptidoglycan, will be described.

## respiratory chemoheterotrophs

The Gram-negative bacteria that depend on respiration for the provision of ATP can be further subdivided on the basis of their motility. The largest group is either motile by flagella, or is immotile. A relatively small group of bacteria displays gliding motility, a trait held in common with the blue-green bacteria, *Chloroflexus,* and some of the chemoautotrophic bacteria.

The great majority of the bacteria belonging to these groups are *strict aerobes,* for which molecular oxygen is always an essential nutrient. However, the *denitrifying bacteria* can use nitrate in place of $O_2$ as a terminal electron acceptor, and thus can also grow anaerobically. Anaerobic growth through denitrification can be readily distinguished from fermentative growth by its strict dependence on the provision of nitrate in quantities sufficient to meet respiratory needs. Denitrification results in vigorous gas formation, since $N_2$ is the major end product of nitrate reduction. Another

form of anaerobic respiration, sulfur reduction, is a strictly anaerobic process.
The principal subgroups of these bacteria are outlined in Table 11.1. They are distinguished principally on the basis of cell shape, mode of flagellar insertion, and biochemical or morphological traits.

### the pseudomonads

Among the many groups of Gram-negative flagellated rods that contain DNA with a base composition in the range of 58 to 70 moles percent G + C (Table 11.1), the organisms known as *aerobic pseudomonads* have received the most extensive study. It must be emphasized that the present limits of this group are somewhat arbitrary, its distinction from such organisms as *Alcaligenes, Agrobacterium, Rhizobium,* and the acetic acid bacteria being based on characters the taxonomic significance of which has not yet been carefully evaluated. The primary criteria for assigning a Gram-negative aerobic chemoheterotroph to this particular group is the mode of flagellar insertion, which is polar (Figure 11.1). However, some rods with polar flagella are excluded (e.g., the Caulobacter group, some acetic acid bacteria, and some members of the Azotobacter group) on the basis of special characters. About 30 species are now recognized among the aerobic pseudomonads; with the exception of yellow-pigmented plant pathogens, assigned to the genus *Xanthomonas,* they are placed in the genus *Pseudomonas.* Nucleic acid hybridization has revealed that the aerobic pseudomonads are a group of considera-

**Table 11.1**
Major Subgroups Among Respiratory Gram-Negative Chemoheterotrophs With Flagellar Motility

| CELL SHAPE | FLAGELLAR INSERTION | MOLE % G + C IN DNA | OTHER DISTINCTIVE PROPERTIES | GROUP |
|---|---|---|---|---|
| Rods | Polar | 58–70 | None | Pseudomonads |
| Rods | Peritrichous or subpolar | 58–70 | Some form nodules or galls on plants | Rhizobium group |
| Rods | Polar or subpolar | 59–65 | Prosthecate; special division cycle | Prosthecate bacteria |
| Rods | Peritrichous, rarely polar | 57–70 | Free-living, aerobic nitrogen fixers | Azotobacter group |
| Rods | Polar or peritrichous | 55–64 | Oxidize organic substrates incompletely | Acetic acid bacteria |
| Rods | Polar or subpolar | 69–70 | Form sheaths | Sphaerotilus group |
| Helical | Polar | 30–65 | None | Spirillum group |
| Cocci or short rods | Nonflagellate | 40–52 | None | Moraxella group |
| Straight or curved rods or helical | Polar or peritrichous | 46–61 | Strictly anaerobic | Sulfur reducers |

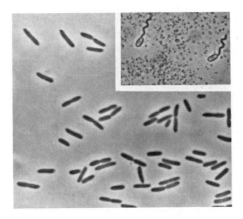

**Figure 11.1**
Phase contrast photomicrograph of cells of *Pseudomonas* (×1,100). Inset (upper right): flagella stain of *Pseudomonas*, showing the polar monotrichous flagellation characteristic of many *Pseudomonas* species (×1,290). Courtesy of N. J. Palleroni.

ble internal heterogeneity, the constituent species being assignable to a total of five major and isolated genetic homology groups. We shall describe here some of the species representative of three of these homology groups; fluorescent pseudomonads, the pseudomallei group, and the Xanthomonas group.

THE FLUORESCENT PSEUDOMONADS. A somewhat variable but distinctive property of fluorescent pseudomonads is the production of a yellow water-soluble pigment, which diffuses into the medium and fluoresces yellow-green under ultraviolet light. This pigment, an iron chelator, is synthesized in response to iron deprivation. It thus increases the availability to the bacteria of iron by solubilizing this mineral from insoluble precipitates such as iron sulfides or hydroxides. In addition to the yellow fluorescent pigments, pyocyanin, a blue pigment, is characteristic of the species *P. aeruginosa*. This species, *P. fluorescens*, and *P. putida* are common members of the microflora of soil and water, and are all nutritionally highly versatile, being able to use 60 to 80 different organic compounds as sole sources of carbon and energy.

P. aeruginosa, which has a considerably higher temperature maximum than *P. fluorescens* and *P. putida*, is sometimes pathogenic for man. It belongs to the category of opportunistic pathogens, which do not normally exist in animal hosts, but which can establish infections in individuals whose natural resistance has been reduced. Thus, *P. aeruginosa* typically causes infections in victims of severe burns and in cancer patients who have been treated with immunosuppressive drugs. An intrinsic resistance to most commonly used antibiotics makes treatment difficult; these infections are thus often fatal.

The fluorescent pseudomonads also include organisms that are pathogenic for plants; the many varieties, which differ in host range, are assigned to one species, *P. syringae*. These plant pathogens are true parasites, readily distinguishable from the free-living soil and water species by their physiological and biochemical properties. They are less versatile nutritionally; and their growth rates, both in synthetic and in complex media, are much lower.

THE PSEUDOMALLEI GROUP.  The pseudomallei group, like the fluorescent group, are mostly nutritionally versatile organisms that do not require growth factors. Although usually pigmented, they never produce a yellow-green diffusible fluorescent pigment. The prototype of this group, *P. pseudomallei*, was originally discovered as the agent of *melioidosis*, a highly fatal tropical disease of man and other mammals. Even in the tropical areas where melioidosis is endemic, it is a relatively rare disease, typically contracted through the contamination of wounds with soil or mud. In fact, *P. pseudomallei* appears to be, like *P. aeruginosa*, an opportunistic pathogen that is a normal member of the microflora of soil and water in the tropics. However, the closely related species *P. mallei* is a true parasite, causing a disease of horses known as *glanders*. *P. mallei* is unable to survive in nature in the absence of its specific animal host. It is the only aerobic pseudomonad that is permanently immotile: its inclusion in the group is based on its close genetic relationship and its phenotypic similarity to *P. pseudomallei*.

The pseudomallei group also contains several species that occur in soil and are occasionally pathogenic for plants; they are exemplified by *P. cepacia*, notable for its extreme nutritional versatility; it can use over 100 different organic compounds as carbon and energy sources.

THE XANTHOMONAS GROUP.  Certain plant pathogenic pseudomonads have long been placed in a special genus, *Xanthomonas*, distinguished by the production of yellow cellular pigments. Nucleic acid homology studies have shown that these xanthomonads are a genetically isolated subgroup, although distantly related to one other aerobic pseudomonad, *Pseudomonas maltophilia*. In contrast to the groups so far discussed, both xanthomonads and *P. maltophilia* require organic growth factors, including methionine and (in the case of xanthomonads) certain of the B vitamins.

It was long assumed that the yellow cellular pigments of xanthomonads are carotenoids, but chemical studies have shown that these pigments are of unique chemical composition. Pigments of this type are not known to occur in any other bacteria; thus, the chemical nature of the cellular pigments is a highly distinctive property of the Xanthomonas group.

### the rhizobium group

The Gram-negative aerobic chemoheterotrophs with rod-shaped cells include many representatives in which flagellar insertion is not polar, and which are hence excluded by definition from the aerobic pseudomonads. These bacteria normally bear few (1 to 4) flagella, and the flagellar insertion is not easy to determine unambiguously. Apart from the practical difficulty of determining this character, there are considerable grounds for doubting whether or not it really possesses the taxonomic importance that has traditionally been ascribed to it. An extension of the study of genetic interrelationships by nucleic acid hybridization, which has been so successful in revealing the internal relationships of the aerobic pseudomonads, to these nonpolarly flagellated rods will probably aid considerably in clarifying their taxonomic status.

At the present time, the members of the genera *Rhizobium* and *Agrobacterium* are distinguished by their special relationships (described below) to plants. The genus *Alcaligenes* is used as a repository for Gram-negative rod-shaped aerobes in which flagellar insertion is nonpolar.

Of these groups, the genus *Rhizobium* has been studied in greatest detail, because of its major agricultural importance. The rhizobia can invade the root hairs of leguminous plants and initiate the formation of root nodules, within which they develop as intracellular symbionts and fix nitrogen (see Chapter 15 for further discussion). Although rhizobia are strict aerobes, their nitrogenase is extremely susceptible to oxygen inactivation within the cell. Hence the enzyme is active only in cultures exposed to very low $O_2$ concentrations, which severely limit growth. The root nodule must therefore provide a special environment that favors the maintenance of high rhizobial nitrogenase activity.

Six species of *Rhizobium* are recognized, primarily on the basis of their host ranges among leguminous plants, these ranges being relatively narrow and specific. The genus is almost certainly a heterogeneous one, being divisible into two groups of species that differ in mode of flagellar insertion and in growth rate on complex media. Although the rhizobia persist for some time in soil, they are probably unable to compete successfully with members of the free-living soil microflora. Hence, when a leguminous plant is introduced into an area where it has not been previously cultivated, inoculation of the seeds with the specific rhizobial symbiont is essential to ensure satisfactory nodulation and nitrogen fixation.

The organisms of the genus *Agrobacterium* are responsible for the formation of galls or tumors on the roots or stems of many different plant families. The principal species is *A. tumefaciens;* the nature of its interaction with the plant host presents a fascinating and still unsolved problem. Although bacterial infection is essential to initiate tumor formation, the bacteria soon disappear from the tumor, which continues to develop in their absence. This suggests that bacterial infection leads to a transmissible genetic modification of the infected plant cells, but the nature and mechanism of the genetic change so induced remain unknown.

### the prosthecate bacteria

The prosthecate bacteria are distinguished from other groups of Gram–negative heterotrophs on the basis of their possession of *prosthecae,* relatively thin extensions of the cell. Some prosthecate bacteria can, by virtue of their metabolic properties, be classified with other bacterial groups. *Hyphomicrobium* and *Rhodomicrobium* both bud from the tip of a prostheca. Both are discussed in Chapter 10, the former as a methane oxidizer and the latter as a purple nonsulfur bacterium. In most prosthecate bacteria, the prosthecae do not serve the reproductive role that they do in *Hyphomicrobium* and *Rhodomicrobium*. The cells, which may bear numerous prosthecae (Figure 11.2), divide by either binary fission or by budding.

In one group, that typified by *Caulobacter,* the cell bears a single prostheca termed a *stalk.* Just prior to division, a flagellum is synthesized at the pole opposite the prostheca. Cell division thus results in two different daughter cells, one prosthecate and the other flagellated. The flagellated *swarmer cell* swims for a period of time, then loses its flagellum and develops a prostheca at the same site. The prosthecate cell is capable of dividing again; the swarmer cell is not—until it has developed a prostheca. This life cycle is illustrated in Figure 11.3.

As a result of the production of an adhesive extracellular holdfast at the tip of the stalk, the caulobacters can attach nonspecifically to solid substrates, including the cell walls of other microorganisms (Figure 11.4). Under natural conditions, it is probable that they mostly grow in attachment to larger microorganisms (algae, protozoa, other bacteria), utiliz-

**Figure 11.2**

A prosthecate, freshwater bacterium, *Ancalomicrobium.* (a) Phase contrast micrograph of a group of cells (×2,180). (b) Electron micrograph of a single cell, showing continuity of prosthecae with the body of the cell (×8,530). From J. T. Staley, "*Prosthecomicrobium* and *Ancalomicrobium:* New prosthecate Freshwater Bacteria," *J. Bacteriol.* **95,** 1921 (1968).

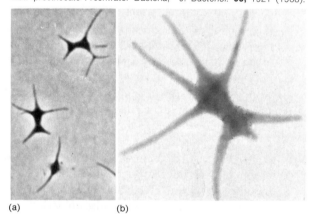

(a)                    (b)

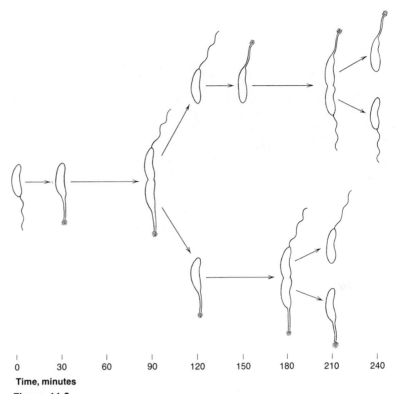

**Figure 11.3**

A diagrammatic representation of clonal growth in *Caulobacter*, based on continuous micro-scopic observations. Note that the time required for the division of a swarmer cell is considera-bly longer than that required for the division of its stalked sibling. After J. L. S. Poindexter. ''Biological Properties and Classification of the Caulobacter group,'' *Bacteriol. Rev.* **28,** 231 (1962).

ing organic materials secreted by the organisms to which they adhere. Rela-tive to other aerobic chemoheterotrophs from the same aquatic habitats (e.g., aerobic pseudomonads), their growth rates (even in complex media) are low. The minimal generation time is never less than 2 hours, and for many species as long as 4 to 6 hours. Most caulobacters require organic growth factors, and their ranges of utilizable substrates are less broad than those of aerobic pseu-domonads. Hence, their capacity for attachment to other microorganisms appears to be an important factor for successful competition with other aero-bic chemoheterotrophs in nature.

The prosthecate bacteria are all aquatic, being found in both fresh and salt water. Most are adapted to quite low nutrient concentrations, and it has been suggested that one role of the prosthecae is to increase the absorbtive surface of the cell. This is supported by the depend-

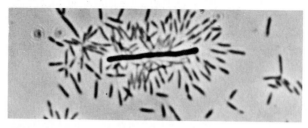

**Figure 11.4**
The attachment of *Caulobacter* to *Bacillus*. From J. L. S. Poindexter, "Biological Properties and Classification of the *Caulobacter* group," *Bact. Rev.* **28,** 231 (1962).

ence of the *Caulobacter* stalk length on the concentration of phosphate, a nutrient that is frequently growth-limiting in aquatic systems. It has also been suggested that the stalk retards sedimentation of these strict aerobes, although some prosthecate bacteria (e.g., *Ancalomicriobium*) regulate their buoyancy by means of gas vacuoles. Prosthecae most probably fill multiple functions, which may differ in different organisms.

### the azotobacter group

Rod-shaped organisms of the Azotobacter group possess a property that does not occur in any other group of Gram-negative chemoheterotrophs: the ability to fix nitrogen under aerobic growth conditions. In view of the extreme sensitivity of nitrogenase to oxygen inactivation, the existence of this ability in bacteria that are strict aerobes appears paradoxical. It is evident that the Azotobacter group must possess special mechanisms for the protection of nitrogenase, since facultatively anaerobic nitrogen-fixing chemoheterotrophs (e.g., *Bacillus polymyxa*) can maintain nitrogenase activity only when growing in the absence of oxygen. The azotobacters have extraordinarily high respiratory rates, far in excess of those of all other aerobic bacteria, and this may prevent $O_2$ penetration to the intracellular sites of nitrogenase activity.

The members of the Azotobacter group have oval to rod-shaped cells, which are large (as much as 2 $\mu$m wide) in most species. Cultures on solid media have a characteristic mucoid appearance (Figure 11.5), since these organisms produce large amounts of extracellular polysaccharide. The members of the genera *Azotobacter* and *Azomonas* are common in temperate regions in neutral or alkaline soils and waters. In tropical regions the far more acid-tolerant members of the genus *Beijerinckia* are the prevalent members of the aerobic, nitrogen-fixing soil microflora; they can grow at pH values as low as 3, and thus are well adapted to the relatively acid soils characteristic of the tropics.

The members of the genus *Azotobacter* (but not of the other two genera) produce distinctive resting cells known as *cysts* (Figure

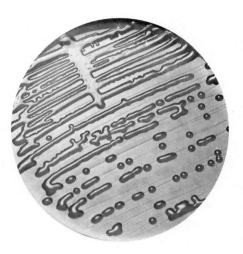

**Figure 11.5**
A streaked plate of *Azotobacter*, showing the
typical smooth, glistening colonies. Courtesy of
O. Wyss; reproduced from his *Elementary Microbiology.* New York: Wiley, 1963.

11.6). These structures, which arise by the deposition of additional outer
layers around the vegetative cell wall, are resistant to desiccation but not to
heat. Although microscopic examination of soil frequently reveals numerous
cysts, laboratory cultures often fail to encyst unless specially treated. Usually,
replacement of the carbon source supporting growth with a short chain fatty
acid or alcohol (such as butanol) will induce the culture to produce cysts.

### the acetic acid bacteria

The rod-shaped acetic acid bacteria are distinguishable from other aerobic Gram-negative chemoheterotrophs by a series
of physiological and metabolic characters, which seem to be at least in part a
reflection of their ecology. Acetic acid bacteria occur on the surface of plants,
particularly flowers and fruits. They develop abundantly as a secondary microflora in decomposing plant material under aerobic conditions, following an
initial alcoholic fermentation of sugars by yeasts. Under these circumstances,
they use ethanol as an oxidizable substrate, converting it to acetic acid. Many
carbohydrates and primary and secondary alcohols can also serve as energy
sources, their oxidation characteristically resulting in the transient or permanent accumulation of partly oxidized organic products. Since most members
of the group have relatively complex growth-factor requirements, they are
usually grown in complex media supplemented with an oxidizable substrate.
These bacteria are markedly acidophilic, growing at pH values as low as 4,
with an optimum between pH 5 and 6.

By virtue of their capacity to convert many organic compounds almost stoichiometrically to partly oxidized organic end
products, the acetic acid bacteria are a group of considerable industrial importance, principally in the manufacture of vinegar from ethanol-containing
materials (e.g., wine, cider).

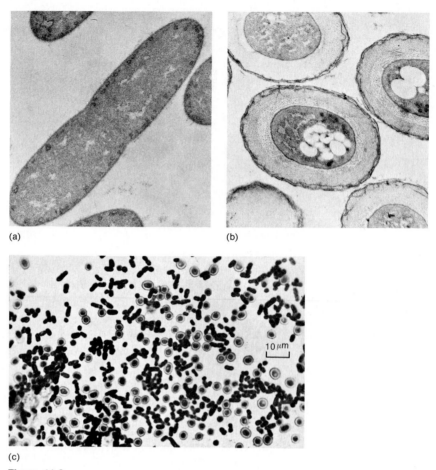

(a)                                                    (b)

(c)

**Figure 11.6**

Electron micrographs of thin sections of vegetative cells and cysts of *Azotobacter*. (a) Vegetative cells (×17,800). (b) Cysts (×10,700). Both from O. Wyss, M. G. Neumann, and M. D. Socolofsky, "Development and Germination of the *Azotobacter* Cyst.," *J. Biophys. Biochem. Cytol.* **10**, 555 (1961). (c) Stained preparation of vegetative cells and cysts of *Azotobacter* (×1,210). The vegetative cells are oval, deeply stained rods; the cysts are spherical, surrounded by thick, lightly stained walls. From S. Winogradsky, *Microbiologie du sol*, p. 780. Paris: Masson, 1949. Reprinted with permission of M. Manigault and the publishers.

The species *Acetobacter xylinum* is able to synthesize the polysaccharide, cellulose. This substance is formed as a result of growth at the expense of glucose and certain other sugars, and is deposited outside the cells in the form of a loose, fibrillar mesh (Figure 11.7). In liquid cultures this organism forms a tough, cellulosic pellicle that encloses the cells and can attain a thickness of several centimeters.

**Figure 11.7**
Extracellular formation of cellulose by *Acetobacter xylinum* (×860). The bacterial cells are entangled in a mesh of cellulose fibrils. From J. Frateur, "Essai sur la systématique des acétobacters," *La Cellule* **53,** 3 (1950).

### the sphaerotilus group

The Sphaerotilus group are relatively large rod-shaped organisms that grow as chains of cells enclosed in tubular sheaths (Figure 11.8), usually attached to solid substrates by basal holdfasts. Reproduction occurs by the liberation of motile cells, bearing polar or subpolar flagella, from the open apex of the sheath. These bacteria are all aquatic. *Sphaerotilus* is found in slowly running streams contaminated with sewage or other organic matter, where it grows as long, slimy, attached white tassels. It also develops in aerobic sewage treatment ponds.

The sheaths contain proteins, polysaccharides, and lipids. New sheath material is laid down only at the apical tip of the filament, presumably by the apical cell in the growing chain.

### the spirillum group

The spirilla are aerobic chemoheterotrophs characterized by the possession of helical cells, bearing bipolar flagellar tufts (Figure 11.9). They are all aquatic bacteria, common in both freshwater and marine environments. A physiological property, shared to some degree by all spirilla, which is uncommon in strictly aerobic bacteria, is a *preference for low oxygen tensions*. The microaerophilic tendencies of these bacteria are shown by their behavior in wet mounts; the highly motile cells accumulate in a dense, narrow band, located at some distance from the edge of the cover slip (Figure 11.10). Despite this, most spirilla can form colonies on the surface of agar media.

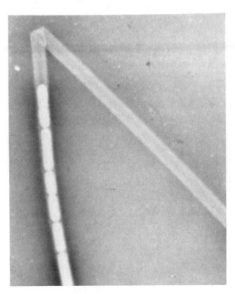

**Figure 11.8**
*Sphaerotilus*, showing a chain of cells enclosed within the sheath. From J. L. Stokes, "Studies on the Filamentous Sheathed Iron Bacterium *Sphaerotilus natans*," *J. Bacteriol.* **67,** 281 (1954).

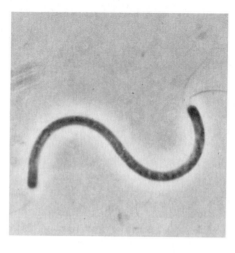

**Figure 11.9**
A single cell of a large spirillum (phase contrast). Note the flagellar tufts at one pole of the cell. From N. R. Krieg, "Cultivation of *Spirillum volutans* in a Bacteria-free Environment," *J. Bacteriol.* **90,** 817 (1965).

*Bdellovibrio* is sometimes placed with the spirilla on the basis of cell shape, although it is not clear that there is any real relationship between the two. *Bdellovibrio* is a minute curved rod (Figure 11.11) that attacks other Gram-negative bacteria. It moves rapidly, being propelled by a single, sheathed polar flagellum, and its impact is capable of knocking a much larger cell, such as a pseudomonad, completely across a microscope field. Following impact, the *Bdellovibrio* penetrates the host cell wall and resides between the cell wall and membrane. Damage to the membrane results in the leakage of cytoplasmic contents, which are used by the parasite. The *Bdellovibrio* grows into a long helix, which then fragments into curved rods,

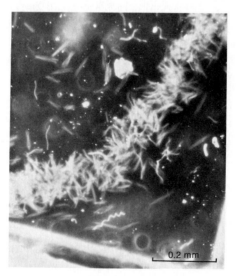

**Figure 11.10**
Typical aerotactic pattern of *Spirillum* in a wet mount, showing the accumation of the cells in a very narrow band some distance from the edge of the cover slip, visible as a white band at the bottom and lower right corner. From N. R. Krieg, "Cultivation of *Spirillum volutans* in a Bacteria-free Environment," *J. Bacteriol.* **90,** 817 (1965).

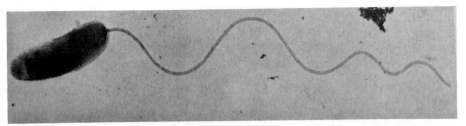

**Figure 11.11**
Electron micrograph of a cell of *Bdellovibrio,* showing the sheathed polar flagellum (uranyl acetate stain; ×29,000). From R. J. Seidler and M. P. Starr, "Structure of the Flagellum of *Bdellovibrio bacteriovorus,*" *J. Bacteriol.* **95,** 1952 (1968).

each of which develops a flagellum. Ultimately, the host cell wall is degraded and the progeny *Bdellovibrio* are released. This life cycle is shown schematically in Figure 11.12.

### the moraxella group

The members of the Moraxella group are non-flagellated short rods or cocci (Figure 11.13), some of which are parasitic on warmblooded animals. An unusual property shared by the parasitic members of the group is their marked penicillin sensitivity; except in strains that have acquired penicillin resistance, growth is inhibited by penicillin at levels of 1 to 10 units/ml, whereas the inhibitory concentration for Gram-negative bacteria usually lies between 100 and 1,000 units/ml.* The coccoid organisms of the genus *Neisseria* are parasites on the mucous membranes of mammals, and include two human pathogens, the agents of gonorrhea and meningitis. The rod-shaped organisms of the genera *Moraxella* and *Acinetobacter,* although similar in many phenotypic respects and in DNA base composition, give no evidence of genetic relationships by nucleic acid hybridization. *Moraxella,* like *Neisseria,* consists of parasites on the mucous membranes of vertebrates, whereas the acinetobacters are nutritionally versatile free-living bacteria with an exceptionally wide natural distribution.

### the sulfur reducers

The respiratory organisms include one highly specialized group of *strict anaerobes,* the sulfur reducers,† which couple the oxidation of organic compounds to the reduction of elemental sulfur or sulfate. This process occurs only under anaerobic conditions, and the bacteria that perform it are incapable (unlike the denitrifying bacteria) of using alternative inorganic electron acceptors such as oxygen or nitrate.

*The activity of penicillins is expressed in arbitrary units; a unit equals approximately one milligram of penicillin G.
†This group has historically been known as the "sulfate reducers," since until quite recently *Desulfuromonas* was unknown.

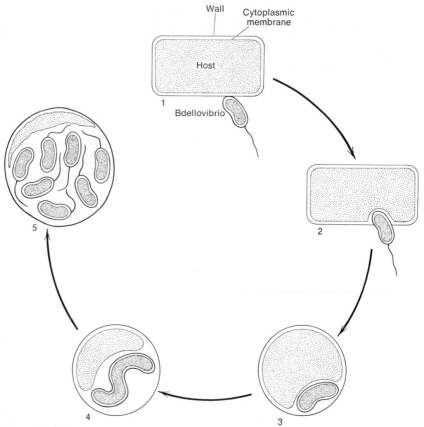

**Figure 11.12**
Life cycle of *Bdellovibrio*. Attachment to the host cell (1) is immediately followed by penetration (2). Growth of the parasite occurs without concommitant cell division, leading to a helical *Bdellovibrio* (3 and 4). Ultimately the helical organism divides into a number of short rods or vibrios (curved rods), and flagella are synthesized (5). The weakened host cell wall (evident from the swelling and loss of shape which occur early in infection) then ruptures and the *Bdellovibrio* daughter cells are released.

The ability to use elemental sulfur as terminal electron acceptor is characteristic of one known genus, *Desulfuromonas*, a rod with several peritrichous flagella. This organism couples the oxidation of a limited range of organic compounds to the reduction of elemental sulfur to $H_2S$. *Desulfuromonas* is incapable of utilizing any other compound, including sulfate, as electron acceptor.

Dissimilatory sulfate reduction is confined to two genera of bacteria: *Desulfotomaculum,* an endospore former that will be considered in Chapter 12, and *Desulfovibrio*. Neither is capable of using elemental sulfur as electron acceptor.

Some sulfate reducers are capable of reducing sulfate at the expense of hydrogen gas. However, they lack the Calvin cycle

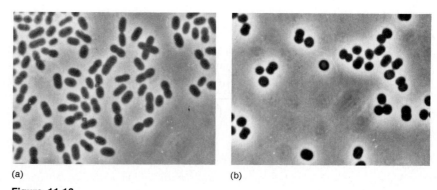

(a)                                                                (b)

**Figure 11.13**
The Moraxella group (phase contrast, ×2,200). (a) *Moraxella osloensis*. (b) *Neisseria catarrhalis*.

and require an organic source of carbon. Since the range of organic compounds that the sulfate reducers can respire is quite limited, it is possible that hydrogen, which is produced in copious amounts in anaerobic environments, is their major respiratory substrate.

The typical habitats of the sulfur reducers are anaerobic sediments that contain organic matter, sulfur, and sulfate; the activities of these organisms result in a massive generation of $H_2S$. This often leads to development in the overlying water layer of purple and green sulfur bacteria, which utilize the $H_2S$ as a photosynthetic electron donor, reoxidizing it under anaerobic conditions in the light to sulfate. The combined activities of the sulfur reducers and of the anoxygenic photosynthetic bacteria therefore drive an anaerobic sulfur cycle (Figure 11.14). In addition, large quantities of $H_2S$ are often sequestered as insoluble metal sulfides (e.g., FeS). It is now generally accepted that these bacteria have been responsible for the formation of metal sulfide ores and, indirectly, for natural deposits of elemental sulfur (formed by a secondary, probably nonbiological, oxidation of sulfides).

In waterlogged (and hence largely anaerobic) soils the production of $H_2S$ by sulfate reducers may cause damage to plants: this is sometimes a serious economic problem in the cultivation of rice, which is sown in flooded fields (see Chapter 15).

**Figure 11.14**
The anaerobic sulfur cycle.

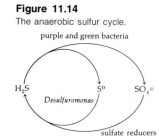

# the gliding bacteria

The Gram-negative chemoheterotrophs include several groups of unicellular or filamentous gliding bacteria. Gliding movement requires contact of the cell with a solid surface. It is considerably slower than flagellar movement, and no locomotor organelles have so far been demonstrated.

## the myxobacteria

The myxobacteria are soil organisms and are usually detected in nature through the development of their fruiting bodies on solid substrates: the bark of trees, decomposing plant material, and in particular the dung of animals. They are differentiated from other gliding bacteria by two properties: their special developmental cycle and the composition of their DNA, which is of very high G + C content (67 to 71 mole percent).

Myxobacterial vegetative cells are small rods, which multiply by binary fission. On solid substrates these organisms form flat, spreading colonies with irregular borders, made up of small groups of advancing cells (Figure 11.15). The migrating cells produce a tough slime layer that underlies and gives coherence to the colony. Under favorable conditions, the vegetative cells aggregate at a number of points in the inner area

**Figure 11.15**

Scanning electron micrographs of the process of fruiting body formation in *Chondromyces*. (a) Aggregation of vegetative cells. Note slime covering most of the colony. (b and c) Early stages of fructification. (d) Composite picture showing progressively later stages, with a mature fruiting body to the right. Courtesy of P. Grilione and J. Pangborn.

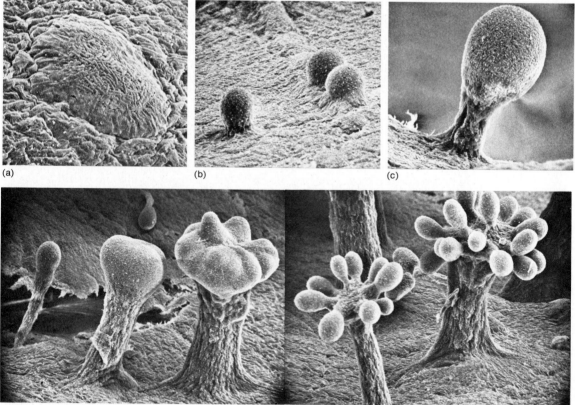

(a)  (b)  (c)

(d)

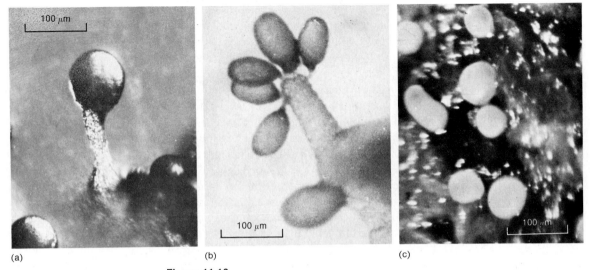

(a)          (b)          (c)

**Figure 11.16**
Light micrographs of fruiting bodies of myxobacteria. (a) *Melittangium lichenicolum* (×232). (b) *Stigmatella aurantiaca* (×318). (c) *Myxococcus xanthus* (×32). (a–b) Courtesy of M. Dworkin and H. Reichenback; (c) courtesy of H. McCurdy.

of the colony, and fruiting bodies then differentiate from these cellular aggregates (Figure 11.16). Each fruiting body is made up of slime and bacterial cells; when mature, it acquires a definite size, form, and color. The pigments of these bacteria are carotenoids, associated with the cells. In some genera, as the fruiting bodies mature, the cells within them become converted to resting cells, known as *myxospores* (Figure 11.17). Upon subsequent germination, each myxospore gives rise to a vegetative rod. In other genera, no myxospores are formed.

**Figure 11.17**
Microcysts and shortened rods from a fruiting body of *Myxococcus* (phase contrast, ×2,210). Courtesy of M. Dworkin and H. Reichenbach.

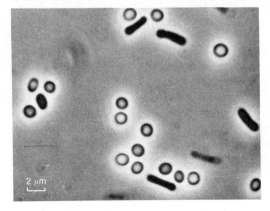

The properties of myxospores have been examined primarily in the genus *Myxococcus*. Although they are more heat-resistant than vegetative cells, the difference in this respect is not great. However, they are much more resistant to desiccation, and can survive for months or years in the dry state. The formation of these resting structures in fruiting bodies that are raised off the surface of the substrate no doubt facilitates their physical dispersion. In the genus *Myxococcus* the myxospore is both the resting structure and the unit of dispersion. However, in some other genera the unit of dispersion is not the individual myxospore, but the cyst that contains many vegetative cells or myxospores. This fact is most evident if one compares the respective modes of germination. The fruiting bodies of most *Myxococcus* spp. are deliquescent, and each of the myxospores liberated by their breakdown can germinate under favorable conditions, giving rise through subsequent growth to a vegetative colony. In cyst-forming genera, germination of myxospores is accompanied by a rupture of the wall of the enclosing cyst, with the release of hundreds of vegetative cells (Figure 11.18). The cells remain in association and give rise to a single vegetative colony. Cyst germination thus permits the establishment of a large vegetative population, prior to the initiation of cell division. Such large initial populations are probably necessary for the rapid accumulation of the extracellular hydrolytic enzymes (see below), upon which these organisms depend to obtain oxidizable carbon compounds.

**Figure 11.18**

Cyst germination in *Chondromyces*. (a) Mature fruiting body, bearing cysts, ×104. (b) Germination of a detached cyst on the surface of an agar plate, showing the emergence of a large population of vegetative cells, derived from the enclosed myxospores; phase contrast, ×120. (c) The empty wall of a germinated cyst; phase contrast, ×485. Courtesy of Dr. Hans Reichenbach.

(a)

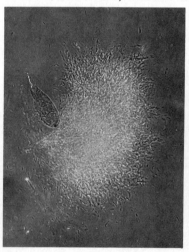

(b)

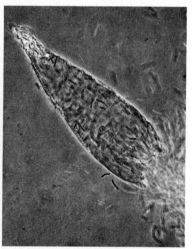

(c)

THE BACTEROLYTIC FRUITING MYXOBACTERIA.  Myxobacteria are strict aerobes that grow poorly (if at all) on conventional complex media. They fall into two nutritional subgroups: bacteriolytic and cellulolytic organisms. The majority of species are bacteriolytic, growing at the expense of bacteria and other microorganisms, living or dead. Living host cells are killed by antibiotics, secreted by the myxobacteria; these substances have not yet been characterized chemically. The host cells are then lysed through the action of extracellular enzymes. Myxobacterial growth occurs at the expense of the soluble hydrolytic products. The exceptionally favorable nature of dung as a substrate is largely attributable to its high bacterial content. The bacteriolytic myxobacteria are most readily cultivated on the surface of mineral agar, over which a heavy suspension of host cells has been spread. Growth in liquid media is often poor. The minimal nutritional requirements of most of these organisms are not well known, since many species will not grow readily except at the expense of microbial cells.

A few species, all of the genus *Polyangium*, have quite different nutrient requirements. They are active cellulose-decomposers, and grow in a medium with a mineral base, supplemented either with cellulose or with the soluble sugars cellobiose and glucose which are the hydrolysis products of cellulose.

Myxobacteria do not readily form fruiting bodies on media that support good vegetative growth. Fructification is generally favored by cultivation on media that are poor in nutrients, but the specific factors that control the process are not understood. Since the classification of these bacteria is very largely based on the structure of their fruiting bodies, the difficulty of obtaining reproducible fructification can be a serious obstacle to their identification.

ALGICIDAL NONFRUITING MYXOBACTERIA.  The death of natural populations of green algae and of blue-green bacteria is often caused by algicidal nonfruiting myxobacteria. Many of these have not been observed to form fruiting bodies, so that their taxonomic position is uncertain; however, their DNA base composition is similar to that of fruiting myxobacteria, lying in the range of 69 to 71 moles percent G + C. Consequently, they are referred to by the noncommittal name of *myxobacters*. In contrast to bacteriolytic fruiting myxobacteria, the myxobacters have simple nutrient requirements; they grow well in liquid media of defined composition, and can use glucose or other carbohydrates as sole sources of carbon energy.

Some of the myxobacters that kill blue-green bacteria produce extracellular enzymes that destroy the peptidoglycan layer of the host cell. The lytic enzymes include a protease of broad substrate specificity, which hydrolyzes peptide bonds of peptidoglycans. Other myxobacters kill blue-green bacteria only by cell-to-cell contact (Figure 11.19); the enzymatic mechanism of the attack is not known.

(a)

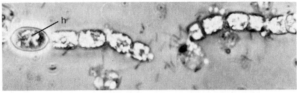

(b)

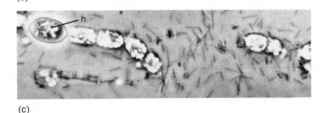

(c)

**Figure 11.19**
Time lapse sequence, showing lysis by the myxobacterium of a filament of *Anabaena*. This blue-green bacterium contains gas vacuoles, which accounts for the phase-bright appearance of the vegetative cells. The filament contains a heterocyst (h), which is not susceptible to attack. In (a) the arrow points to a lysed vegetative cell, the destruction of which has caused a break in the filament. All ×1,260. From M. J. Daft, and W. D. P. Stewart, "Light and Electron Microscope Observations on Algal Lysis by Bacterium CP-1," *New Phytol.* **72,** 799 (1973).

## the cytophaga group

The gliding bacteria of the Cytophaga group do not form fruiting bodies, and they differ markedly from the myxobacteria (including the nonfruiting myxobacters) in their DNA base composition, which lies in the range of 30 to 50 moles percent G + C. Some of these bacteria produce chains of cells, 100 μm or more in length, a character that does not occur in myxobacteria. However, other cytophagas grow as single, slender rods, and cannot be distinguished by vegetative cell structure from myxobacteria. Resting cells are formed only in the genus *Sporocytophaga*, which produces spherical, refractile microcysts, similar in structure and development to the myxospores of *Myxococcus*. The base composition of the DNA is therefore a character of primary importance to distinguish the members of the Cytophaga group from the myxobacteria, and in particular from the nonfruiting myxobacters.

The principal genera of the Cytophaga group are distinguished primarily by their nutritional properties. *Cytophaga* and *Sporo-*

*cytophaga* spp. can hydrolyze and grow at the expense of complex polysaccharides, whereas *Flexibacter* spp. are not polysaccharide decomposers. Most of these bacteria are strict aerobes; a few *Cytophaga* and *Flexibacter* spp. are facultative anaerobes and can ferment carbohydrates.

The most active aerobic cellulose-decomposing bacteria in soil are certain species of *Cytophaga* and *Sporocytophaga*. They can be readily enriched from soil in a medium with a mineral base containing filter paper as the sole source of carbon and energy. Upon initial isolation, these bacteria are unable to grow at the expense of any other organic substrate; by subsequent selection, mutants able to grow with the soluble sugars, glucose and cellobiose, can be obtained. The cellulolytic ability of these organisms is remarkable: when streaked on a sheet of filter paper placed on the surface of a mineral agar plate, they completely destroy its fiber structure, the attacked areas being converted into slimy colored patches filled with bacterial cells (Figure 11.20). Another distinctive property of the cellulose-decomposing soil cytophagas is the necessity for direct contact with the cellulosic substrate; this behavior suggests that the primary attack on cellulose is mediated by an exoenzyme that is nondiffusible, remaining bound to the cell surface. As a consequence, the cells in a culture containing cellulose adhere closely to the cellulose fibers, often in a very regular alignment, the rod-shaped cells being oriented parallel to the polysaccharide fibrils (Figure 11.21).

Other species of *Cytophaga* grow at the expense of the polysaccharides *chitin* and *agar*. Chitin, a polymer of N-acetylglucosamine, is a component of insect and crustacean exoskeletons. It is consequently common in both marine and terrestrial environments. Agar is a complex mixture of polysaccharides that serve as structural polymers in certain marine algae but do not occur in terrestrial environments. The agarolytic cytophagas

**Figure 11.20**
Agar plate covered with a layer of filter paper and streaked with a culture of *Cytophaga*. Note that the filter paper has been completely dissolved where growth has occurred.

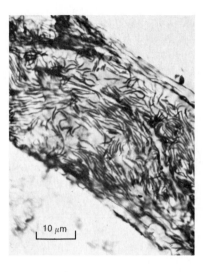

**Figure 11.21**
Cellulose fiber heavily attacked by *Cytophaga* (stained preparation). Note the characteristic regular arrangement of the cells. From S. Winogradsky, *Microbiologie du Sol*. Paris: Masson, 1949. Reprinted with permission of M. Manigault and the publisher.

are, in consequence, all organisms of marine origin, whereas chitinolytic members of the group occur both in soil and in sea water.

The hydrolysis of chitin and agar is mediated by inducible, extracellular enzymes, and hence the cells need not be in direct contact with the substrate. The chitinolytic and agarolytic cytophagas are also much less specialized nutritionally than the cellulose-decomposing species. They can all use a wide range of soluble sugars as carbon and energy sources, and most can grow in complex nitrogenous media (peptone or yeast extract) in the absence of carbohydrates.

The members of the genus *Flexibacter* are widely distributed in soil, freshwater, and marine environments. They include several species specifically pathogenic for fish. Infections caused by *Flexibacter columnaris* often assume epidemic proportions with massive mortalities in fish hatcheries during the warmer months of the year.

### filamentous, gliding chemoheterotrophs

In addition to the Cytophaga group, the gliding bacteria with DNA of low G + C content include a number of filamentous chemoheterotrophs.

*Saprospira* and *Vitreoscilla* are organisms that develop as flexible, gliding filaments as much as 500 μm long, made up of cells 2 to 5 μm in length. The filaments of *Saprospira* are helical (Figure 11.22), those of *Vitreoscilla* straight. Both groups occur largely in aquatic environments. The genus *Simonsiella* is distinguished by the formation of flattened, ribbon-shaped filaments (Figure 11.23), which are motile only when the broad surface of the filament is in contact with the substrate. *Simonsiella* is an aerobic member of the microflora of the oral cavity of man and other animals. Its nutri-

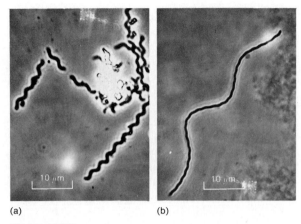

(a)                                    (b)

**Figure 11.22**
Phase contrast photomicrographs of *Saprospira* (×451): (a) *S. albida;*
(b) *S. grandis.* Courtesy of Ralph Lewin.

tional requirements are not precisely known; it develops best in complex
media, supplemented with serum or blood.

The most highly differentiated of the gliding
chemoheterotrophs is *Leucothrix,* a marine organism that grows as an epiphyte
on seaweeds, and is also found in decomposing algal material. The very long
filaments are immotile and are attached to substrates by an inconspicuous
basal holdfast. Reproduction occurs not by random fragmentation of the filament, as in *Saprospira, Vitreoscilla,* and *Simonsiella,* but by the breaking off of

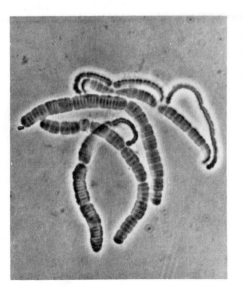

**Figure 11.23**
*Simonsiella,* a gliding bacterium which forms
ribbon-shaped filaments of flattened cells
(phase contrast, ×440). Some of the filaments
are viewed on edge, and they appear much
thinner than the filaments that lie flat. Courtesy
of Mrs. P. D. M. Glaister.

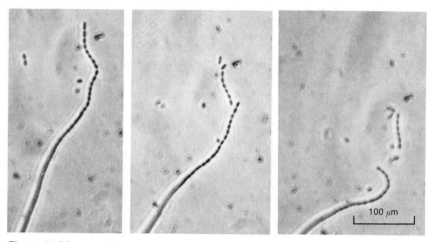

**Figure 11.24**
Successive pictures of a *Leucothrix* filament showing liberation of reproductive cells (phase contrast, ×309). From Ruth Harold and R. Y. Stanier, ''The Genera *Leucothrix* and *Thiothrix*,'' *Bact. Rev.* **19,** 49 (1955).

ovoid cells, singly or in short chains, from the apical end of the filament (Figure 11.24). These reproductive cells are capable of gliding movement.

## the enteric group

The organisms treated in this section constitute one of the largest well-defined groups among the Gram-negative, nonphotosynthetic bacteria. They have small, rod-shaped cells, either straight or curved, not exceeding 0.5 $\mu$m in width. Some are permanently immotile; motile representatives include organisms with peritrichous flagella, with polar flagella, and with ''mixed'' (sheathed polar-unsheathed peritrichous) flagellation (Figure 11.25). They can be distinguished from all other Gram-negative bacteria of similar structure by the property of *facultative anaerobiosis.* Under anaerobic conditions, energy is provided by fermentation of carbohydrates; under aerobic conditions, a wide range of organic compounds can serve as substrates for respiration.

The classical representative is *Escherichia coli,* one of the most characteristic members of the normal intestinal flora of mammals. Closely related to *Escherichia* are the genera *Salmonella* and *Shigella.* They are pathogens, responsible for such intestinal infections as bacterial dysentery, typhoid fever, and some bacterial food poisoning.

Clearly related to these bacteria, but of different ecology, are the genera *Enterobacter, Serratia,* and *Proteus,* which occur primarily in soil and water; also included are the plant pathogens of the genus *Erwinia.*

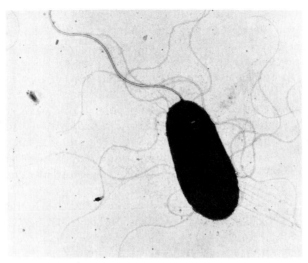

**Figure 11.25**
Electron micrograph of a *Beneckia* with "mixed" polar peritrichous flagel-lation. The cell bears a single sheathed polar flagellum, together with numerous laterally inserted unsheathed flagella (×16,600). From R. D. Allen and P. Baumann, *J. Bacteriol.* **107,** 295 (1971).

Together with the *E. coli* subgroup, these genera constitute the enteric group, as classically defined.

The bacteria so far discussed are either immotile or peritrichously flagellated. The primary importance that was for so long accorded to the mode of flagellar insertion as a taxonomic character impeded the recognition that some polarly flagellated bacteria are also allied to the enteric group, and most appropriately treated as members of it. These are all aquatic bacteria, which occur either in fresh water (*Vibrio, Aeromonas*) or in the marine environment (*Beneckea, Photobacterium*). Many of the marine forms show mixed flagellation under certain conditions of growth and polar flagel-lation under others. Some are animal pathogens; these include two species that cause intestinal diseases (*Vibrio cholerae,* and *Beneckea parahemolytica*).

As yet, there is no generally accepted collective name for this entire assemblage. It will be termed here the *enteric group,* but it must be emphasized that this designation includes genera (the polarly flagel-lated group) that fall outside the traditional confines of the group.

The mean DNA base composition for members of the enteric group is rather wide, extending from 37 to 63 mole percent G + C. With the exception of *Proteus,* in which one species differs widely from the rest in DNA base composition, the ranges within each genus are narrow. The values for the closely related organisms of the three genera *Escherichia, Salmonella,* and *Shigella* are not significantly different. The total span of base composition for the "classical" peritrichously flagellated enteric bacteria (37 to 59 mole percent G + C) corresponds rather closely to that for the polarly flagellated members (39 to 63 mole percent G + C).

Under anaerobic conditions, all bacteria of the enteric group utilize substantially the same biochemical pathways for the fermentative breakdown of sugars. Differences in detail, however, result in quantitative and qualitative differences in the end products of the fermentation, accounting for description of two major variations on the fermentation pattern of this group. The *mixed-acid fermentation* is characterized by the accumulation of large amounts of organic acids, as a consequence of which the pH of the medium may drop substantially. In the *butanediol* fermentation, the neutral end product, butanediol, replaces a significant amount of the acid end products; thus there is substantially less pH change. These two fermentations are used as primary diagnostic characters in this group.

Another biochemical trait used frequently in the identification of enteric bacteria is the fermentative production of gas. In fermentations by the enteric bacteria, gas, if formed, is produced from formic acid (an end product of both mixed acid and butanediol fermentations) by the enzyme formic hydrogenlyase. This enzyme produces equal amounts of $CO_2$ and $H_2$ from formic acid.

Another character of considerable diagnostic importance in the enteric group is the ability to ferment the disaccharide, lactose. Lactose fermentation is characteristic of *Escherichia* and *Enterobacter* and is absent from *Shigella, Salmonella,* and *Proteus.*

The members of the genera *Beneckea* and *Photobacterium* can be readily distinguished from all other members of the enteric group by their *salt requirements.* Indigenous marine bacteria have an absolute requirement for sodium ions, no growth occurring if sodium salts are omitted from the medium (Chapter 5).

A scheme for the subdivision of the enteric group is shown in Tables 11.2 and 11.3. It must be noted that the specialists who have been concerned with the taxonomy and epidemiology of enteric bacteria have created a very large number of genera among these organisms, for the most part distinguished by phenotypic differences so minor that other bacterial taxonomists would employ them (at best) to distinguish species. A few examples will illustrate this problem. The maintenance of the three genera *Escherichia, Salmonella,* and *Shigella* cannot really be justified in view of the close genetic relationships now known to exist among them. The generic distinction is based on pathogenic properties alone. Indeed, additional genera (*Arizona, Citrobacter, Edwardsiella*) have been proposed for certain members of this complex. The differences between *Enterobacter* and *Serratia* similarly do not justify a generic separation; and two additional genera (*Klebsiella* and *Hafnia*) have been proposed for organisms very similar to *Enterobacter.* Finally, the genus *Erwinia* is united solely by the dubious character of plant pathogenicity. It is internally heterogeneous; some of the species included resemble the *Enterobacter-Serratia* group, while others clearly do not.

**Table 11.2**
Taxonomic Subdivision of the Peritrichously Flagellated Enteric Bacteria and Related Immotile Forms
(All Straight Rods)

| MAJOR SUBGROUP | DNA BASE COMPOSITION, MOLE % G + C | MOTILITY | PRODUCTION OF | | | CONSTITUENT GENERA |
|---|---|---|---|---|---|---|
| | | | BUTANEDIOL | $H_2 + CO_2$ | UREASE | |
| I | 50–53 | V | – | V | – | *Escherichia, Salmonella, Shigella* |
| II | 50–59 | V | + | V | – | *Enterobacter, Serratia, Erwinia* |
| III | 37–50 | + | – | + | + | *Proteus* |
| IV | 46–47 | V | – | – | + | *Yersinia* |

V denotes variable within group.

The second major group of enteric bacteria consists of organisms that bear polar flagella or display mixed polar-peritrichous flagellation (Table 11.3). A primary separation within this group can be made on the basis of ionic requirements, which distinguish *Beneckea-Photobacterium* from *Vibrio-Aeromonas*. One character of considerable utility in this group is flagellar structure: the polar flagella of *Vibrio* and *Beneckea* spp. are relatively thick, being enclosed by a sheath that is made up of an extension of the outer membrane. The polar flagella of *Aeromonas* and *Photobacterium* are not sheathed.

## group I:
### *Escherichia-Salmonella-Shigella*
The members of group I are all inhabitants of the intestinal tract of man and other vertebrates. Highly detailed, intraspecific subdivisions among the species of group I have been made on the basis of the immunological analysis of the surface structures of the cell. The extreme specificity of antigen-antibody reactions makes it possible to recognize differences in these respects between strains of a bacterial species that are indis-

**Table 11.3**
Taxonomic Subdivision of Polarly or Mixed Flagellated
Enteric Bacteria (Straight or Curved Rods)

| GENUS | DNA BASE COMPOSITION, MOLE % G + C | SHEATHED POLAR FLAGELLA | NA+ REQUIREMENT |
|---|---|---|---|
| *Aeromonas* | 57–63 | – | – |
| *Vibrio* | 45–49 | + | – |
| *Beneckea* | 45–47 | + | + |
| *Photobacterium* | 39–43 | – | + |

tinguishable on the basis of other phenotypic criteria. In the genus *Salmonella* the detailed analysis of the surface antigenic structure has made it possible to distinguish many hundreds of different *serotypes;* comparable analyses of *Escherichia* and *Shigella* are less extensive. The principal utility of these systems of antigenic classification is not taxonomic, but *epidemiological.* The serotype of a pathogenic *Salmonella* strain is a marker that permits its recognition (and hence allows one to follow the course of an epidemic) where other phenotypic characters do not.

Escherichia coli is a component of the normal intestinal flora and gives rise to disease only under exceptional conditions. The genera *Salmonella* and *Shigella* comprise pathogens that cause a wide variety of enteric diseases in man and other animals. In all cases, entry occurs through the mouth; and the small intestine is the primary locus of infection, although some of these pathogens may subsequently invade other body tissues and cause more generalized damage in the infected host. The members of the genus *Shigella* are the agents of a specifically human enteric disease, bacterial dysentery. In the genus *Salmonella,* both the host range and the variety of diseases produced are much broader than in *Shigella. Salmonella typhi* and *S. schottmülleri,* the agents of typhoid and paratyphoid fever, are specific pathogens of man, whereas certain other species are specific pathogens of other mammals or of birds. The great majority of the *Salmonella* group, however, have a low host specificity. They exist, often without causing disease symptoms, in the intestine and in certain tissues of animals or birds. If these forms gain access to and develop in foods, their subsequent ingestion by man can give rise to food poisoning. This can be defined as a gastrointestinal infection that is usually acute but transient, although in some cases it may be more serious. Outbreaks of food poisoning often have an epidemic character because food preparation on a large scale may provide favorable opportunities for the growth of these organisms.

### group II:
### Enterobacter-Serratia-Erwinia

Enterobacter aerogenes, the prototype of group II, is common in soil and water, and sometimes also occurs in the intestinal tract of animals. Similar bacteria, distinguished from *E. aerogenes* by permanent immotility and the presence of capsules, occur in the respiratory tract. Although such pathogens are customarily classified in a separate genus, *Klebsiella,* it is questionable whether or not a specific distinction, let along a generic one, is justified. A biochemical property that distinguishes some (though not all) *Enterobacter* strains from other enteric bacteria is the ability to fix nitrogen. This property can be expressed only under anaerobic growth conditions.

Serratia marcescens, also a common soil and water organism, differs from *Enterobacter* principally by its failure to produce formic hydrogenlyase (little or no visible gas formed during sugar fermentation) and by its inability or weak ability to ferment lactose. Many (but by no means all) strains of *Serratia* produce a characteristic red cellular pigment, *prodigiosin*.

Relative to the enteric bacteria so far discussed, the representatives of the genus *Erwinia* constitute a very heterogeneous group, in which three principal subgroups are now recognized, exemplified by the species *Erwinia amylovora, E. carotovora,* and *E. herbicola.*

*E. amylovora* is the agent of fire blight, a necrotic disease of pears and related plants. This species is notable for its limited range of utilizable sugars and its requirement for organic growth factors, characters absent from other erwinias. *E. carotovora* causes soft rots of the storage tissues of many plants, an action attributable in part to its ability to produce pectolytic enzymes, which destroy the pectic substances that serve as intracellular cementing materials in plant tissues. *E. herbicola,* which produces yellow cellular pigments, occurs commonly on the leaf surfaces of healthy plants; some strains are plant pathogens. Similar pigmented enteric bacteria have occasionally been isolated from human sources, although their pathogenicity for man remains uncertain.

### group III:
### *Proteus*

The members of the Proteus group are probably soil inhabitants, although they are found in particular abundance in decomposing animal materials. The relatively low G + C content of their DNA distinguishes most species from the groups so far discussed, from which they are also distinguishable by certain physiological properties. These include strong proteolytic activity (gelatin is rapidly liquefied) and ability to hydrolyze urea. Most members of the Proteus group are very actively motile and can spread rapidly over the surface of a moist agar plate, a phenomenon known as *swarming*. A curious feature of the swarming phenomenon is its periodicity: it occurs in successive waves, separated by periods of quiescence and growth. This produces a characteristic zonate pattern of development on an agar plate (Figure 11.26).

### group IV:
### *Yersinia*

The genus *Yersinia* contains two or three species that are agents of disease in rodents. *Yersinia pestis* can be transmitted by fleas from its normal rodent hosts to man; it is the cause of human bubonic plague, a disease that has been responsible throughout human history for massive epidemics with a very high mortality rate. In man the disease can also be transmitted through the respiratory route. Both in their mode of transmission and in their symptoms, the diseases caused by *Yersinia* species are entirely different from the major enteric diseases, such as dysentery and typhoid fever.

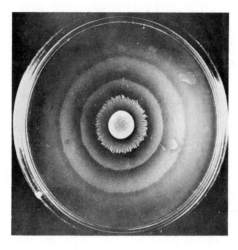

**Figure 11.26**
Swarming of *Proteus* on the surface of nutrient agar plate. The plate was inoculated in the center with a drop of a bacterial suspension and was photographed after incubation at 37°C for 20 hours. From H. E. Jones and R. W. A. Park, "The Influence of Medium Composition on the Growth and Swarming of *Proteus*," *J. Gen. Microbiol.* **47**, 369 (1967).

The members of the genus *Yersinia* carry out a mixed-acid fermentation without production of $H_2$ and $CO_2$; in this respect, they resemble the Shigella group. They ferment lactose and have a powerful urease. Motility is a variable character; *Y. pestis* is permanently immotile. The G + C content of the DNA is significantly lower than that of the members of Group I. A cultural character that distinguishes them is their relatively slow growth on complex media.

### the polar flagellates

The polarly flagellated facultative anaerobes have been studied far less intensively and systematically than the classical enteric bacteria, and their classification is still uncertain. Provisionally, four groups may be recognized (Table 11.3).

The genus *Aeromonas* includes organisms that perform a butanediol fermentation, accompanied by $H_2$ and $CO_2$ production, similar to that of *Enterobacter*, and ones that perform a mixed-acid fermentation without gas production, similar to that of *Shigella*. Some strains are capable of causing disease in both frogs and fish; others have been implicated in human gastroenteritis.

The limits of the genus *Vibrio* are somewhat controversial. Several of the marine organisms now assigned to *Beneckea* and *Photobacterium* have been included in it, largely on the basis of the curvature of their cells. However, the significance of this structural character seems increasingly doubtful; and if these strains are excluded, the genus *Vibrio* becomes essentially reduced to one species: *V. cholerae*, a water-transmitted pathogen that causes the gastrointestinal disease known as *cholera*.

The bacteria of the genera *Beneckea* and *Photobacterium* are among the most abundant chemoheterotrophs in marine environments. They occur in sea water, and in the intestinal tract and on the body surfaces of marine animals. Many of these bacteria can decompose chitin. *Beneckea parahemolytica* is a frequent cause of human gastroenteritis in Japan, where raw fish is commonly consumed. The property of bioluminescence (Figure 11.27), common to all members of the genus *Photobacterium*, also occurs in one *Beneckea* species. Bioluminescence occurs only under aerobic conditions.

### enteric bacteria in sanitary analysis

The intestinal diseases caused by the enteric bacteria are transmitted almost exclusively by the fecal contamination of water and food materials. Transmission through contaminated water supplies is by far the most serious source of infection and is responsible for the massive epidemics of the more serious enteric diseases (particularly typhoid fever and cholera). Today these diseases are almost unknown in most parts of the Western world, although cholera has recently reappeared in some countries bordering on the Mediterranean. Control of these diseases was achieved primarily by appropriate sanitary methods. An essential part of this operation was *the development of bacteriological methods for ascertaining the occurrence of fecal contamination of water and foodstuffs.*

It is seldom possible to isolate enteric pathogens directly from contaminated water because they are usually present in small numbers. To demonstrate the fact of fecal contamination, it is sufficient to

**Figure 11.27**
Luminous bacteria photographed by their own light: left, a streaked plate of *Photobacterium;* right, two flasks containing a suspension of the same organism in a sugar medium. A stream of air was passed continuously through the flask on the right during the photographic exposure. The bacteria in the unaerated flask on the left had exhausted the dissolved oxygen and had ceased to luminesce except at the surface, where organisms were exposed to the air.

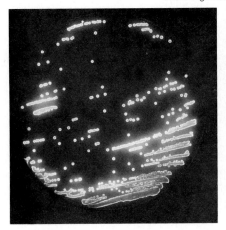 

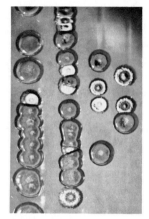

**Figure 11.28**
A plate of EMB agar streaked with a mixture of *Escherichia coli* and *Enterobacter aerogenes*. The colonies of *E. coli* are relatively small and appear light as a result of their metallic sheen. Courtesy of N. J. Palleroni.

show that the sample under examination contains bacteria known to be specific inhabitants of the intestinal tract, even though these may not be agents of disease. The bacteria that have principally served as indices of such contamination are the fecal streptococci (discussed in Chapter 12) and *E. coli*. The methods of sanitary analysis developed by bacteriologists differ somewhat from country to country.

One method for detecting *E. coli* is to inoculate dilutions of the sample under test into tubes of lactose broth, which are then incubated at 37°C, and examined after 1 and 2 days for acid and gas production. Cultures showing acid and gas formation are then streaked on a special medium, with a composition that facilitates recognition of *E. coli* colonies. One of the media most commonly used is a lactose-peptone agar containing two dyes, eosin and methylene blue (EMB agar). On this medium, *E. coli* produces blue-black colonies with a metallic sheen, whereas the other principal member of the group capable of fermenting lactose with acid and gas production, *Enterobacter aerogenes* (not necessarily indicative of fecal contamination) produces pale pink mucoid colonies without a sheen (Figure 11.28). For a final distinction between these two organisms, a series of physiological tests, known as the *IMViC tests,** can be performed on material from an isolated colony. The typical results obtained with the two species are shown in Table 11.4. Of these four tests, the Methyl Red (M) and Voges-Proskauer (V) tests are the most significant, since they indirectly reveal the mode of fermentative sugar breakdown. Both are performed on cultures grown in a glucose-peptone medium. The Methyl Red test affords a measure of the final pH: this indicator is yellow at a pH of 4.5 or higher and red at lower pH values. A positive test (red color) is therefore indicative of substantial acid production, characteristic of a mixed-acid fermentation. The Voges-Proskauer test is a color test for acetoin, an intermediate in the butanediol fermentation. The test for indole production (I) from tryptophan, performed on a culture grown in a peptone medium rich in tryptophan, is a test for the presence of the enzyme tryptophanase, which splits tryptophan to indole, pyruvate, and ammonia. This enzyme is present in many bacteria of the enteric group (including *E. coli*) but is not found in *E. aerogenes*. The citrate utilization test (C) determines

---

*IMViC is an acronym of the tests: Indole production (I), Methyl red color change (M), Voges-Proskauer (V) and Citrate utilization (C).

**Table 11.4**
IMViC Tests for the Differentiation Between *Escherichia coli* and *Enterobacter aerogenes*

| | | TYPICAL REACTIONS | | |
| --- | --- | --- | --- | --- |
| | INDOLE | METHYL RED | VOGES-PROSKAUER | CITRATE |
| *Escherichia coli* | + | + | − | − |
| *Enterobacter aerogenes* | − | − | + | + |

ability to grow in a synthetic medium containing citrate as the sole carbon source. This ability is lacking in most strains of *E. coli,* as a result of the absence of a citrate permease.

The first step of the analytical procedure described above is relatively nonspecific, since many bacteria, not members of the enteric group, also can grow at 37°C in lactose broth with acid and gas production. A more specific enrichment can be achieved by incubation at 44°C.

## fermentative chemoheterotrophs

Many of the strictly fermentative bacteria are unable to tolerate even brief exposure to air, thus rendering their isolation and cultivation quite difficult. As a consequence, it has only been quite recently that this group has become known, principally through the development of new techniques by R. E. Hungate during work on the bacterial flora of the rumen. Nevertheless, this group is still one of the least well studied.

Most of these strict anaerobes inhabit the body cavities of animals, which provide a variety of anaerobic niches. In appropriate habitats, the numbers of these bacteria can be enormous; representatives of the genus *Bacteroides,* for example, outnumber *Escherichia* in the large intestine by an order of magnitude and comprise the bulk of mammalian fecal material. Some strains have been implicated in human or animal infections (e.g., *Bacteroides, Fusobacterium, Veillonella*), but such pathogenesis is probably quite rare. In general, these bacteria appear to be benign components of the normal animal microflora.

Properties of some representative genera are shown in Table 11.5. Differentiation between genera depends heavily on determination of the fermentable substrates. Most strains ferment sugars or

**Table 11.5**
Strictly Anaerobic Gram-Negative With a Fermentative Mode of Metabolism

| | *BACTEROIDES* | *FUSOBACTERIUM* | *SELENOMONAS* | *VEILLONELLA* |
|---|---|---|---|---|
| Structure | Straight rods | Straight rods | Crescent-shaped | Cocci |
| Flagellar insertion | Peritrichous or immotile | Peritrichous or immotile | Tuft on concave side | Immotile |
| Fermentable substrates | | | | |
| Sugars | + | + | + | − |
| Amino acids | + | + | − | − |
| Lactic acid | − | − | − | + |
| DNA base composition | | | | |
| (mole % G + C) | 40–55 | 26–34 | 53–61 | 40–44 |
| Source | Intestinal tract, mouth, rumen, | Mouth, intestinal tract | Rumen, mouth | Mouth, intestinal tract |

amino acids (or, by virtue of the possession of extracellular hydrolases, poly-saccharides or proteins). *Veillonella,* however, has the relatively rare ability to ferment lactic acid, a common end product of many sugar fermentations. Most strains are relatively undistinguished morphologically, although *Seleno-monas* has a unique site of flagellar insertion; a tuft of flagella is inserted laterally on the concave side of the crescent-shaped cell (Figure 11.29).

## spirochetes

One group of Gram-negative bacteria, of distinct cell structure, contains aerobic, anaerobic, and facultatively anaerobic strains, and thus does not fit the physiological classification given above. These are the spirochetes, helical cells that, with one exception, are extremely thin (Figure 11.30). Indeed, some are barely resolvable in the light microscope.

Spirochetes are all motile, appearing very active in wet mounts. They frequently thrash about, swim vigorously through the liquid, or creep slowly across a solid substratum. These movements are thought to be mediated by an organelle known as the *axial filament,* composed of two sets of fibers that originate from the cytoplasmic membrane—one set at each pole of the cell—and then penetrate the peptidoglycan layer and wrap around the cell within the outer membrane (Figure 11.31). The two sets overlap in the middle. The individual fibers appear to be homologous to bacterial flagella, and although the mechanism is unclear, it is probable that spirochetal motility constitutes a special modification of flagellar motility.

Many spirochetes are parasites of vertebrate animals, and include the causative agents of a number of diseases (e.g., *Trepo-*

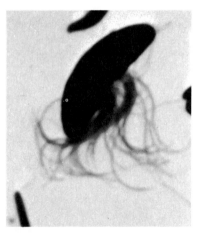

**Figure 11.29**
*Selenomonas:* flagella stain, showing the crescent-shaped cell and the tuft of flagella inserted laterally on the concave surface of the cell (×3,600). Photo courtesy of C. F. Robinow.

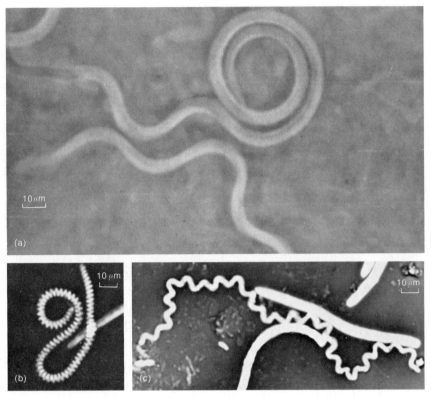

**Figure 11.30**
Some of the larger spirochetes. (a) Phase contrast of living *Cristispira*, a form found in the crystalline style of clams (×380). Courtesy of S. W. Watson. (b) Nigrosin mount of an unidentified spirochete from water (×341). (c) *Spirochaeta*, a large spirochete common in water (×341). The preparation also contains cells of rod-shaped bacteria. (b) and (c) courtesy of C. F. Robinow.

*nema pallidum* and *T. pertenue* cause syphilis and yaws, respectively). A number of these parasites occur as part of the normal flora of animals; the human mouth, for example, contains great numbers of anaerobic spirochetes (principally nonpathogenic *Treponema* species) in the crevices between the teeth and the gums. Other strains are normally relatively benign parasites, occasionally erupting into a fulminating infection. *Leptospira*, an obligate aerobe, is apparently capable of infecting the kidney, living on the glomerular filtrate and causing no overt problems for the host for years. If these parasites gain access to the bloodstream, however, they may cause a fatal septicemia.

Spirochetes are also found free-living. The most common habitat is anaerobic mud, which is frequently teeming with spirochetes. The bulk of these are strict anaerobes, although one species of *Spirochaeta* is a facultative anaerobe. Many free-living strains have a relatively high

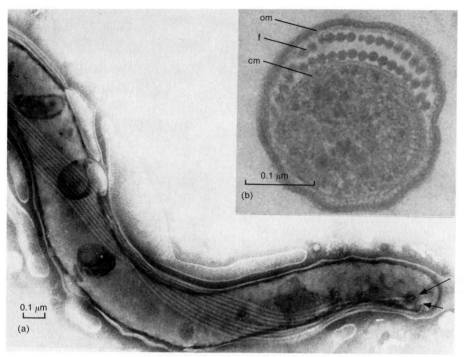

**Figure 11.31**
The structure of the spirochetal cell as shown by electron micrographs of spirochetes from the mouth. (a) End of a cell, negatively stained with phosphotungstic acid, showing the relationship of the multifibrillar axial filament to the protoplast ($\times$51,000). The insertion points of two fibrils from the axial filament are just visible at the pole of the cell (arrows). (b) Cross section of a large spirochete, showing the location of the fibrils (f) of the axial filament between the cell membrane (cm) and the outer membrane (om) ($\times$183,000). From M. A. Listgarten and S. S. Socransky, "Electron Microscopy of Axial Fibrils, Outer Envelope and Cell Division of Certain Oral Spirochetes," *J. Bacteriol.* **88,** 1087 (1964).

tolerance to the toxic effects of $H_2S$ and can therefore live in habitats in which vigorous sulfur reduction is occurring.

Although united by their unique morphology, the spirochetes are obviously a genetically and metabolically diverse group. The G + C content of their DNA spans the range from 32 to 66 mole percent. The full range of oxygen relationships is observed within the group, and they are found in a great diversity of habitats.

## obligate intracellular parasites

### rickettsias

The rickettsias are obligate intracellular parasites in certain groups of arthropods (notably fleas, lice, and ticks). They do not appear to produce symptoms of disease in their arthropod hosts, but if they are transmitted by bite to a vertebrate host, a severe and often fatal infection may result. The major rickettsial disease of man is epidemic typhus, the biography of which has been recounted in popular form in one of the few literary classics written by a microbiologist, Hans Zinsser's *Rats, Lice and History*.* Other human rickettsial diseases of minor importance are Rocky Mountain spotted fever, transmitted by ticks, and scrub typhus, normally transmitted by mites to field mice, but also transmissible to man.

The rickettsias are small rods, about 0.3 by 1.0 $\mu$m (Figure 11.32), which multiply by binary fission. The explanation of their obligate intracellular parasitism is still far from clear. Rickettsias have an autonomous system of energy-yielding metabolism. However, isolated cells

*Hans Zinsser, *Rats, Lice and History*, Little, Brown and Co. Inc., Boston, 1935.

**Figure 11.32**
Electron micrographs of thin sections through the tissues of chicken embryos infected with Rickettsia. (a) A single embryonic cell in the later stage of rickettsial infection ($\times$10,026). The whole cytoplasmic region is filled with the rod-shaped rickettsial cells (r); the centrally located nucleus, although abnormal in appearance, is not infected. (b) Portion of an infected embryonic cell at a high magnification ($\times$57,031). Several rickettsial cells (r) are to be seen in longitudinal or transverse section; also a mitochondrion (m), which is roughly of the same dimensions as the rickettsial cells. From S. L. Wissig, L. G. Caro, E. B. Jackson, and J. E. Smadel, "Electron Microscopic Observations on Intracellular Rickettsiae," *Am. J. Pathol.* **32,** 1117 (1956).

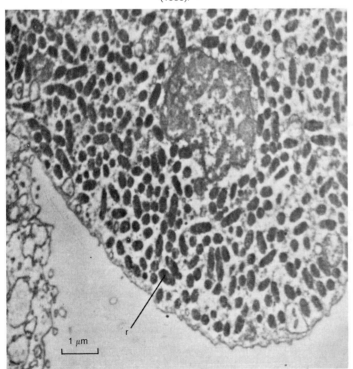

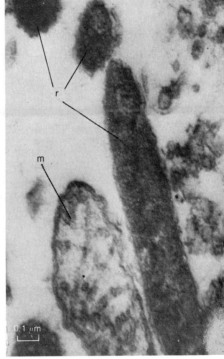

(a)                                                    (b)

rapidly become inviable. Loss of viability can be retarded by adding certain coenzymes to the suspending medium. Since these are substances to which intact cells are usually impermeable, it has been suggested that the adaptations of the rickettsias to an intracellular mode of life include changes in membrane permeability that enable them to use complex metabolites produced by the host. Such changes in membrane properties would, however, make the rickettsias extremely vulnerable in an extracellular environment.

### chlamydias

The chlamydias are the agents of a number of diseases of birds and of mammals. In man, these include *trachoma*, a serious eye infection that is a major world health problem, and *lymphogranuloma venereum*, an increasingly common venereal disease. In contrast to rickettsias, these bacteria are not transmitted through invertebrate hosts, but pass directly between their vertebrate hosts.

The cells (Figure 11.33) are more or less spherical and slightly smaller than those of rickettsias, being 0.2 to 0.7 μm in diameter. As a result of their very small size, it is difficult to establish with certainty the mode of growth within the host cell; and there is some dispute about whether multiplication occurs by binary fission or by budding. Although biochemical studies show that the chlamydias possess extensive enzymatic capacities, it

**Figure 11.33**

Electron micrograph of a thin section of part of a mammalian cell growing in tissue culture, and infected by many chlamydia cells (×31,185). Each of the small round-to-ovoid chlamydia cells is surrounded by a unit membrane; many are in the course of binary fission. Courtesy of R. R. Friis, Department of Microbiology, University of Chicago.

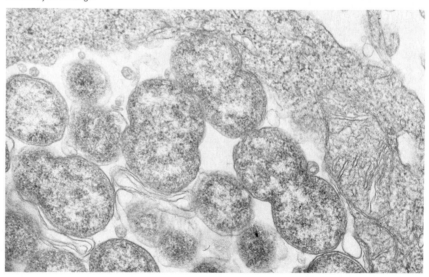

has not yet been demonstrated that they are capable of performing ATP yielding metabolism. This has led to the interesting hypothesis that they may be "energy parasites," the growth of which is strictly dependent on the provision of energy-rich materials derived from the ATP-yielding metabolism of the host cell.

## the mycoplasmas

The mycoplasmas are a group of procaryotes distinguished by their lack of a peptidoglycan cell wall. Many of them appear to be bounded only by the cell membrane; others have variable amounts of chemically unidentified material external to the cell membrane. Due to the lack of peptidoglycan, the cells are quite plastic, frequently exhibiting a considerable variation in shape—coccoidal, pear-shaped, and filamentous forms being most common (Figure 11.34). Many strains are capable of passing through filters with pore size smaller than the cell diameter, a feature that may make them a nuisance as contaminants in filter-sterilized media.

Although the layers sometimes found external to the cell membrane may confer structural strength, the cell membrane of mycoplasmas is itself significantly stronger than that of other bacteria. The basis for this extra resilience is not known, although for some members of the group it may be a consequence of the presence in their membranes of sterols, a class of lipid characteristic of eucaryotic membranes. Although some other procaryotes have traces of sterols, the mycoplasmas are the only ones in which sterols are quantitatively important. Even these bacteria, however, are incapable of synthesizing sterols; sterols are consequently required growth factors for these strains of mycoplasmas. Strains that do not require sterols for growth will incorporate them if they are available in the medium.

In addition to sterols, most mycoplasmas require numerous other growth factors that are usually supplied to the culture media in animal serum or other complex material. An exception is *Thermoplasma*, which can be cultivated on a relatively simple defined medium.

On solid media, the mycoplasmas characteristically produce quite small colonies, rarely larger than a few millimeters in diameter; they have a nippled or "fried-egg" appearance (Figure 11.35). The raised center is a nearly spherical mass of cells, which penetrates into the agar; it is surrounded by a thin zone of surface growth.

Two types of motility (gliding and swimming) are found in the group, both of unknown mechanism. Gliding motility is shown by some *Mycoplasma* strains in contact with a solid substratum. *Spiroplasma* is capable of translational movements in liquid, but does not possess either flagella or axial filaments.

**Figure 11.34**
Electron micrograph of cells of a member of the *Mycoplasma* group, the agent of bronchopneumonia in the rat (×1,230). From E. Klieneberger-Nobel and F. W. Cuckow, "A Study of Organisms of the Pleuropneumonia Group by Electron Microscopy," *J. Gen. Microbiol.* **12**, 99 (1955).

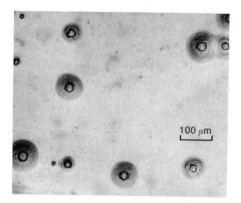

**Figure 11.35**
Characteristic colony structure of organisms of the *Mycoplasma* group (×79). Courtesy of M. Shifrine.

Four genera are recognized on the basis of cell shape, sterol requirement, and habitat (Table 11.6). Members of the genus *Mycoplasma* are parasitic on animals, principally on moist mucosal surfaces such as the respiratory and urogenital tracts. They are frequently pathogenic, causing a variety of animal diseases, including some forms of arthritis and respiratory infections such as sinusitis and some pneumonias. *Acholeplasma* contains both parasitic and free-living forms, characterized by the absence of a sterol requirement. Members of both genera are fermentative, but can grow in the presence of air. Some *Mycoplasma* strains are also capable of respiration.

*Thermoplasma*, a strict aerobe, is characteristically found in piles of refuse from coal mines, which are rich in iron pyrite. Oxidation of this iron and sulfide by chemoautotrophs generates acid and heat, thus providing the high temperatures (optimum about 60°C) and low pH (optimum about pH 2) required for growth of *Thermoplasma*. In addition, the complex organic components of the coal presumably satisfy the unusual polypeptide growth factor requirement of this organism. This requirement may explain the rarity of *Thermoplasma* in acidic hot springs.

**Table 11.6**
Distinguishing Characteristics of the Genera Comprising the Mycoplasmas

| GENUS | CELL SHAPE | G + C (MOLES PERCENT) IN DNA | STEROL REQUIREMENT | MOTILITY | HABITAT |
|---|---|---|---|---|---|
| *Mycoplasma* | Variable | 23–41 | + | ± (Gliding) | Parasitic |
| *Acholeplasma* | Variable | 20–33 | − | − | Parasitic or free-living |
| *Thermoplasma* | Variable | 46 | − | − | Free-living thermophile |
| *Spiroplasma* | Variable (solid media) Helical (liquid media) | 25–27 | + | + (Swimming) | Parasitic |

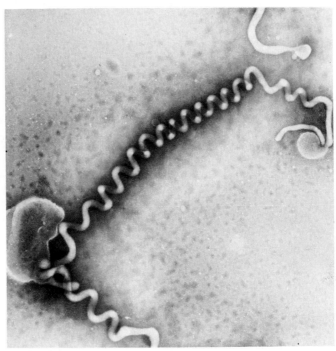

**Figure 11.36**
Electron micrograph of a *Spiroplasma* cell, showing the helical shape. From R. M. Cole, J. C. Tully, T. J. Popkin, and J. M. Bové, "Morphology, Ultrastructure, and Bacteriophage Infection of the Helical Mycoplasma-Like Organism (*Spiroplasma citri*, gen. nov., sp. nov.) Cultured from 'Stubborn' Disease of Citrus," *J. Bact.* **115,** 367 (1973).

*Spiroplasma* is unusual among mycoplasmas in that cells growing in liquid culture have a well defined helical shape (Figure 11.36); however, stationary phase cultures, or cultures grown on solid media, show the morphology typical of other mycoplasmas. Strains of *Spiroplasma* are parasitic on a wide variety of plants, insects, and animals, causing disease in some cases.

Since the mycoplasmas lack a peptidoglycan-containing cell wall, the Gram stain is of dubious systematic value. Despite the lack of peptidoglycan, *Spiroplasma* stains Gram-positive and *Thermoplasma* Gram-variable (some strains retain the dye, others do not). The basis for and significance of retention of the dye complex in these strains is unknown.

# 12

# GRAM-POSITIVE BACTERIA

Although the Gram-positive bacteria, like the Gram-negative ones, contain aerobes, anaerobes, and facultative anaerobes, primary division of the group is usually not made on this basis. The Gram-positive bacteria are instead normally divided into two major assemblages on the basis of morphological characteristics: the first group—the endospore-formers—form the resting cell termed an *endospore,* and the second—the actinomycete line—consists of a continuum of morphological complexity that culminates in the formation of a true mycelium. Each group shows considerable physiological heterogeneity.

In addition to these two major groups, a small and morphologically diverse group of Gram-positive anaerobes are united by their unique metabolism, in which methane is produced. A miscellaneous assortment of other anaerobes completes the list of known Gram-positive bacteria.

## the endosporeformers

Endospores (Figure 12.1) can be readily recognized microscopically by their intracellular site of formation, their extreme refractility, and their resistance to staining by dyes that readily stain vegetative cells. They are not normally formed during active growth; their differentiation begins when a population of vegetative cells passes out of the exponential growth phase as a consequence of nutrient limitation. Typically, one endospore is formed in each vegetative cell. The mature spore is liberated by lysis of the vegetative cell in which it has developed. Free endospores have no detectable metabolism, but for many years (often decades) they retain the

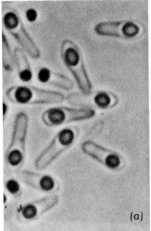
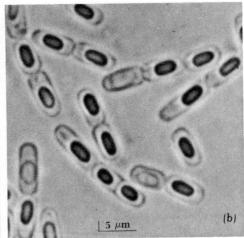
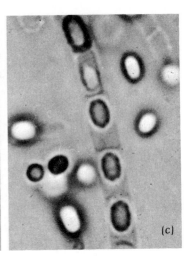

**Figure 12.1**
Sporulating cells of *Bacillus* species: (a) unidentified bacillus from soil; (b) *B. cereus;* (c) *B. megaterium.* From C. F. Robinow, in *The Bacteria* (I. C. Gunsalus and R. Y. Stanier, editors), **1,** p. 208. New York: Academic Press, 1960.

potential capacity to germinate and develop into vegetative cells. Endospores are in addition highly resistant to heat, ultraviolet and ionizing radiations, and many toxic chemicals. Their heat resistance is frequently taken advantage of in the isolation of spore-forming bacteria; these organisms can be selected by subjecting suspensions of source materials to a thermal pretreatment sufficient to kill all vegetative cells.

### endospore formation

The structural events associated with spore formation have been elucidated by a combination of light and electron microscopic observations. A synthesis based on both types of observations will be presented here (Figures 12.2 and 12.3).

At the end of the logarithmic phase of growth, just prior to the onset of sporulation, each cell contains two nuclear bodies. The coalescence of these into a single nuclear area is the first microscopically detectable stage of sporulation. A transverse septum is then formed near one cell pole, which separates the cytoplasm and the DNA of the smaller cell (destined to become the spore) from the rest of the cell contents. Septum formation is not accompanied, as in normal cell division, by the development of a transverse wall; instead, the membrane of the larger cell rapidly grows around the smaller cell, which thus becomes completely engulfed within the cytoplasm of the larger cell, to produce a so-called *forespore*. In effect, the forespore is a protoplast enclosed by two concentric sets of unit membranes: its own bounding membrane, and the membrane of the mother cell that has grown around it. At this stage, the development process becomes irreversible: the cell is said to be "committed" to undergo sporulation. By light microscopy, the forespore appears as a small, dark, nonrefractile area that is free of granular inclusions.

Once the forespore has been engulfed by the mother cell, there is a rapid synthesis and deposition of new structures that enclose it. The first to appear is the *cortex,* which develops between the two membranes; shortly afterward, a more electron-dense layer, the *spore coat,* begins to form exterior to the outer unit membrane surrounding the cortex. In the *B. cereus* group an additional, looser and thinner layer, the *exosporium,* forms outside the spore coat (Figure 12.4). Once the spore coat is synthesized, the maturing spore begins to become refractile, although it is not yet heat-resistant. The development of heat resistance closely follows two major chemical changes: a massive uptake of $Ca^{2+}$ ions by the sporulating cell, and the synthesis in large amounts of dipicolinic acid (Figure 12.5), a compound absent from vegetative cells. This compound represents 10 to 15 percent of the spore dry weight, and it is located within the spore protoplast. The time sequence of the development of refractility and heat resistance and the synthesis of dipicolinic acid are shown in Figure 12.6.

**Figure 12.2**

A schematic representation of endospore formation. (a) Vegetative cell showing cell wall, C.W., cell membrane, C.M., and two nuclear areas, N. (b) Condensation of nuclear material. (c–d) Septum formation. (e–f) Engulfment. (g–h) Cortex and coat synthesis. (i) Release.

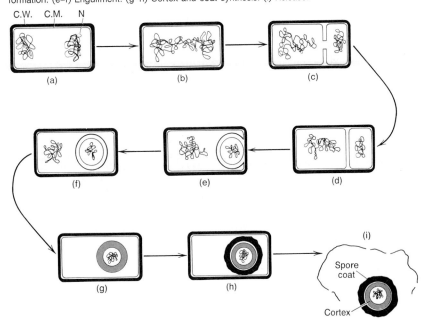

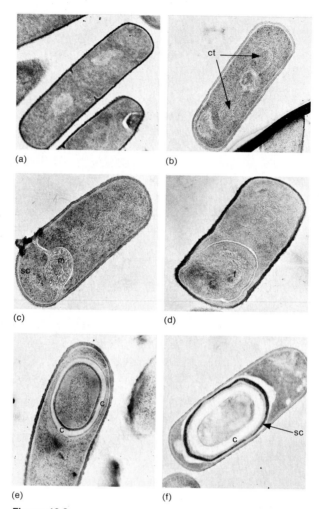

**Figure 12.3**

Electron micrographs of thin sections of *Bacillus*, showing the sequence of structural changes associated with endospore formation. (a) Vegetative cell in course of exponential growth. (b) Coalescence of the nuclei, to form an axial chromatin thread, ct. (c) Formation of transverse septum, containing a mesosome, m, near one cell pole, delimiting the future spore cell, sc, from the rest of the cell. (d) Formation of a forespore f, completely enclosed by the cytoplasm of the mother cell. (e) The developing spore is surrounded by the cortex, c. (f) The terminal stage of spore development: the mature spore, still enclosed by the mother cell, is now surrounded by both cortex, c, and spore coat, sc. From A. Ryter, ''Étude Morphologique de la Sporulation de *Bacillus subtilis*,'' *Ann. Inst. Pasteur* **108,** 40 (1965).

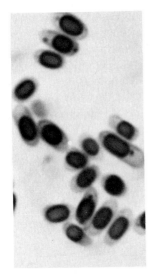

**Figure 12.4**
Mature endospores of *Bacillus cereus* (×3,600). Each stained spore is surrounded by a less deeply stained exosporium. Courtesy of C. F. Robinow.

HOOC—N—COOH

**Figure 12.5**
Dipicolinic acid.

The cortex is composed of a chemically modified peptidoglycan, in which relatively few of the muramic acid residues bear a tetrapeptide side chain. Cross-linking is thus substantially reduced, and the cortex is thought to be more elastic than rigid.

The spore coat, which represents 30 to 60 percent of the dry weight of the spore, is largely composed of protein and accounts for about 80 percent of the total spore protein. The spore coat proteins are highly resistant to treatments that solubilize most proteins.

After the completion of spore development, the spore protoplast, accordingly, contains an extremely high content of $Ca^{+2}$ and dipicolinic acid and it is enclosed by newly synthesized outer layers of unique chemical structure (the cortex and the spore coat, sometimes also an exosporium), which account for a very large fraction of the spore dry weight. When liberated by autolysis of the mother cell, the mature endospore is highly dehydrated, shows no detectable metabolic activity, and is highly resistant to heat and radiation damage and to attack by either enzymatic or chemical agents. It remains in this state until a series of environmental triggers initiate its conversion into a new vegetative cell.

In many sporeformers, both aerobic and anaerobic, the onset of sporulation is accompanied by the synthesis of a distinctive class of antimicrobial substances: small peptides. Many of these peptide

**Figure 12.6**
The increases in the refractility, thermostability of the cells, and dipicolinate content of the population that occur during sporulation in a culture of *Bacillus cereus*. All values are plotted against the age of the culture in hours. After T. Hasimoto, S. H. Black, and P. Gerhardt, "Development of fine structure, thermostability, and dipicolinate during sporogenesis in a bacillus." *Can. J. Microbiol.* **6**, 203 (1960).

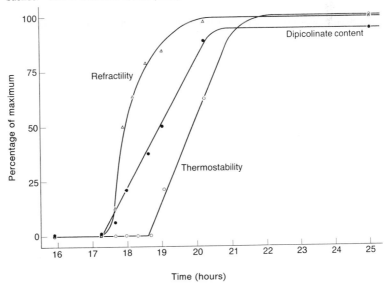

antibiotics have been characterized chemically and functionally. They can be assigned to three classes: *edeines,* linear basic peptides that inhibit DNA synthesis; *bacitracins,* cyclic peptides that inhibit cell wall synthesis; and the *gram-icidin-polymyxin-tyrocidin group,* linear or cyclic peptides that modify membrane structure or function. Many of these peptides contain amino acids of D-configuration, a form of amino acid that does not occur in proteins, and even constituents that are not amino acids. The production of peptide antibiotics occurs rather early in the sporulation process. The biosynthesis of these compounds involves a novel assembly mechanism, in which the amino acid sequence is determined by a protein (enzyme); neither tRNAs nor ribosomes participate. The role of these compounds in sporulation is unknown, but it has been suggested that they may control various stages of the differentiation process.

### activation, germination, and outgrowth of endospores

Freshly formed endospores will remain largely dormant even if placed in optimal conditions for germination. The state of dormancy can be broken by a variety of treatments, collectively termed *activation.* Perhaps the most general mechanism for activating spores is *heat shock:* exposure for several hours to an elevated, but sublethal, temperature (e.g., 65°C). Heat activation is not accompanied by any detectable change in the appearance of the spores: it simply enables them to germinate when subsequently placed in a favorable environment. A much slower activation takes place upon the storage of spores, even at relatively low temperatures (5°C) or under dry conditions.

When activated spores are placed under favorable conditions, germination can take place. This process is very rapid and is characterized by a loss of refractility, by a loss of resistance to heat and other deleterious agents, and by the sudden onset of respiration. These processes are accompanied by the liberation, as soluble materials, of about 30 percent of the spore dry weight. This spore exudate consists largely of $Ca^{+2}$ and dipicolinic acid (derived from the spore protoplast) and peptidoglycan fragments (derived from the cortex). The cortex is rapidly destroyed, only the outer spore coat remaining.

The process of germination does not appear to be accompanied by significant synthesis of biopolymers; the appearance of metabolic activity is a consequence of the unmasking of preexisting but inactive enzymes in the spore protoplast. If germination occurs in a medium that does not contain the nutrients required for vegetative growth, no further change will occur. In a complete medium, the germinated spore will proceed to grow out into a vegetative cell. Outgrowth involves an initial swelling of the spore within its spore coat, accompanied by a rapid synthesis of a vegetative cell wall. The newly formed vegetative cell then emerges from the spore coat (Figure 12.7), elongates, and proceeds to undergo the first vegetative division.

**Figure 12.7**
Spore outgrowth in (a) *Bacillus polymyxa* and (b) *B. circulans* (stained preparations, ×5,260). Courtesy of C. F. Robinow and C. L. Hannay.

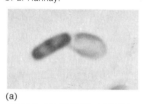

(a)

(b)

## classification of
## the endosporeformers

Morphologically and biochemically, the endosporeformers are diverse. They are united only by their ability to form endospores, a trait of sufficient biochemical and structural complexity to serve as a useful taxonomic characteristic. In other respects, however, they undoubtedly constitute a heterogenous group.

Many species fail to react uniformly to the Gram stain and are termed *gram-variable*; i.e., cultures contain a mixture of Gram-positive and Gram-negative cells, and may become completely Gram-negative in older cultures. This loss of the Gram-positive staining characteristic as the culture ages is presumably due to the accumulation of enzymes that modify the chemical structure of the cell wall. Such enzymes have a role ultimately in the release of the endospore from the mother cell. Despite their variable reaction to the Gram stain, most endosporeformers have a wall that in ultrastructural terms is typical of Gram-positive bacteria. One possible exception is a genus of an endospore-producing sulfate-reducer physiologically identical to *Desulfovibrio*; even young cells of this organism stain Gram-negative. Electron microscopic analysis of the cell wall will thus be necessary to determine whether this organism should be placed with the Gram-positive or Gram-negative bacteria.

A number of types of sporeformers have been observed but never isolated in pure culture. These are typically inhabitants of the gut of herbivorous animals, and are presumably anaerobic. They include spiral, rod, and filamentous forms, and some appear to produce more than one endospore per cell (Figure 12.8).

The primary subdivision of the endosporeformers is usually based on their oxygen relationships. *Bacillus* is the principal genus of the aerobic (either strict or facultative) endosporeformer, while *Clostridium* is strictly fermentative. The G + C content of their DNA is typically low, with that of *Bacillus* being 33 to 51 mole percent; and *Clostridium*, with 22 to 28 mole percent, containing organisms with the lowest known G + C content.

**Figure 12.8**
*Metabacterium polyspora:* smear from intestinal tract of guinea pig, showing three sporulating cells, each containing two or more rod-shaped endospores (×1,700). Courtesy of C. F. Robinow.

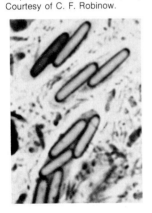

### Bacillus

Most aerobic sporeformers are versatile chemoheterotrophs capable of utilizing a considerable range of simple organic compounds (sugars, amino acids, organic acids) as respiratory substrates, and in some cases also capable of fermenting carbohydrates. A few species require no organic growth factors; others may require amino acids, B vitamins, or both. The majority are mesophiles, with temperature optima in the range of 30 to 45°C. Many strains are motile by means of peritrichous flagella.

A special physiological group consists of the extreme thermophiles, which are capable of growing at temperatures as high as 65°C and usually fail to grow at temperatures below 45°C. There are probably a number of species with this attribute, but their taxonomy has not been studied in detail. The characteristic environment for these organisms is decomposing plant material, in which the heat generated by microbial metabolic activity cannot be readily dissipated. The classical example is a moist haystack, in which the rise of temperature is so marked that it sometimes leads to spontaneous combustion. As the temperature rises, in the first place through the metabolic activities of mesophilic microorganisms, the primary microbial population is displaced by extreme thermophiles, principally bacilli and actinomycetes (p. 312).

A distinctive species cluster of mesophilic bacilli consists of *B. cereus*, one of the most abundant aerobic sporeformers in soil, and two related pathogens, *B. anthracis* and *B. thuringiensis*. In all three species the endospores are enclosed by an *exosporium*, a loose outer coat that is not formed by other bacilli. Certain strains of *B. cereus* produce distinctive loosely spreading colonies, superficially resembling those of fungi (Figure 12.9). *B. anthracis* is the agent of anthrax, a disease of cattle and sheep that is also transmissible to man. It is one of the few spore-forming bacteria that is a true parasite, in the sense that it is able to develop massively within the body of the animal host. Apart from its pathogenic properties, the biochemical mechanisms of which have been extensively studied (see Chapter 16), *B. anthracis* differs from *B. cereus* by its permanent immotility.

*B. thuringiensis* is the causal agent of paralytic disease of the caterpillars of many lepidopterous insects, paralysis resulting from the ingestion of plant materials that carry on their surface spores or sporulating cells of the bacterium. Each sporulating cell of *B. thuringiensis* produces, adjacent to the spore, a regular bipyrimidal protein crystal (Figure 12.10), which is liberated, along with the spore, by autolysis of the parent cell (Figure 12.11). The protein, of which the crystal is composed, is toxic for insects;

**Figure 12.9**
A colony of *Bacillus cereus var. mycoides* (×1.29). Courtesy of David Cornelius and C. F. Robinow.

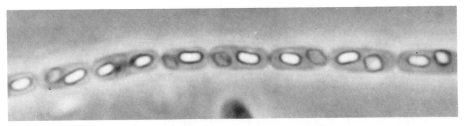

**Figure 12.10**

A chain of sporulating cells of *Bacillus thuringiensis* (phase contrast, ×3,900). Each cell contains, in addition to the bright, refractile spore, a less refractile bipyramidal crystalline inclusion. Courtesy of P. FitzJames.

after ingestion, it dissolves in the alkaline gut contents of the caterpillar and causes a loosening of the epithelial gut wall, with a consequent diffusion of liquid from the gut into the blood. This leads to rapid paralysis. Germination of the accompanying endospores is followed by extensive bacterial growth in the host tissues, leading to death of the host. Since the parasporal protein of *B. thuringiensis* is toxic for a wide range of lepidopterous larvae, but nontoxic for vertebrates or predatory insects, preparations of sporulating cells of this organism have found an extensive application in agriculture as a biological insecticide (see Chapter 18).

### *Clostridium*

Anaerobic sporeformers with a fermentative mode of metabolism (genus *Clostridium*) were discovered by Pasteur in the middle of the nineteenth century when he demonstrated that some of these organisms carry out a fermentation of sugars accompanied by the formation of butyric acid. Shortly afterward it was recognized that clostridia are also the principal agents of the anaerobic decomposition of proteins ("putrefaction"). Toward the end of the nineteenth century it became evident that some clos-

**Figure 12.11**

Electron micrograph of free crystals and a free spore surrounded by a exosporium of a crystal-forming *Bacillus* (metal-shadowed preparation, ×7,500). From C. L. Hannay and P. FitzJames, "The protein crystals of *B. thuringiensis*." *Can. J. Microbiol.* **1,** 694 (1955).

tridia are agents of human or animal disease. Like other members of the group, the pathogenic clostridia are normal soil inhabitants, with little or no invasive power; the diseases they produce result from the production of a variety of highly toxic proteins (exotoxins). Indeed, botulism (caused by *C. botulinum*) and less serious types of clostridial food poisoning (caused by *C. perfringens*) are usually pure intoxications, resulting from the ingestion of foods in which these organisms have previously developed and formed exotoxins. The other principal clostridial diseases, tetanus (cause by *C. tetani*) and gas gangrene (caused by several other species), are the results of wound infections; tissue damage leads to the development of an anaerobic environment that permits localized growth and toxin formation by these organisms. Some clostridial toxins (those responsible for botulism and tetanus) are potent inhibitors for nerve functions. Others (those responsible for gas gangrene) are enzymes that cause tissue destruction (see Chapter 16).

Clostridia are common soil organisms, and are especially common in the anaerobic sediments of both fresh- and saltwater systems. In these places, they are the principal organisms responsible for the initial attack on organic molecules, including proteins and complex polysaccharides such as cellulose. The clostridial fermentations result in the formation of various gases (especially $H_2$ and $CO_2$) and organic end products that are used as energy sources by methanogenic bacteria (see p. 315) and by organisms capable of anaerobic respiration, principally the sulfur-reducers. Due to the high clostridial population densities often achieved in anaerobic lake and estuary sediment, clostridial intoxication (botulism) is a major cause of death for bottom-feeding waterfowl.

Clostridial cells are rod-shaped, typically long and thin. The spore is formed terminally or subterminally, and is larger in diameter than the vegetative cell. The sporulating cell is thus swollen at the end, giving rise to characteristic club or tennis-racket shapes (Figure 12.12). Many strains are motile by means of peritrichous flagella. Some species fix nitrogen.

**Figure 12.12**
Bacterial endospores. Stained wet mount of an anaerobic spore-former (*Clostridium*) in the course of sporulation. The rod-shaped vegetative cells are swollen at one end by the presence of oval, highly refractile spores. Photo courtesy of C. F. Robinow.

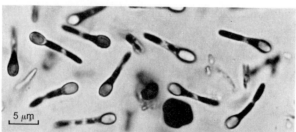

5 μm

Some clostridia are particularly sensitive to oxygen, their vegetative cells being killed by even momentary exposure to air; their spores, however, are not oxygen-sensitive. The ability of clostridial spores to withstand prolonged exposure to air is undoubtedly important to the ubiquity of the genus in nature.

## the actinomycete line

Among the procaryotes, a mycelial* growth habit is confined to Gram-positive bacteria, being characteristic of the organisms known as *actinomycetes*. Some of these organisms, the *euactinomycetes*, develop only in the mycelial state, and they reproduce through the formation of unicellular spores, differentiated either singly or in chains at the tips of the hyphae. This group is a large and complex one, containing many genera that are distinguished primarily by their structural and developmental properties.

In a second group, known as the *proactinomycetes*, mycelial development is transitory and often limited; specialized spores are not produced, and reproduction occurs primarily by mycelial fragmentation into short, usually rod-shaped cells. The proactinomycetes intergrade with a large group of organisms known collectively as *coryneform bacteria*, which are unicellular and reproduce by binary fission. These organisms are characterized by a marked irregularity and variability of cell form; the cells are commonly tapered and club-shaped, and in some cases undergo a transition to a coccoid stage during the developmental cycle.

Lastly, certain Gram-positive, unicellular bacteria that do not form endospores, and are distinguished from the coryneform group by their regular cell shape, appear to be associated with the actinomycete line. They include the lactic acid bacteria and the micrococci.

If motile, members of the actinomycete line bear flagella. Motility is rather rare, although it occurs sporadically in some members of all the groups described above. Most euactinomycetes are permanently immotile, but flagellated spores are formed in a few genera.

For the purposes of the ensuing discussion, it is convenient to divide the members of the actinomycete line into three principal groups: group I comprises the unicellular lactic acid bacteria and the micrococci; group II, the coryneform bacteria and proactinomycetes; group III, the euactinomycetes. An enormous number of genera have been described in this assemblage; only a few representatives are discussed here.

The overall range of DNA base composition in

* A *mycelium* is a multinucleate mass of cytoplasm contained within a branched tubular cell wall. Individual tubes are termed *hyphae* (singular *hypha*). Although septa may occur, they are normally widely spaced. Consequently the individual cells possess many nuclei contained within a common mass of cytoplasm, a condition termed *coenocytic*.

the actinomycete line is wide (Table 12.1). However, within each group, the ranges are, with a few notable exceptions, considerably narrower. Most members of group I have DNA of low G + C content, except for *Micrococcus*, in which the values lie near the top of the G + C scale. In group III, the G + C content is characteristically very high. The members of group II have DNA of relatively high G + C content, although the span is somewhat wider than that characteristic of group III.

### group I

THE LACTIC ACID BACTERIA. The lactic acid bacteria are immotile, rod-shaped or spherical organisms (Figure 12.13), united by an unusual constellation of metabolic and nutritional properties. The name derives from the fact that ATP is synthesized through fermentations of carbohydrates, which yield lactic acid as a major (and sometimes as virtually the sole) end product.

The lactic acid bacteria are all facultative anaerobes, which grow readily on the surface of solid media exposed to air. However, they are unable to synthesize ATP by respiratory means, a reflection of their failure to synthesize cytochromes and other heme-containing enzymes. The growth yields of lactic acid bacteria are, accordingly, unaffected by the presence or absence of air, the fermentative dissimilation of sugars being the source of ATP under both conditions.

One consequence of the failure to synthesize heme proteins is that the lactic acid bacteria are catalase-negative, and hence cannot mediate the decomposition of $H_2O_2$ to water and oxygen gas. The absence of catalase activity, readily demonstrated by the absence of gas bubble formation when cells are mixed with a drop of dilute $H_2O_2$, is one of the most useful diagnostic tests for the recognition of these organisms, since they are virtually the only bacteria devoid of catalase that can grow in the presence of air.

**Table 12.1**
DNA Base Composition of the Actinomycete Line

|  | MOLE % G + C |
|---|---|
| **Group I** |  |
| Lactic acid bacteria | 33–51 |
| Micrococci: |  |
|   *Staphylococcus* and *Sarcina* | 30–40 |
|   *Micrococcus* | 66–75 |
| **Group II** |  |
| Coryneforms | 57–72 |
| Proactinomycetes | 57–72 |
| **Group III** |  |
| Euactinomycetes | 69–76 |

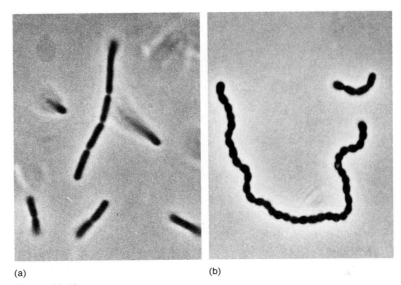

(a)                                                    (b)

**Figure 12.13**
The form and arrangement of cells in two genera of lactic acid bacteria: (a) *Lactobacillus*; (b) *Streptococcus* (phase contrast, ×2,180).

The inability of lactic acid bacteria to synthesize heme proteins is correlated with an inability to synthesize hemin, the porphyrin that is associated with these enzymes. However, certain lactic acid bacteria acquire catalase activity when grown in the presence of a source of hemin (e.g., on media containing red blood cells). Such species appear to synthesize a protein that can combine with exogenously supplied hemin to produce an enzyme with the properties of catalase.

The inability to synthesize hemin is only one manifestation of the extremely limited synthetic abilities characteristic of the lactic acid bacteria. All these organisms have complex growth-factor requirements; they invariably require B vitamins and, with one exception (*Streptococcus bovis*), a considerable number of amino acids. The amino acid requirements of many lactic acid bacteria are considerably more extensive than those of higher animals. As a result of their complex nutritional requirements, lactic acid bacteria are usually cultivated on media containing peptone, yeast extract, or other digests of plant or animal material. These media must be supplemented with a fermentable carbohydrate to provide an energy source.

Even when growing on very rich media, the colonies of lactic acid bacteria (Figure 12.14) always remain relatively small (at most, a few millimeters in diameter). They are never pigmented; as a result of the absence of cytochromes, the growth has a chalky white appearance that is very characteristic. The small colony size of these bacteria is attributable

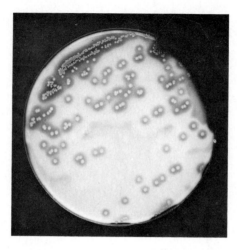

**Figure 12.14**
Colonies of *Streptococcus* growing on an agar medium containing a suspension of calcium carbonate. Lactic acid has dissolved the calcium carbonate, producing clear zones around each colony.

primarily to low growth yields, a consequence of their exclusively fermentative metabolism. Some species can produce unusually large colonies when grown on sucrose-containing media (Figure 12.15), as a result of the massive synthesis of extracellular polysaccharides (either the polymer of glucose units termed *dextran* or the polymer of fructose units termed *levan*) at the expense of this disaccharide; in this special case, much of the volume of the colony consists of polysaccharide. Since dextrans and levans are synthesized only from sucrose, the species in question form typical small colonies on media containing any other utilizable sugar. In the isolation of the spherical lactic acid bacteria, which can grow in media that have an initial pH of 7 or above, the incorporation of finely divided $CaCO_3$ in the plating medium is useful, since the colonies can be readily recognized by the surrounding zones of clearing, caused by the acid they produce (Figure 12.14).

Another distinctive physiological feature of lactic acid bacteria is their high tolerance of acid, a necessary consequence of their mode of energy-yielding metabolism. Although the spherical lactic acid bacteria can initiate growth in neutral or alkaline media, most of the rod-shaped forms cannot grow in media with an initial pH greater than 6. Growth of all lactic acid bacteria continues until the pH has fallen, through fermentative production of lactic acid, to a value of 5 or less.

The capacity of lactic acid bacteria to produce and tolerate a relatively high concentration of lactic acid is of great selective value, since it enables them to eliminate competition from most other bacteria in environments that are rich in nutrients. This is shown by the fact that lactic acid bacteria can be readily enriched from natural sources through the use of complex media with a high sugar content. Such media can, of course, support the growth of many other chemoorganotrophic bacteria, but the competing organisms are almost completely eliminated as growth proceeds by the accumulation of lactic acid, formed by the lactic acid bacteria.

As a result of their extreme physiological specialization, the lactic acid bacteria are confined to a few characteristic natural environments. Some live in association with plants and grow at the expense of the nutrients liberated through the death and decomposition of plant tissues. They occur in foods and beverages prepared from plant materials: pickles, sauerkraut, ensilaged cattle fodder, wine, and beer. A lactic fermentation of the sugar initially present occurs during the preparation of pickles, sauerkraut, and ensilage. In beer the lactic acid bacteria are potential spoilage agents, which sometimes produce an undesirable acidity and odor. In wine they sometimes cause desirable changes (p. 434).

Other lactic acid bacteria constitute part of the normal flora of the animal body and occur in considerable numbers in the nasopharynx, the intestinal tract, and the vagina. These forms include a number of important pathogens of man and other mammals; all pathogenic lactic acid bacteria belong to the genus *Streptococcus*.

A third characteristic habitat of the lactic acid bacteria is milk, to which they gain access either from the body of the cow or from plant materials. The normal souring of milk is caused by certain strep-

**Figure 12.15**
The formation of extracellular polysaccharides by lactic acid bacteria. Two plates of *Leuconostoc*, streaked on glucose medium (a) and sucrose medium (b). The large size and mucoid appearance of the colonies on sucrose are caused by the massive synthesis and deposition around the cells of dextran.

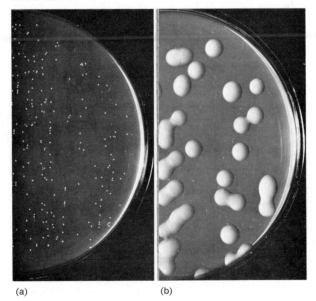

(a)                              (b)

tococci; both rod-shaped and spherical lactic acid bacteria play important roles in the preparation of the fermented milk products: butter, cheeses, buttermilk, yogurt (p. 440).

Because of their activities in the preparation of foods and as agents of human and animal disease, the lactic acid bacteria are a group of major economic importance.

THE MICROCOCCI. In addition to lactic acid bacteria with spherical cells, the Gram-positive cocci include three other genera, distinguished by a combination of physiological, metabolic, and structural characters. (1) The cells of *Staphylococcus* usually occur in irregular clusters and are relatively small (approximately 1 μm in diameter). (2) *Micrococcus* cells are of similar size, occurring as single cells, pairs, irregular clusters, or cubical packets. (3) The cells of *Sarcina* are much larger (2 to 3 μm in diameter), and occur in regular, cubical packets (Figure 12.16).

The members of the genus *Staphylococcus* are facultative anaerobes and ferment sugars with the formation of lactic acid as one of the major end products. Their DNA base composition (30 to 40 mole percent G + C) is in the same range as that of many spherical lactic acid bacteria. However, they can be readily distinguished from these organisms by several criteria: possession of catalase and other heme pigments; capacity for respiratory metabolism; and much less restricted requirements for carbon and energy (growth will occur on complex media in the absence of carbohydrates). Many also produce carotenoid pigments, which are absent from all lactic acid bacteria. Staphylococci are typical members of the normal microflora of the skin, and some are potential pathogens causing either infections or food poisoning.

*Micrococcus* closely resembles *Staphylococcus* in cell structure, but its members are all strict aerobes and have a wholly different DNA base composition (66 to 72 mole percent G + C). The normal habitat of these bacteria is obscure; they are common in air, and also occur in milk.

The two species of *Sarcina* are relatively oxygen-tolerant anaerobes and vigorous sugar fermenters. An unusual structural property of *S. ventriculi* is the synthesis of a cell wall which has a thick outer layer of cellulose.

**Figure 12.16**

*Sarcina maxima* (phase contrast, ×1,630). From S. Holt and E. Canale-Parola, "Fine Structure of *Sarcina maxima* and *Sarcina ventriculi*," *J. Bacteriol.* **93**, 399 (1967).

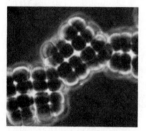

group II:
coryneforms and proactinomycetes

The members of group II (coryneform bacteria and proactinomycetes) present many difficult taxonomic problems; the classification of this group is more unclear and much more controversial than than of almost any other major bacterial group.

These organisms include several major human or animal pathogens: e.g., *Corynebacterium diphtheriae,* the causal agent of diphtheria, and *Mycobacterium tuberculosis,* the causal agent of tuberculosis.

*Corynebacterium diphtheriae* is a normal inhabitant of the respiratory tract. Most strains are nonpathogenic, but they acquire the ability to cause diphtheria when infected by a specific phage that confers toxigenicity on the host cell (see Chapter 16).

*C. diphtheriae* and related animal parasites or pathogens are unicellular, immotile organisms with a characteristic cell shape and arrangement. Just after division, the daughter cells (which are club-shaped, tapering toward the outer poles) undergo a sudden "snapping" movement, which brings them into a characteristic angular relationship between pairs of cells (Figure 12.17).

Corynebacteria are facultative anaerobes capable of both respiration and fermentation. The nutritional requirements are complex, and are not known in detail for most species: *C. diphtheriae* requires several vitamins.

The Mycobacterium group can form a rudimentary mycelium, which is unstable, and fragments early in growth into slender, immotile rods, sometimes branched. In contrast to corynebacteria, the cells are not tapered and do not show snapping postfission movements. A distinctive though often variable property of mycobacteria, which also occurs in some nocardias, is *acid-fastness*. Cells that have been stained with a hot phenolic solution of red dye (basic fuchsin) retain it through subsequent treatment with a mineral acid (dilute $H_2SO_4$ or HCl); all other bacteria stained in this manner are rapidly decolorized by the acid treatment. The property of acid-fastness is associated with a very high content of complex lipids in the cell wall, which makes the cells of mycobacteria and some nocardias waxy and strongly hydrophobic. As a result, colonies have a dry, rough wrinkled surface (Figure 12.18); and in liquid cultures (unless they are grown in the

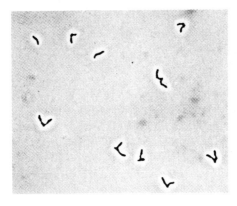

**Figure 12.17**
The typical angular arrangement of dividing cells of a coryneform bacterium, brought about by "snapping" post-fission movement. Phase contrast, ×1,400. From T. A. Krulwich and J. L. Pate, "Ultrastructural Explanation for Snapping Post-Fission Movements in *Arthrobacter crystallopoietes,*" *J. Bacteriol.* **105,** 408 (1971).

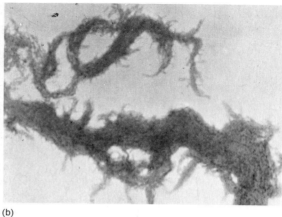

(a)                                                    (b)

**Figure 12.18**
The characteristic appearance of cultures of *Mycobacterium tuberculosis*. (a) Colony growing on the surface of an agar plate, ×7. (b) Cordlike aggregations of cells from a liquid culture, ×345. Courtesy of Professor N. Rist, Institut Pasteur, Paris.

presence of a detergent), the cells cohere to form a tough surface pellicle and a film of growth attached to the wall of the culture vessel.

The mycobacteria are strict aerobes, with a purely respiratory mode of metabolism. Growth, particularly of pathogenic species, is slow. Nonpathogenic mycobacteria, which occur in soil, have higher growth rates. They do not require growth factors, whereas the nutritional requirements of the pathogenic species are complex.

Many mycobacteria produce yellow or orange carotenoid pigments; and in certain of these species the synthesis of carotenoids is specifically induced by exposure of cultures to light.

The Nocardia group differs from mycobacteria primarily by its greater tendency to mycelial growth. In the initial stages of development, an abundant mycelium is produced, and subsequently fragments into short rods (Figure 12.19). Some of the nocardias are animal pathogens; others are nonpathogenic soil organisms. In nutritional and metabolic respects, they resemble the mycobacteria.

Recent chemical studies have shown that the members of the genera *Corynebacterium, Mycobacterium,* and *Nocardia* share a distinctive cell wall composition unique to the members of these three genera. The peptidoglycan of the wall is covalently linked to a polysaccharide made up of arabinose. Furthermore, the wall in all three genera has a high content of lipid, including a distinctive class of lipids known as *mycolic acids.* These are high molecular weight branched chain fatty acids containing hydroxyl groups. This unique wall structure shared by these three genera suggest that, despite their structural and physiological differences, they comprise an interrelated taxomic cluster.

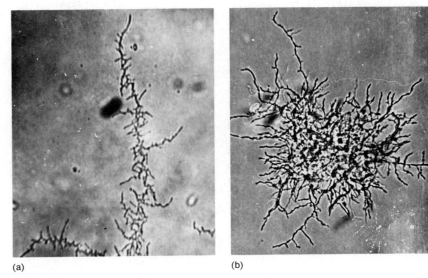

**Figure 12.19**
Young surface colonies on agar plates of (a) *Mycobacterium fortuitum* and (b) *Nocardia asteroides* (×648). Courtesy of Ruth Gordon and H. Lechevalier.

The genus *Corynebacterium* is a relatively small and well-defined assemblage, if it is considered to contain only *C. diphtheriae* and the related animal parasites and pathogens that are facultative anaerobes and possess cell walls of the structure described above. However, many other nonspore-forming Gram-positive bacteria display the cell shape and the ability to undergo snapping postfission movements that are characteristic of *Corynebacterium*. These so-called *coryneform bacteria* are diverse with respect to physiological, nutritional, and metabolic properties, and they occur in a wide variety of habitats.

Strictly aerobic coryneform bacteria are abundant in soil and in milk products; some are pathogenic for plants. Some of these organisms have been attached to the genus *Corynebacterium*, others placed in a variety of special genera. The best-characterized representatives are the soil coryneforms of the genus *Arthrobacter*. These organisms constitute a large fraction of the aerobic chemoheterotrophic population of soil bacteria, and are important agents for the mineralization of organic matter in soil. The most distinctive property of *Arthrobacter* ssp. is the succession of changes in cell form that accompany growth (Figure 12.20). In cultures that have entered the stationary phase, the cells are spherical and of uniform size, resembling micrococci. When growth is reinitiated, these cells elongate into rods, which undergo binary fission, accompanied by typical snapping postfission movements. From these rods, thinner outgrowths may develop near one or both poles of the cell, producing branched forms that resemble early developmen-

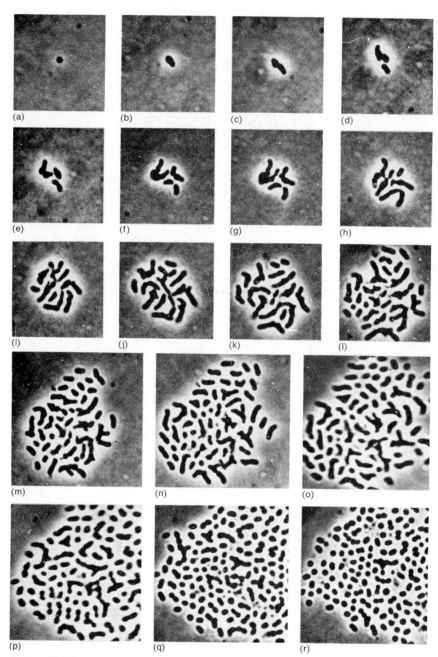

**Figure 12.20**

The cellular life cycle of *Anthrobacter*, shown by successive photomicrographs of the growth of a microcolony from a single coccoid cell on agar over a period of 40 hours. Phase contrast, ×1,020. From H. Veldkamp, G. Van den Berg, and L. P. T. M. Zevenhuizen, "Glutamic Acid Production by *Arthrobacter globiformis*," *Antonie van Leeuwenhoek* **29,** 35 (1963).

tal stages of mycobacteria or proactinomycetes. Return to the coccoid state may occur either by multiple fragmentation (as in nocardias) or by a progressive shortening of the rods through successive binary fissions. This remarkable growth cycle thus includes features suggestive not only of corynebacteria, but also of mycobacteria and nocardias.

In their nutritional properties the *Arthrobacter* group show interesting analogies to aerobic pseudomonads; most species can utilize a wide and varied range of simple organic compounds as principal carbon and energy sources. The majority require growth factors.

Among the coryneform bacteria and proactinomycetes, several genera are made up of oxygen-tolerant anaerobes, which perform distinctive fermentations of carbohydrates. These include *Propionibacterium, Bifidobacterium,* and *Actinomyces.*

The coryneform bacteria of the genus *Propionibacterium* were first isolated from Swiss cheese (they play an important role in its ripening). They develop as a secondary microflora, fermenting the lactate initially produced in the curd by lactic acid bacteria, with formation of propionate, acetate, and $CO_2$. The two fatty acids give this cheese its distinctive flavor, and the $CO_2$ produces the characteristic holes. Subsequent work has shown that the primary natural habitat of propionic acid bacteria is the rumen of herbivores, where they ferment the lactate produced by other members of the rumen population. In addition to fermenting lactate, these organisms can ferment a variety of sugars. Although they cannot grow in the presence of air, requiring anaerobic conditions or low tensions of $O_2$, they contain heme pigments, both cytochromes and catalase. Their metabolism is fermentative.

The cells of bifidobacteria are typically swollen, irregular, and branched. The complex nutritional requirements of these organisms include a requirement for N-acetyl sugars. These compounds are present in milk, which accounts for the fact that this is the most favorable medium for bifidobacteria and probably explains their predominance in the intestinal flora of breast-fed babies. When cultivated in a medium containing an excess of N-acetylglucosamine, the cells of bifidobacteria assume a much more regular rod form. Hence the branched, swollen cells characteristic of these organisms probably reflect the fact that they are usually grown with a limiting supply of N-acetylglucosamine, an essential precursor of peptidoglycan, the structural determinant of the cell wall.

The members of the genus *Actinomyces* include organisms that are members of the normal flora of the mouth and throat, as well as several pathogenic species that produce infections in man and cattle. Like bifidobacteria, they are catalase-negative aeroduric anaerobes, and have complex nutritional requirements. $CO_2$ is often required for good growth. They also show a greater tendency than bifidobacteria to mycelial growth in young cultures, although the mycelium is fragile and breaks up readily into rod-shaped or branched fragments (Figure 12.21).

**Figure 12.21**
*Actinomyces israelii* from a broth culture, showing branched cells and short mycelial fragments (dark field illumination, ×1,120). From J. M. Slack, S. Landfried, and M. A. Gerencer, "Morphological, Biochemical and Serological Studies on 64 Strains of *Actinomyces israelii,*" *J. Bacteriol.* **97,** 873 (1969).

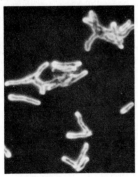

### group III:
### the euactinomycetes

The true actinomycetes are characterized by a mycelial growth habit throughout the period of active growth, and by reproduction through formation of specialized spores rather than by mycelial fragmentation. Taxonomy of the actinomycetes is based heavily upon the morphology of the reproductive structures, which vary from single spores borne on short lateral branches as in *Micromonospora* (Figure 12.22), through the chains of spores made by *Streptomyces* (Figure 12.23); to motile spores produced within a sporangium by *Actinoplanes* (Figure 12.24). The spores function both as a mechanism of dispersal and as resting cells. Dispersal is accomplished by special mechanisms in some genera, such as the motility associated with *Actinoplanes* spores, or the hydrophobic sheath found on *Streptomyces* spores, which makes these spores spread on the surface of a film of water. Although actinomycete spores respire slowly and are not especially resistant to heat or ultraviolet light, they are resistant to dessication, and can remain viable for months in the dry state.

One thermophilic actinomycete, *Thermoactinomyces*, produces true endospores, chemically and ultrastructurally identical to those of *Bacillus* and *Clostridium*. This distinctive property, and the G + C content (about 50 mole percent), indicate that *Thermoactinomyces* is probably best classified with the sporeformers. Its habitat appears to be the same as that of thermophilic bacilli.

Spore formation in many genera is accompanied by the production of an *aerial mycelium* that arises from the *substrate mycelium* (Figure 12.25); an aid, presumably, to spore dispersal. Other genera, such as *Actinoplanes* and *Micromonospora*, do not produce an aerial mycelium. On solid media, colonies of actinomycetes that produce aerial mycelia can be recog-

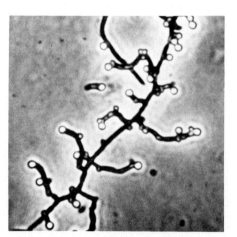

**Figure 12.22**
*Micromonospora chalcea*, showing spherical spores borne singly at the tips of hyphae (phase contrast, ×2,300). Courtesy of G. M. Luedemann and the Schering Corporation.

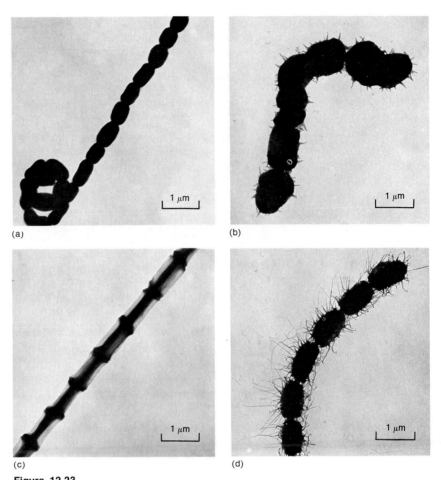

(a)

(b)

(c)

(d)

**Figure 12.23**
Electron micrographs of the spores of four different *Streptomyces* species, which illustrate various types of surface structure and ornamentation: (a) *S. cacaoi,* showing smooth spores; (b) *S. hirsutus,* showing spiny spores with obtuse spines; (c) *S. aureofaciens,* of a smooth but special ''phalangiform'' type; (d) *S. flavoviridis,* showing hairy spores. Courtesy of H. D. Tresner, Lederle Laboratories; reproduced in part from E. B. Shirling and D. Gottlieb, ''Cooperative Description of Type Cultures of *Streptomyces. II.* Species Descriptions from First Study,'' *Intern. J. Syst. Bacteriol.* **18,** 69–189 (1968).

nized readily by their fluffy or chalky appearance; before the onset of sporulation, the colony is smooth and consists entirely of substrate mycelium. Actinomycetes that do not produce an aerial mycelium retain their smooth appearance throughout the life cycle.

Actinomycetes are common soil and water organisms. In soil, they are often the most common component of the microflora, and may frequently be isolated without enrichment. The characteristic odor of damp earth is, in fact, largely attributable to volatile compounds produced

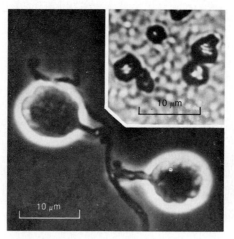

**Figure 12.24**
Spore vesicles of *Actinoplanes*: (a) (inset) A group of mature spore vesicles viewed on the surface of a colony (bright field illumination). (b) Two mature spore vesicles attached to a hypha, mounted in water (phase contrast illumination). From H. Lechevalier and P. E. Holbert, ''Electron Microscopic Observation of the Sporangial Structure of a Strain of *Actinoplanes*,'' *J. Bacteriol.* **89,** 217 (1965).

by *Streptomyces*. These organisms appear to be nutritionally versatile and to be important in the mineralization of organic compounds, especially complex polymers such as peptidoglycan and rubber.

One organism, named *Frankia* but not yet obtained in pure culture, forms nitrogen-fixing symbiotic associations with a variety of nonleguminous plants. By virtue of this symbiosis, such plants as alder are frequently among the first vascular plants to invade nitrogen-poor soils, such as those exposed by glacier retreat.

Most euactinomycetes are strict aerobes, although a few anaerobic, cellulose-fermenting strains of *Micromonospora* have been isolated from the termite gut.

**Figure 12.25**
Diagram of a *Streptomyces* colony growing on agar medium, showing the spore that initiated the colony (center), hyphae of the substrate mycelium (horizontal and below the medium surface), aerial hyphae (vertical above the medium surface), and spores. The right half of the colony is sectioned to show nuclear regions and septa. After Hopwood and Sermonti by permission of Academic Press, Inc.

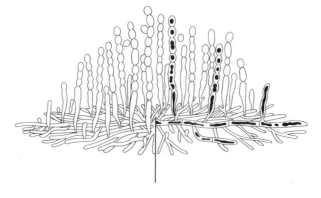

A number of clinically important antibiotics are produced by actinomycetes, especially by *Streptomyces* (e.g., streptomycin, tetracycline, chloramphenicol). As in the case of *Bacillus* antibiotics, these antibiotics are produced principally during sporulation, and their function in nature is not clear. In soil they are produced in relatively small amounts, but pharmaceutical firms have selected a variety of mutant strains that produce large quantities of these drugs when cultivated in special media. The industrial importance of these organisms, coupled with patent laws, has led to the description of hundreds of species.

## gram-positive strict anaerobes

Although the actinomycete line contains a number of strictly fermentative organisms, most are relatively insensitive to $O_2$, some even being able to grow in its presence. There are, however, a few strict anaerobes among the nonsporeforming Gram-positive bacteria. These include several genera of fermentative cocci that inhabit the animal intestine or rumen, and the more widespread methanogenic bacteria.

### the methanogenic bacteria

The biological formation of methane ($CH_4$) is a geochemically important process that occurs in all anaerobic environments in which organic matter undergoes decomposition: swamps, lake sediments, and the digestive tract of animals. It results from the metabolic activity of a small and highly specialized bacterial group, which are terminal members of the food chain in these environments; they convert fermentation products formed by other anaerobes (notably $CO_2$, $H_2$, and formate) to methane. Since this gas has a low solubility in water, it escapes from the anaerobic environment and is eventually reoxidized under aerobic conditions by methylotrophs (see Chapter 10). Methanogenesis in the rumen of cattle and sheep has been intensively studied, since it is an important component of the complex microbial activities associated with ruminant digestion (see Chapter 15). In waste disposal plants where sewage is subjected to anaerobic treatment, the growth of these organisms results in the formation of large amounts of methane, which is often collected and used to generate electricity.

The methanogenic bacteria stain Gram-variable. Although preliminary electron microscopic examination indicates a Gram-positive type of wall, chemical analysis reveals no peptidoglycan.

These bacteria are divided into several genera on the basis of cell shape, which can be either coccoidal (unicellular or in packets), rod-shaped, or helical (e.g., Figure 12.26).

All strains utilize $CO_2$ as electron acceptor, coupling its reduction to the oxidation of $H_2$. Most are also able to oxidize for-

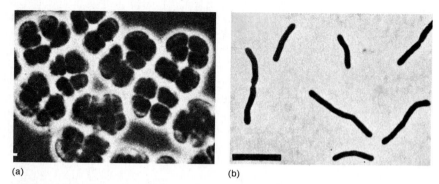

(a)                                                    (b)

**Figure 12.26**
Phase-contrast photomicrographs of several methanogenic bacteria. (a) *Methanosarcina*. (b) *Methano-bacterium*. In each case, the bar indicates 5 $\mu$m. From J. G. Zeikus and V. G. Bowen, "Comparative Ultrastructure of Methanogenic Bacteria," *Can. J. Microbiol.* **21,** 121 (1975).

mate, and a few can use methanol or acetate. Growth factors are generally not required, although some strains from the rumen have fatty acid and cofactor requirements.

# 13

# THE PROTISTS

Among protists (eucaryotic microorganisms), three major groups can be recognized: *algae, protozoa,* and *fungi.* Each of these groups is very large and internally diverse. The more highly specialized representatives can be readily assigned to an appropriate group. However, there are many protists for which the assignment is arbitrary: numerous forms exist intermediate between algae and protozoa and between protozoa and fungi. For this reason, the three major groups of protists cannot be sharply distinguished in terms of simple sets of clear-cut differences. Broadly speaking, the algae may be defined as protists that perform photosynthesis and hence possess chloroplasts. Some of them are unicellular microorganisms; some are filamentous, colonial, or coenocytic; and some have a plantlike structure that is formed through extensive multicellular development, with little or no differentiation of cells and tissues. In organismal terms, accordingly, the algae are highly diverse, and by no means all fall into the category of microorganisms. The brown algae known as *kelps* may attain a total length of as much as 50 m. The protozoa and fungi are nonphotosynthetic organisms, and the difference between them is essentially one of organismal structure; protozoa are predominantly unicellular, whereas fungi are predominantly coenocytic and grow in the form of a filamentous, branched structure known as a *mycelium.*

For historical reasons, the algae and fungi were traditionally regarded as plants and have been largely studied by botanists, while the protozoa were traditionally regarded as animals and have been largely studied by zoologists. As a result of this specialization, the many interconnections between the three groups have been largely overlooked. We shall attempt in this chapter to provide a unified account of the properties of protists that emphasizes possible evolutionary interrelationships.

# the algae

The primary classification of algae is based on cellular, not organismal, properties: the chemical nature of the wall, if present; the organic reserve materials produced by the cell; the nature of the photosynthetic pigments; and the nature and arrangement of the flagella borne by motile cells. In terms of these characters, the algae are arranged in a series of major groups, summarized in Table 13.1.

The groups are not equivalent to one another in terms of the range of organismal structure of their members. For example, the Euglenophyta (euglenoid algae) consist entirely of unicellular or simple colonial organisms, while the Phaeophyta (brown algae) consist only of plantlike, multicellular organisms. The largest and most varied group, the Chlorophyta (green algae), from which the higher plants probably originated, span the full range of organismal diversity, from unicellular organisms to multicellular representatives with a plantlike structure.

The common cellular properties of each algal group suggest that its members, however varied their organismal structure may be, are representatives of a single major evolutionary line. Evolution among the algae thus in general appears to have involved *a progressive increase in organismal complexity in the framework of a particular variety of eucaryotic cellular organization*. Although it is possible to perceive these evolutionary progressions *within* each algal group, the relationships *between* groups are completely obscure. The primary origin of the algae as a whole is accordingly an unsolved problem.

## the photosynthetic flagellates

In many algal groups, the simplest representatives are motile, unicellular organisms, known collectively as *flagellates*. The cell of a typical flagellate, illustrated by *Euglena* (Figure 13.1), has a very marked polarity: it is elongated and leaf-shaped, the flagella usually being inserted at the anterior end. In the Euglenophyta, to which *Euglena* belongs, there are two flagella of unequal length, which originate from a small cavity at the anterior end of the cell. Many chloroplasts and mitochondria are dispersed throughout the cytoplasm. Near the base of the flagellar apparatus is a specialized organelle, the *eyespot*, which is red, owing to its content of special carotenoid pigments; the eyespot serves as a photoreceptor to govern the active movement of the cell in response to the direction and intensity of illumination. The cell of *Euglena*, unlike that of many other flagellates, is not enclosed within a rigid wall; its outer layer is an elastic *pellicle*, which permits considerable changes of shape. Cell division occurs by *longitudinal fission* (Figure 13.2). About the time of the onset of mitosis, there is a duplication of the anterior organelles of the cell, including the flagella and their basal apparatus; cleavage subsequently occurs through the long axis, so that the duplicated organelles are equally

**Table 13.1**
**Major Groups of Algae**

| GROUP NAME | PIGMENT SYSTEM | | COMPOSITION OF CELL WALL | NATURE OF RESERVE MATERIALS | NUMBERS AND TYPE OF FLAGELLA | RANGE OF STRUCTURE |
|---|---|---|---|---|---|---|
| | CHLOROPHYLLS* | OTHER SPECIAL PIGMENTS | | | | |
| Green algae: division Chlorophyta | a + b | — | Cellulose | Starch† | Generally two identical flagella per cell | Unicellular, coenocytic, filamentous multicellular forms |
| Euglenoids: division Euglenophyta | a + b | — | No wall | Paramylum† and fats | One, two, or three flagella per cell | All unicellular |
| Dinoflagellates and related forms: division Pyrrophyta | a + c | Special carotenoids | Cellulose | Starch† and oils | Two flagella, dissimilar in form and position on cell | Mostly unicellular, a few filamentous forms |
| Chrysophytes and diatoms: division Chrysophyta | a ± c | Special carotenoids | Wall composed of two overlapping halves, often containing silica (some have no walls) | Leucosin† and oils | Two flagella, arrangement variable; or no flagella | Unicellular, coenocytic, filamentous |
| Brown algae: division Phaeophyta | a + c | Special carotenoids | Cellulose and algin | Laminarin† and fats | Two flagella, of unequal length | Plantlike multicellular forms |
| Red algae: division Rhodophyta | a | Phycobilins | Cellulose | Starch† | No flagella | Unicellular, plantlike multicellular forms |

* Like bacteriochlorophylls, eucaryotic chlorophylls are of several types, depending on the substituents on the tetrapyrrole ring system.
† Specific classes of polysaccharides.

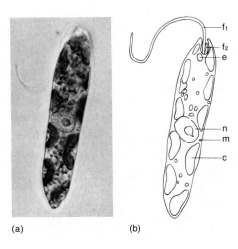

**Figure 13.1**

*Euglena gracilis.* (a) Photomicrograph of fixed cell (×1,000). Courtesy of Gordon F. Leedale. (b) Schematic drawing of the same cell, to show principal structural features: n, nucleus; c, chloroplast; m, mitochondrion; e, eyespot; $f_1$, $f_2$, the two flagella of unequal length, originating within a small cavity of the anterior end of the cell.

(a)                    (b)

partitioned between the two daughter cells. This mode of cell division is characteristic of all flagellates except those belonging to the Chlorophyta, such as *Chlorogonium*, where each cell undergoes *two or more fissions* to produce a number of smaller daughter cells, liberated by rupture of the parental cell wall (Figure 13.2). As we shall see in a subsequent section, longitudinal division also occurs in the nonphotosynthetic flagellate protozoa and is one of the primary characters that distinguish these organisms from the other major group of protozoa that possess flagellalike locomotor organelles, the ciliates.

Most multicellular algae are immotile in the mature state. However, their reproduction frequently involves the formation and liberation of motile cells, either asexual reproductive cells (*zoospores*) or gametes. Figure 13.3 shows the liberation of zoospores from a cell of a filamentous member of the Chlorophyta, *Ulothrix;* it can be seen that these zoospores have a structure very similar to that of the *Chlorogonium* cell, illustrated in Figure 13.2(c). The structure of the motile reproductive cells of multicellular algae thus often reveals their relatedness to a particular group of unicellular flagellates.

### the nonflagellate unicellular algae

By no means are all unicellar algae flagellates; several algal groups also contain unicellular members that are either immotile or possess other means of movement. Many of these unicellular nonflagellate algae possess strikingly specialized and elaborate cells, which may be illustrated by considering the *diatoms.*

The diatoms (Figure 13.4), members of the Chrysophyta, have organic walls impregnated with silica. The architecture of the diatom wall is exceedingly complex; it always consists of two overlapping halves, like the halves of a petri dish. Division is longitudinal, each daughter cell retaining half of the old wall and synthesizing a new half.

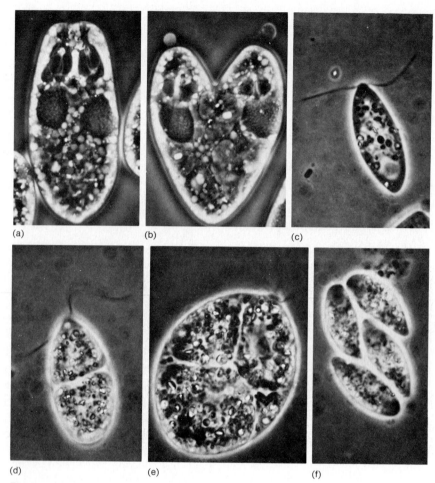

**Figure 13.2**

Longitudinal and multiple fission in flagellate algae. (a, b) Two cells of *Euglena* in the course of longitudinal fission (phase contrast, ×1,240). [Reproduced from G. F. Leedale, in *The Biology of Euglena*, D. E. Buetow, ed. (New York: Academic Press, 1968).] In (a), division of the nucleus and of the locomotor apparatus at the anterior end of the cell is complete. In (b), cell cleavage has begun. (c, d, e, f) Four steps in the cellular life cycle of *Chlorogonium*, a green alga that reproduces by multiple fission (phase contrast, ×1,430). (c) Newly liberated daughter cell. (d) Two-celled stage. (e) Four-celled stage. (f) Four daughter cells just after liberation from the mother cell. Original photomicrographs of material provided by Paul Kugrens, Department of Botany, University of California, Berkeley.

**Figure 13.3**

The filamentous green alga, *Ulothrix* (×1,250). At left, the formation and liberation of biflagellate zoospores.

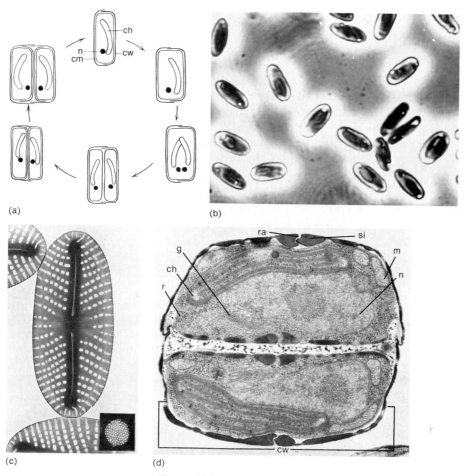

**Figure 13.4**

The diatom *Navicula*. (a) Diagrammatic representation of the division cycle. (b) Living cells, phase contrast illumination (×1,320). (c) Electron micrograph of the wall (×9,800). Insert depicts fine structure of one of the wall pores (×56,000). (d) Transverse section of a dividing cell (×23,800): ch, chloroplast; g, Golgi apparatus; n, nucleus; m, mitochondrion; r, ribosomes; ra, raphe; si, silica in wall; cw, cell wall; cm. Courtesy of M. L. Chiappino and B. E. Volcani, University of California, San Diego.

Although devoid of flagella, some diatoms can move slowly over solid substrates by a special modification of ameboid movement. In motile diatoms, there is a narrow longitudinal slot in the wall, known as *raphe*, through which the protoplast can make direct contact with the substrate. Movement is brought about by directed cytoplasmic streaming in the canal of the raphe, which pushes the cell over the substrate.

Many fossil diatoms are known, because the siliceous skeleton of the wall is practically indestructible, and as diatoms are one of the major groups of algae in the oceans, large fossil deposits of diatom walls have accumulated in many areas. These deposits, known as *diatomaceous earth*, have industrial uses as abrasives and filtering agents.

### the natural distribution of algae

Most algae are aquatic organisms that inhabit either fresh water or the oceans. These aquatic forms are principally free-living, but certain unicellular marine algae have established durable symbiotic relationships with specific marine invertebrate animals (e.g., sponges, corals, various groups of marine worms) and grow within the cells of the host animal. Some terrestrial algae grow in soil or on the bark of trees. Others have established symbiotic relationships with fungi, to produce the curious, two-membered natural associations termed *lichens*, which form slowly growing colonies in arid inhospitable environments, notably on the surface of rocks.

The marine algae play a very important role in the cycles of matter on earth, since their total mass (and consequently their gross photosynthetic activity) is approximately equal to that of all land plants combined. This role is by no means evident, because the most conspicuous of marine algae, the seaweeds, occupy a very limited area of the oceans, being attached to rocks in the intertidal zone and in the shallow coastal waters of the continental shelves. The great bulk of marine algae are unicellular floating (*planktonic*) organisms, predominantly diatoms and dinoflagellates, distributed through the surface waters of the oceans. Although they sometimes become abundant enough to impart a definite brown or red color to local areas of the sea, their density is usually so low that there is no gross sign of their presence. It is the large total surface area of the earth's oceans that accounts for their tremendous quantities.

### nutritional versatility of algae

The ability to perform photosynthesis confers on many algae very simple nutrient requirements; in the light they can grow in a completely inorganic medium. This is not the case for all algae, however, because many have specific vitamin requirements, a requirement for vitamin $B_{12}$ being particularly common. In nature the source of these vitamins is probably bacteria that inhabit the same environment. Thus, the ability to perform photosynthesis does not necessarily preclude the utilization of organic compounds as the principal source of carbon and energy, and many algae have a mixed type of metabolism.

Even when growing in the light, certain algae cannot use $CO_2$ as their principal carbon source and are therefore dependent on the presence of acetate or some other suitable organic compound to fulfill their carbon requirements. This is due to their defective photosynthetic machinery: although these algae can obtain energy from their photosynthetic activity, they cannot reduce pyridine nucleotides, which are required to convert $CO_2$ to organic cell materials.

Many algae that perform normal photosynthesis in the light, using $CO_2$ as the carbon source, can grow well in the dark at the expense of a variety of organic compounds; such forms can thus shift from photosynthetic to respiratory metabolism, the shift being determined primarily by the presence or absence of light. Algae completely enclosed by cell walls are dependent on dissolved organic substrates as energy sources for dark growth. However, a considerable number of unicellular algae that lack a cell wall, or are not completely enclosed by it, can phagocytize bacteria or other smaller microorganisms. It is not correct, accordingly, to regard the algae as an *exclusively* photosynthetic group; on the contrary, many of their unicellular members possess and can use the nutritional capacities characteristic of the two major subgroups of nonphotosynthetic eucaryotic protists, the protozoa and fungi.

### the leucophytic algae

Loss of the chloroplast is an *irreversible event*, which results in a *permanent loss of photosynthetic ability*. Such a change appears to have taken place many times among unicellular algal groups with a mixed mode of nutrition, to yield nonpigmented counterparts, which can be clearly recognized on the basis of other cellular characters as *nonphotosynthetic derivatives of algae*. Such organisms, known collectively as *leucophytes*, exist in many flagellate groups and also in diatoms and in nonmotile groups among the green algae. Leucophytes often may preserve a virtually complete structural identity with a particular photosynthetic counterpart. In some cases, this structural near-identity may include the preservation of vestigial, nonpigmented chloroplasts, as well as a pigmented eyespot. There can be little doubt accordingly that these nonphotosynthetic organisms are close relatives of their structural counterparts among the algae and have arisen from them by a loss of photosynthetic ability in the recent evolutionary past. Indeed, the transition can be demonstrated experimentally in certain strains of *Euglena*, which yield stable, colorless strains when treated with the antibiotic streptomycin or when exposed to small doses of ultraviolet irradiation or to high temperatures (Figure 13.5). These colorless strains cannot be distinguished from the naturally occurring nonphotosynthetic euglenoid flagellates which belong to the genus *Astasia*.

The classification of the leucophytes raises a difficult problem. In terms of cell structure, they can be easily assigned to a particular division of algae, as nonphotosynthetic representatives, and this classification is no doubt the most satisfactory one. However, since they are nonphotosynthetic, unicellular eucaryotic protists, they can alternatively be regarded as protozoa, and they are, in fact, included among the protozoa by

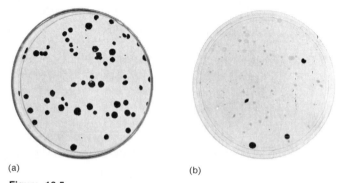

(a)                                        (b)

**Figure 13.5**
The loss of chloroplasts in *Euglena gracilis* as a result of ultraviolet irradiation. (a) A light-grown plate culture of *E. gracilis*. (b) A light-grown culture of the same organism, after exposure to brief ultraviolet irradiation. Most of the cells have given rise to clones devoid of chloroplasts (pale colonies). Courtesy of Jerome A. Schiff.

zoologists. The leucophytes accordingly provide the first and by far the most striking case of a group, or rather a whole series of groups, that are clearly transitional between two major assemblages of eucaryotic protists.

### the origins of the protozoa

The protozoa are a highly diverse group of unicellular, nonphotosynthetic protists, most of which show no obvious resemblances to the various divisions of algae. Nevertheless, the various kinds of leucophytes, which are recognizably of algal origin, provide a plausible clue concerning the evolutionary origin of many groups among the protozoa. The loss of photosynthetic function abruptly reduces the nutritional potentialities of an organism; leucophytes are therefore immediately confined to a more restricted range of environments than their photosynthetic ancestors. Specific features of cellular construction that possessed adaptive value in the context of photosynthetic metabolism become superfluous; the eyespot is the most obvious example. Hence, one could expect that loss of photosynthetic ability would be followed by a series of evolutionary changes in the structure of the cell that would better fit the organism for a heterotrophic mode of life. Beyond a certain point, these changes would make the algal origin of the organism unrecognizable, and it would then be classified as a protozoon.

One group of protists, the dinoflagellates, have several features of cell structure that permit the biologist to recognize a close relationships between photosynthetic and nonphotosynthetic representatives (Figure 13.6). The motile cell of a dinoflagellate has two flagella, which differ in structure and arrangement. One lies in a groove or girdle around the equator of the cell; the other extends away from the cell in a posterior direction.

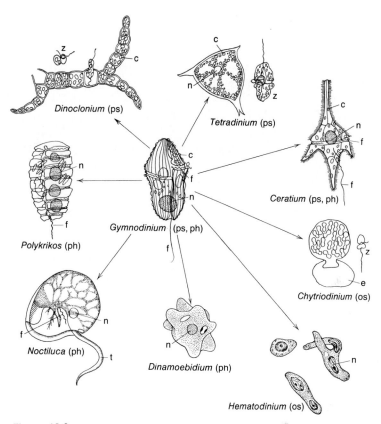

**Figure 13.6**
The different evolutionary trends that are represented among dinoflagellates. *Gymnodinium* is a relatively unspecialized photosynthetic dinoflagellate, which is both photosynthetic (ps) and phagotrophic (ph). *Ceratium* is a more specialized photosynthetic dinoflagellate, characterized by a very complex wall with spiny extensions, comprised of many plates. *Tetradinium* and *Dinoclonium* are nonmotile, strictly photosynthetic organisms, which reproduce by multiple cleavage to form typical dinoflagellate zoospores. *Polykrikos, Noctiluca,* and *Dinamoebidium* are three free-living phagotrophic dinoflagellates. *Polykrikos* is a coenocytic, multinucleate organism, the cell of which bears a series of pairs of flagella. *Noctiluca* has one small flagellum, and bears a large and conspicuous tentacle. *Dinamoebidium* is an ameboid organism. *Chytriodinium* and *Hematodinium* are parasitic dinoflagellates whose nutrition is osmotrophic (os). *Chytriodinium* parasitizes invertebrate eggs and reproduces by cleavage of a large sac-like structure into dinoflagellate zoospores. *Hematodinium* is a blood parasite in crabs: n, nucleus; f, flagellum; c, chloroplast; z, zoospore; e, parasitized invertebrate egg; t, tentacle.

The dinoflagellate nucleus is also unusual; its division is highly specialized, and its chromosomes remain visible in interphase.

Most photosynthetic dinoflagellates are unicellular planktonic organisms, widely distributed in the oceans, and characteristically brown or yellow in color as a result of the possession of a distinctive set of photosynthetic pigments. Many (the so-called "armored" dinoflagellates) possess very elaborate cell walls, composed of a series of plates, which do not completely enclose the protoplast. There is a very pronounced tendency to phagotrophic nutrition among these photosynthetic members of the group, because the wall structure permits pseudopodial extension and the engulfment of small prey. A few filamentous algae, completely enclosed by walls, can be recognized as of dinoflagellate origin, since they form zoospores with the characteristic flagellar arrangement.

A much more extensive series of specialized forms can be traced among the nonphotosynthetic members of this flagellate group. Many of the free-living unicellular dinoflagellates are nonphotosynthetic phagotrophic organisms. Some preserve close structural similarities to photosynthetic members of the group; others, such as the large marine organism, *Noctiluca*, have a highly specialized cellular organization not found in any photosynthetic member of the group. However, the most far-reaching modifications of cell structure within the dinoflagellates are to be found among its parasitic members, most of which occur in marine invertebrates. *Hematodinium*, which occurs in the blood of certain crabs, in completely devoid of flagella. *Chytriodinium*, which parasitizes the eggs of copepods, develops as a large, saclike structure within the egg, subsequently giving rise by multiple internal cleavage to numerous motile spores with a typical dinoflagellate structure. Were it not for the retention of the distinctive nuclear organization (and, in the case of *Chytriodinium*, the flagellar structure of the spores), neither of these parasitic protists could be recognized as belonging to the same group as the photosynthetic dinoflagellates. *Hematodinium* could be classified with the sporozoan protozoa and *Chytriodinium* with the primitive group of fungi known as *chytrids*.

Accordingly, within this one small flagellate group, it is possible to reconstruct some major patterns of evolutionary radiation that were probably characteristic of protists as a whole.

## the protozoa

In the light of the preceding discussion, the protozoa can best be regarded as comprising a number of groups of nonphotosynthetic, typically motile, unicellular protists, which have probably derived at various times in the evolutionary past from one or another group among the unicellular algae (see Table 13.2).

### the flagellate protozoa:
### the Mastigophora

The Mastigophora are protozoa that always bear flagella as the locomotor organelles. In contrast to the ciliates, in which cell division is transverse, flagellate protozoa undergo longitudinal division, preceded by duplication of the flagellar apparatus at the anterior end of the cell.

**Table 13.2**
**Primary Subdivisions of the Protozoa**

| | |
|---|---|
| I. *Class Mastigophora:* | The flagellate protozoa. Motile by means of one or more flagella. Cell division always longitudinal. |
| | Included in this class are the "phytoflagellates" (i.e., unicellular motile representatives of the various algal divisions) as well as the "zooflagellates," nonphotosynthetic organisms not recognizable as leucophytes. These forms are in the main osmotrophic (utilize soluble nutrients). |
| II. *Class Rhizopoda:* | The ameboid protozoa. Motile by means of pseudopodia. It should be noted that the distinction from class I on the basis of locomotion is not absolute, since many of the *Rhizopoda* can also form flagella. Reproduction by binary fission. Phagotrophic (utilize particulate nutrients). |
| III. *Class Sporozoa:* | A very diverse group of parasitic protozoa. Immotile or showing gliding movement. Reproduction by multiple fission. Osmotrophic. Some examples are discussed in Chapter 17. |
| IV. *Class Ciliata:* | The ciliates. Motile by means of numerous cilia, organized into a coordinated locomotor system. The cell has two nuclei, differing in structure and function. Division always transverse. Phagotrophic. |

This mode of division has already been described for a photosynthetic flagellate, *Euglena*. In addition to leucophytes, this protozoan group includes many representatives that show no resemblance to photosynthetic flagellates and are for the most part parasites of animals.

The trypanosomes are frequently parasitic in vertebrates, where they develop in the bloodstream, being transmitted from host to host by the bite of insects. They include important agents of disease, such as the agent of African sleeping sickness, transmitted by the tsetse fly. The cell is slender and leaf-shaped, its single flagellum being directed posteriorly and attached through part of its length to the body of the cell, to form an undulating membrane [Figure 13.7(a)]. The trypanosomes are osmotrophic protozoa, which absorb their nutrients from the blood of the host.

Other parasitic flagellates inhabit the gut of vertebrates or invertebrates. The trichomonads, which have four to six flagella [Figure 13.7(b)] are harmless inhabitants of the gut of vertebrates. Several very highly specialized groups of flagellate protozoa inhabit the gut of termites; one of the most striking of these organisms, the cellulolytic genus *Trichonympha*, is illustrated in Figure 13.7(c).

### the ameboid protozoa:
### the Rhizopoda

The Rhizopoda are protozoa in which ameboid locomotion is the predominant mode of cell movement, although some of them are able to produce flagella as well. The simplest members of this group

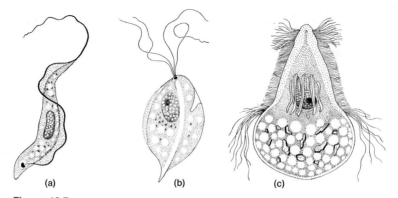

**Figure 13.7**
Some nonphotosynthetic flagellate protozoa (Mastigophora). (a) A trypanosome. The leaf-shaped cell and the long undulating membrane, to which the flagellum is attached, are characteristic of this organism. (b) *Trichomonas*. (c) *Trichonympha*.

are amebas, which have characteristically amorphous cells as a result of the continuous changes of shape brought about by the extension of pseudopodia. Most amebas are free-living soil or water organisms that phagocytize smaller prey. A few inhabit the animal gut, including forms that cause disease (amebic dysentery). Other members of the Rhizopoda have a well-defined cell form, as the result of the formation of an exoskeleton or shell (typical of the foraminifera) or an endoskeleton (typical of the heliozoa and radiolaria). Several members of the Rhizopoda are illustrated in Figure 13.8.

**Figure 13.8**
Some ameboid protozoa (*Sarcodina*). (a) An ameba. (b) A foraminiferan. Note the many-chambered shell, from which the pseudopodia extend. (c) A heliozoan.

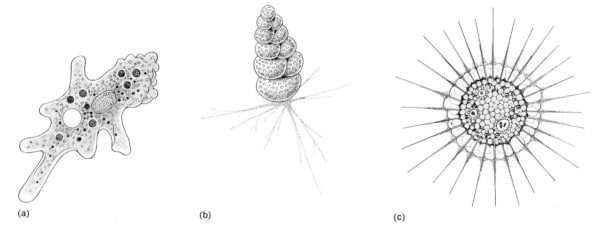

(a)　　　　　(b)　　　　　(c)

## the ciliate protozoa:
## the Ciliata

The ciliate protozoa are a very large and varied group of aquatic, phagotrophic organisms that are particularly widely distributed in fresh water. The ciliates share a number of fundamental cellular characters that distinguish them sharply from all other protists. This suggests that despite the very great internal diversity of this group, it is the one class of protozoa that may have had a single common evolutionary origin.

The common characters of ciliates can be summarized as follows:

1. At some time in the life history, the cell is motile by means of numerous short, hairlike projections, structurally homologous with flagella, which are termed *cilia*.
2. Each cilium arises from a basal structure, the kinetosome, which is homologous with the basal body of a flagellum (centriole—see p. 54); however, in ciliates the kinetosomes are interconnected by rows of fibrils called *kinetodesmata* to form very elaborate compound locomotor structures termed *kineties*. This internal system persists, even when the cell is devoid of cilia.
3. Cell division is transverse, not longitudinal, as in flagellates. Ciliates show a marked polarity, with posterior and anterior differentiation of the cell, so the transverse mode of cell division necessarily entails an elaborate process of morphogenesis each time division occurs, during which the anterior daughter cell resynthesizes posterior structures, while the posterior daughter cell resynthesizes anterior structures. The morphogenetic transformations are generally almost complete when the two daughter cells separate.
4. Each individual contains two dissimilar nuclei, a large *macronucleus* and a much smaller *micronucleus*, which differ in function as well as in structure.

We may illustrate the distinctive character of the ciliates by considering the properties of a simple member of the group, *Tetrahymena* (Figure 13.9). It has a pear-shaped body about 50 $\mu$m long, enclosed by a semirigid pellicle. The surface is covered with hundreds of cilia, arranged in longitudinal rows. The beating of the cilia, which propels the organism, is rhythmic and coordinated.

Near the narrow anterior end of the cell is the mouth or *cytopharynx*. It consists of an oral aperture, a mouth cavity that extends some distance into the cell, an undulating membrane, and three membranelles. The undulating membrane and membranelles are composed of specialized, adherent cilia, the movements of which sweep food particles into the mouth cavity. Captured food enters the cytoplasm by being enclosed in food vacuoles that are formed in succession at the base of the mouth cavity. These food vacuoles then circulate within the cell as a result of cytoplasmic streaming until the food material has been digested and the soluble products absorbed; undigested material is ejected from the cell by a posteriorly located pore known as the *cytoproct*. In nature *Tetrahymena* is normally a predator and feeds on smaller microorganisms. However, in the laboratory it can be grown

(a)

(b)

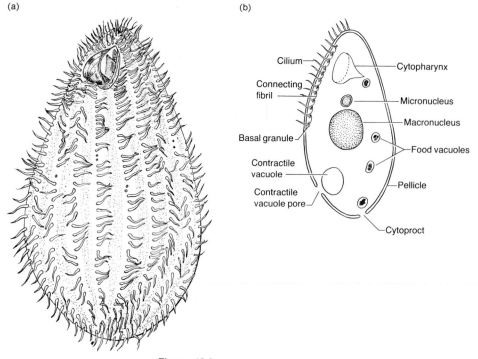

**Figure 13.9**

The ciliate protozoon, *Tetrahymena*. (a) A general view, showing external appearance. (b) Diagrammatic cross section, showing main structural features of the cell.

in pure culture on a medium that contains only soluble nutrients. Under such conditions, the liquid nutrients must still be taken in through the mouth, in the form of pinocytic vacuoles.

Although its natural environment is a dilute one, with an osmotic pressure far below that of the contents of the cell, *Tetrahymena* is able to maintain water balance by the operation of a *contractile vacuole*. This structure, located near the posterior end of the cell, is formed by the coalescence of smaller vacuoles in the cytoplasm; when it reaches a certain critical size, it discharges its liquid contents into the environment through a pore in the pellicle and then starts to grow in volume again.

As mentioned above, typical ciliates have two dissimilar nuclei in the cell. The larger *macronucleus*, which is polyploid, is associated with normal cell division and growth and is therefore sometimes referred to as the "vegetative nucleus." Some strains of *Tetrahymena* have only this kind of nucleus; they can reproduce indefinitely by binary fission but cannot undergo sexual reproduction. Other strains possess also a small, diploid *micronucleus*, which plays an essential role in sexual reproduction. The macronucleus can be derived after conjugation from a micronucleus; hence, strains of *Tetrahymena* having only a macronucleus can be regarded as deficient cell lines that have probably lost their micronucleus by an accident of vegetative growth. In *Tetrahymena* the first step in cell division is an elongation of the macronucleus parallel to the long axis of the cell. At the same time, a structural reorganization of the cytoplasm begins. Its principal feature is the formation of a *second cytostome* just posterior to the future plane of cell division. A furrow then develops across the center of the cell, which becomes dumb-bell-shaped. If a micronucleus is present, it divides mitotically, and the daughter nuclei migrate to the anterior and posterior portions of the cell. Finally, the elongated macronucleus divides, and the two daughter cells separate.

*Tetrahymena* is among the simplest of ciliates. The foregoing account suffices to show what an extraordinarily elaborate and complex biological organization has been evolved in this protozoan group within the framework of unicellularity. The ciliates represent the apex of biological differentiation on the unicellular level, but they appear to be a terminal evolutionary group. The development of more complex biological systems took place through the establishment of multicellularity and involved the differentiation of specialized cell types during the growth of the individual organism, characteristic of all plants and animals.

## the fungi

Like the protozoa, the fungi are nonphotosynthetic. Although some of the more primitive aquatic fungi show resemblances to flagellate protozoa, the fungi as a whole have developed a highly distinctive biological organization that can be regarded as an adaptation to life in their most common habitat, the soil. We shall start out by considering the main features of this type of biological organization.

Most fungi are coenocytic mycelial organisms (Figure 13.10). Fungal growth is characteristically confined to the tips of the hyphae; as the mycelium extends, the cytoplasmic contents may disappear from the older, central regions. The size of a single mycelium is not fixed; as long as nutrients are available, outward growth by hyphal extension can continue, and in some fungi a single mycelium may be as much as 15m in diameter. Usually, asexual reproduction occurs by the formation of uninucleate or multinucleate spores, which are pinched off at the tips of hyphae. Neither the spores nor the mycelium of higher fungi are capable of movement. However, the internal contents of a mycelium show streaming movements, which cannot be translated into progression over the substrate because the cytoplasm is completely enclosed within its wall. In fact, the simplest brief definition of the structure of a higher fungus is: *a multinucleate mass of cytoplasm, mobile within a much-branched enclosing system of tubes.*

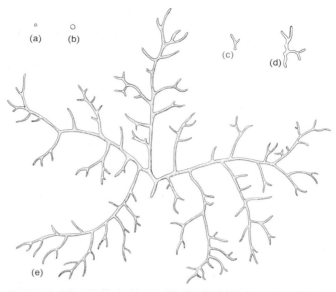

**Figure 13.10**
Successive stages in the development of a fungal mycelium from a reproductive cell
or conidium (×85). After C. T. Ingold, *The Biology of Fungi*. London: Hutchinson,
1961.

Since a mycelium is capable of almost indefinite
growth, it frequently attains macroscopic dimensions. In nature, however, the
vegetative mycelium of fungi is rarely seen because it is normally embedded
in soil or other opaque substrates. Many fungi (the mushrooms) form specialized, spore-bearing fruiting structures, however, which project above soil
level and are readily visible as macroscopic objects. Such structures were
known long before the beginning of scientific biology, although their nature
and mode of formation were not clearly understood until the nineteenth
century. The superficial resemblance of these fruiting structures to plants was
undoubtedly a very important factor in the decision of the early biologists to
assign fungi to the plant kingdom, despite their nonphotosynthetic nature.

Since fungi are always enclosed by a rigid wall,
they are unable to engulf smaller microorganisms. Most fungi are free-living
in soil or water and obtain their energy by the respiration or fermentation of
soluble organic materials present in these environments. Some are parasitic
on plants or animals. A number of soil forms are predators and have developed ingenious traps and snares, composed of specialized hyphae, which
permit them to capture and kill protozoa and small invertebrate animals such
as the soil-inhabiting nematode worms. After the death of their prey, such
fungi invade the body of the animal by hyphal growth and absorb the nutrients contained in it.

The fungi comprise three major groups: the Phycomycetes, the Ascomycetes, and the Basidiomycetes. A fourth group, the Fungi Imperfecti, has been set aside to include those species for which the sexual stage, and hence the correct classification, is not yet known.

### aquatic Phycomycetes

Although soil is by far the most common habitat of the fungi, some fungal groups are aquatic. These fungi are known collectively as *water molds* or *aquatic Phycomycetes*. They occur on the surface of decaying plant or animal materials in ponds and streams; some are parasitic and attack algae or protozoa. These fungi show the closest resemblances to protozoa; they produce motile spores or gametes, furnished with flagella, and in the simpler forms the vegetative structure is not mycelial. This description applies, for example, to many of the fungi known as *chytrids*.

The developmental cycle of a typical simple chytrid, which occurs in ponds on decaying leaves, is shown in Figure 13.11. The mature vegetative structure consists of a sac about 100 μm in diameter, which is anchored to the solid substrate by a number of fine, branched threads known as *rhizoids*. The sac is a *sporangium*, within which reproductive cells, or spores, are produced. The enclosed cytoplasm contains many nuclei, formed by repeated nuclear division during vegetative growth. Each nucleus, surrounded with a distinct volume of cytoplasm, eventually becomes bounded

**Figure 13.11**

The life cycle of a primitive fungus, a chytrid. The flagellated zoospore (a) settles down on a solid surface. As development begins (b), a branching system of rhizoids is formed, anchoring the fungus to the surface. Growth results in the formation of a spherical zoosporangium, which cleaves internally to produce many zoospores (c). The zoosporangium ruptures to liberate a fresh crop of zoospores (d).

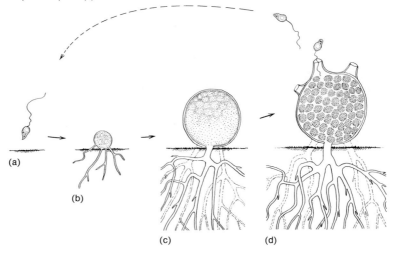

(a)

(b)

(c)

(d)

by a membrane. The sporangium then ruptures to release uninucleate flagellated zoospores, each of which can settle down and grow into a new organism. The rhizoids serve to anchor the developing sporangium to the substrate and to absorb the nutrients required for its growth.

The aquatic Phycomycetes are a varied group with respect to their mechanisms of reproduction and life cycles. The range of this variation can be well illustrated by comparing a chytrid with another aquatic phycomycete, *Allomyces*. *Allomyces* shows a well-marked alternation of haploid and diploid generations (Figure 13.12). We shall describe first the diploid *sporophyte*. When mature it looks like a microscopic tree, with a basal system of anchoring rhizoids from which springs a much-branched mycelium bearing two different kinds of sporangia. The mitosporangia have thin, smooth, colorless walls, whereas the meiosporangia have brown thick, dark-pitted walls. Upon maturation, both kinds of sporangia liberate flagellated spores, but the subsequent development of these spores is very different. The mitospores derived from mitosporangia are diploid and germinate into sphorophytes. The meiospores derived from meiosporangia are haploid, because meiosis takes place during the maturation of the meiosporangium; they give rise to haploid or *gametophytic* individuals.

**Figure 13.12**

The life cycle of *Allomyces*, an aquatic phycomycete with a well-marked alternation of haploid and diploid generations. From a drawing made by Raphael Rodriguez and reprinted by permission of Arthur T. Brice.

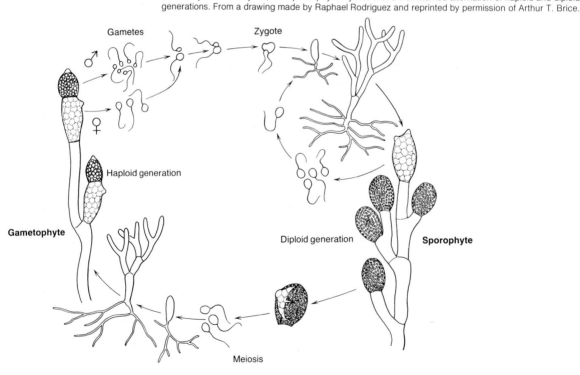

The gametophyte is grossly similar in structure to the sporophyte, but instead of bearing meiosporangia and mitsoporangia, it produces male and female gametangia, which are generally borne in pairs. The female gametangium looks very much like a mitosporangium, whereas the male gametangium is distinguished by its brilliant orange color. The gametangia rupture to liberate male and female gametes, a considerable number arising from each gametangium. Both male and female gametes are motile, moving by means of flagella, but they can be readily distinguished from one another by size and color. The female gamete is larger than the male and is colorless, and the male has an orange oil droplet at the anterior end. The gametes fuse in pairs to form biflagellate zygotes, which eventually settle down and develop once more into sporophytes.

### the terrestrial Phycomycetes

The Phycomycetes also include a group known as the *terrestrial Phycomycetes*, which are inhabitants of soil. These organisms differ from all aquatic Phycomycetes in not possessing motile flagellated reproductive cells. *They are thus permanently immotile.* This is a property they share with all the higher groups of fungi. The absence of motility characteristic of the higher fungi is understandable in terms of their ecology: motile reproductive cells are of value only when dispersion occurs through water. The reproductive cells of soil-inhabiting fungi are dispersed in the main through the air.

As a typical example of a terrestrial phycomycete, we may take *Rhizopus*. The mycelium is differentiated into branched rhizoids that penetrate the substrate, horizontal hyphae known as *stolons* that spread over the surface of the substrate, bending down at intervals to form tufts of rhizoids, and erect sporangiophores that emerge from the stolons in tufts [Figure 13.13(a)]. The unbranched sporangiophore enlarges at the tip to form a rounded sporangium, which becomes separated from the rest of the sporangiophore by a cross wall. Within this sporangium, large numbers of spherical spores are formed. These asexual *sporangiospores* are eventually released by rupture of the surrounding wall and are dispersed by air currents. They give rise on germination to new vegetative mycelia.

*Rhizopus* also reproduces sexually, but sexual reproduction can occur only when two mycelia of opposite sex come into contact with one another. Fungi that show this phenomenon are known as *heterothallic fungi* in contrast to *homothallic fungi* (such as *Allomyces*) that can produce both kinds of sex cells on a single mycelium. In *Rhizopus* the two kinds of mycelia between which sexual reproduction can take place are known as + and − strains, because there are no morphological indications of maleness and femaleness. As the hyphae from a + and − mycelium meet, each produces a short side branch at the point of contact. This side branch then divides to form a *gametangium*. The two gametangia, which are in direct contact with one another, fuse to form a large zygospore, surrounded by a thick, dark wall.

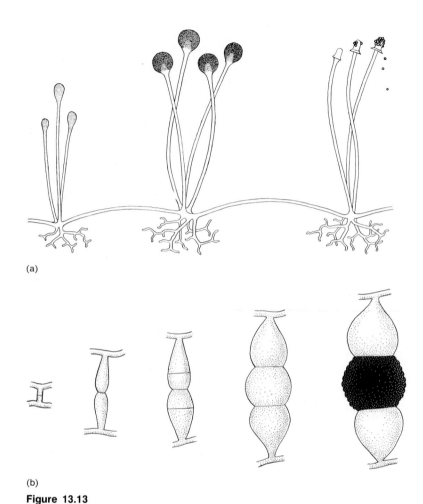

(a)

(b)

**Figure 13.13**

(a) The vegetative stage of *Rhizopus,* a terrestrial phycomycete. (b) Sexuality in *Rhizopus.* Successive stages of sexual fusion and the formation of a zygospore.

This sequence of events is shown in Figure 13.13(b). It can be seen that the behavior of both partners in the sexual act is identical; hence, there is no basis for designation as "male" or "female." Upon germination of the zygospore, meiosis occurs, and a hypha emerges and produces a sporangium. The haploid spores from this sporangium in turn develop into the typical vegetative mycelium.

### distinctions between Phycomycetes and other fungi

Despite the considerable differences among them, all Phycomycetes share two properties that readily distinguish them from the remaining classes of fungi (Ascomycetes, Basidiomycetes, and Fungi

Imperfecti). First, *their asexual spores are always endogenous*, formed inside a saclike structure, the zoosporangium of the aquatic types or the sporangium containing immotile sporangiospores of the terrestrial types. In the other groups of fungi, the asexual spores are always exogenous, being formed free at the tips of hyphae (Figure 13.14). Second, *the mycelium in Phycomycetes shows no cross walls* except in regions where a specialized cell, such as a sporangium or gametangium, is formed from a hyphal tip. Such a mycleium is known as a *nonseptate mycelium*. In the remaining groups of fungi, distinct cross walls occur at regular intervals along the hyphae. Thus, on the basis of these two simple criteria, one can readily distinguish a phycomycete from the other major classes of fungi.

Since the mycelium of Phycomycetes is nonseptate, it is clear that these organisms are coenocytic. The regular occurrence of cross walls in the mycelium of other groups of fungi suggests, in contrast, that they are cellular organisms. This is not true, however. The cross walls do not divide the cytoplasm into a number of separate cells: each cross wall has a central pore, through which both cytoplasm and nuclei can move freely. There is thus just as much cytoplasmic continuity in the septate fungi as in the Phycomycetes, and both groups are, in fact, coenocytic.

### the Ascomycetes and Basidiomycetes

The fungi with septate mycelia and exogenous asexual spores are broadly classified into two groups, Ascomycetes and Basidiomycetes, on the basis of their sexual development. Following zygote formation in these fungi, there is usually an immediate division, followed by the formation of four or eight haploid sexual spores, which are borne in or on

**Figure 13.14**

Penicillium. Left, edge of a colony at relatively low magnification, showing spore heads. Right, conidiophore at high magnification, showing branched structure and terminal chains of spherical conidia. Courtesy of Dr. K. B. Raper.

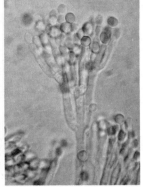

structures known as *asci* and *basidia*, characteristic of Ascomycetes and Basidiomycetes respectively. In Ascomycetes, the zygote develops into a saclike structure, the *ascus*, while the nucleus undergoes two meiotic divisions, often followed by one or more mitotic divisions. A wall is formed around each daughter nucleus and the neighboring cytoplasm to produce four, eight, or more ascospores within the ascus (Figure 13.15). Eventually the ascus ruptures, and the enclosed spores are liberated.

In Basidiomycetes, the zygote enlarges to form a club-shaped cell, the *basidium*; at the same time, the diploid nucleus undergoes meiosis. The subsequent course of events is strikingly different from that which occurs in an ascus. No spores are formed within the basidium; instead, a slender projection known as a *sterigma* develops at its upper end, and a nucleus migrates into this sterigma as the latter enlarges. Eventually, a cross wall is formed near the base of the sterigma, the cell thus cut off being a basidiospore. The same process is repeated for the remaining three nuclei in the basidium, so that a mature basidium bears on its surface four basidiospores (Figure 13.16). Basidiospore discharge is a remarkable phenomenon. After the basidiospore has matured, a minute droplet of liquid appears at the point of its attachment to the basidium. This droplet grows rapidly until it is about one-fifth the size of the spore, and then, quite suddenly, both spore and droplet are shot away from the basidium.

### the Fungi Imperfecti

The classification of the septate fungi into Ascomycetes and Basidiomycetes has one practical disadvantage. Obviously, the assignment of a fungus to its correct class is possible only if one has observed the sexual stage of its life cycle. If one happens to deal with a fungus that is incapable of sexual reproduction, or in which the sexual stage is unknown, it cannot be assigned either to the Ascomycetes or to the Basidiomycetes. Since heterothallism is very common in the higher fungi, it often happens that a

**Figure 13.15**
Successive stages in the formation of an ascus. (a) Binucleate fusion cell. (b) Nuclear fusion. (c), (d), (e), Nuclear divisions. (f) Ascospore formation.

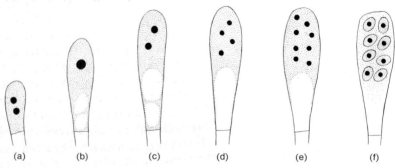

(a)      (b)      (c)      (d)      (e)      (f)

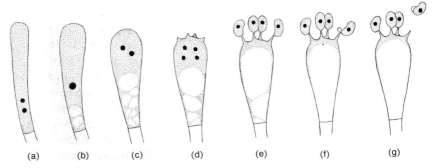

**Figure 13.16**
Successive stages in basidium formation and basidiospore discharge. (a) Binucleate cell. (b) Nuclear fusion.
(c), (d) Nuclear division. (e) Formation of basidiospores. (f), (g) Basidiospore discharge.

single isolate of an ascomycete or basidiomycete will never undergo sexual reproduction, which requires the presence of another strain of opposite mating type. Accordingly, it has been necessary to create a third class, the Fungi Imperfecti, for those kinds in which a sexual stage has not so far been observed. It should be realized that the Fungi Imperfecti are essentially a provisional taxonomic group; from time to time the sexual stage is discovered in a fungus originally assigned to this group, and the organism in question is then transferred to either the Ascomycetes or the Basidiomycetes.

### the development of an ascomycete

As a typical ascomycete, we may consider a mold of the genus *Neurospora*. The vegetative stage of *Neurospora* consists of a mycelium, on the surface of which there develop special hyphae, the *conidiophores*, carrying chains of exogenous asexual spores, the *conidia*. The conidia are pigmented and are responsible for the characteristic pink to orange color of a *Neurospora* colony. When mature, conidia are easily dislodged and float through the air. When they come into contact with a substrate favorable for development, they germinate and give rise once more to the development of a mycelium.

*Neurospora* is a heterothallic ascomycete, which also has a sexual reproductive cycle. The haploid mycelia form immature fruiting bodies, termed *protoperithecia*, each consisting of a coiled hypha lying within a hollow sphere formed from a compact mass of ordinary hyphae. When a hypha from a mycelium of opposite mating type comes into contact with a protoperithecium, it fuses with the coiled hypha within the protoperithecium, and the nuclei of the two haploid strains mingle in a common cytoplasm. Each type of nucleus divides repeatedly, giving rise to many haploid nuclei of opposite mating types; these eventually fuse in pairs, to form many diploid nuclei, which then undergo immediate meiosis. An ascus develops at the site of each meiosis, around the four haploid nuclei. Meanwhile, the wall of the protoperithecium thickens and becomes pigmented, to form a mature *perithecium* that contains several dozen asci. The maturation of the asci is completed by the formation of ascospores, each delimited within the ascus by a resistant spore wall that surrounds one of the haploid nuclei and the adjacent cytoplasm. A mature ascus may contain four or eight ascospores, the number depending on whether or not meiosis is followed by a further mitotic division of the four haploid nuclei.

The mature perithecium is roughly spherical in shape, with a short protruding neck. At maturity, a pore forms at the tip of the neck, through which the ascospores are violently discharged. Upon germination, ascospores (like conidia) produce a haploid mycelium. This life cycle is shown in Figure 13.17.

### the development of a Basidiomycete

The most conspicuous members of the Basidiomycetes are the mushrooms.* The portion of a mushroom that is seen by the casual observer is a small part of the whole organism, being the specialized fruiting structure that bears the basidia. The vegetative portion of the organism is entirely concealed from view and consists of a loose mycelium, spreading, often for many meters, under the soil (Figure 13.18). The vegetative mycelium

*Some mushrooms (e.g., the morels) are ascomycetes.

**Figure 13.17**

The life cycle of *Neurospora*, a heterothallic ascomycete. Asexual reproduction occurs by the formation of conidia from a haploid mycelium of each mating type, A and a. These mycelia also bear protoperithecia which, when fertilized by conidia or hyphae of the opposite mating type, develop into perithecia, within which numerous zygotes are formed. Each zygote undergoes two meiotic and one mitotic division to form an ascus containing eight ascospores, four of mating type A and four of mating type a. Germination of the ascospores gives rise once more to haploid mycelia. (N) indicates a haploid stage; (2N) indicates a diploid stage.

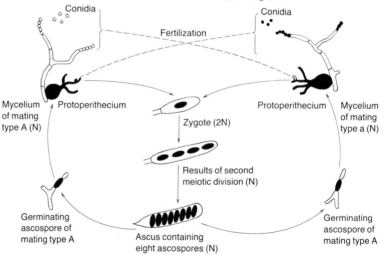

**Figure 13.18**
Cross section of a mushroom, showing the subterranean vegetative mycelium and the fruiting structure. In the mature fruiting body at left, large numbers of basidiospores are being discharged from the gills on the underside of the cap and are being dispersed by the wind. A second, immature fruiting body is shown at right, just emerging from the soil. From A. H. R. Buller, *Researches on Fungi*, Vol. 1, p. 219. New York: Longmans Green, 1909.

grows more or less continuously. When conditions are favorable (generally following a spell of wet, warm weather), fruiting bodies are formed at various points on its surface and push up through the soil to become the parts of the fungus visible to the observer.

In the common field mushroom (Figure 13.18), the fruiting body consists of a stalk surmounted by a cap, both composed of closely packed hyphae. The underside of the cap consists of rows of radiating gills, each gill being lined with thousands of basidia. The basidia project horizontally from the vertical walls of the gill, and consequently when the basidiospores are ejected, they pass into the air space between adjacent gills, and from there fall to the ground below. When a mushroom is mature, basidiospore discharge is a massive phenomenon. If one places a ripe cap on a piece of paper for a few hours, a "negative" of the gill structure will be formed on the paper by the deposition of millions of basidiospores.

Heterothallism is widespread among the Basidiomycetes. Consequently, isolated haploid basidiospores give rise on germination to haploid mycelia that are incapable of fructifying. However, if two such haploid mycelia of compatible mating types come into contact, hyphal fusion followed by nuclear exchange takes place, to produce a still haploid *dicaryon;* the two kinds of nuclei become associated in a very regular fashion, one pair occurring in each compartment of the septate mycelium. During growth of the dicaryon, the two kinds of nuclei divide synchronously.

A dicaryotic mycelium may continue to grow vegetatively for a long time, and during such vegetative growth, fusion between the paired nuclei never occurs. Fusion takes place only at the time of fructification, when the basidia are produced, and is followed in each basidium by immediate meiosis to produce the four haploid nuclei destined to enter the basidiospores.

### the yeasts

Among the Ascomycetes, Basidiomycetes, and Fungi Imperfecti, the characteristic vegetative structure is the coenocytic mycelium. Nonetheless, there are a few groups in these classes that have largely lost the mycelial habit of growth and have become unicellular. Such organisms are known collectively as *yeasts*. A typical yeast consists of small, oval cells that multiply by forming buds. The buds enlarge until they are almost equal in size to the mother cell, nuclear division occurs, and then a cross wall is formed between the two cells (Figure 13.19). Although the yeasts constitute a minor branch of the higher fungi in terms of number of species, they are very important microbiologically. Most yeasts do not live in soil but have instead become adapted to environments with a high sugar content, such as the nectar of flowers and the surface of fruits. Many yeasts (the fermentative ones) perform an alcoholic fermentation of sugars and have been long exploited by man (see Chapter 18).

Yeasts are classified in all three classes of higher fungi: Ascomycetes, Basidiomycetes, and Fungi Imperfecti. The principal agent of alcoholic fermentation, *Saccharomyces cerevisiae*, is an ascomycetous yeast. Budding ceases at a certain stage of its growth, and the vegetative cells become transformed into asci, each containing four ascospores. For a long time it was believed that ascospore formation in *S. cerevisiae* was not preceded by zygote formation, because pairing of vegetative cells prior to the formation of ascospores could never be observed. Eventually, however, it was discov-

**Figure 13.19**

A sequence of photomicrographs of a budding cell of the ascomycetous yeast, *Wickerhamia*, showing nuclear division and transverse wall formation (phase contrast, ×1,770): n, nucleus; v, vacuole; tw, transverse wall. From P. Matile, H. Moore, and C. F. Robinow, p. 219 in *The Yeasts*, Vol. 1, A. N. Rose and J. S. Harrison, eds. New York: Academic Press, 1969.

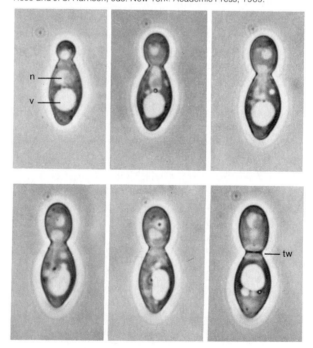

ered that zygote formation takes place at an unexpected stage of life cycle—immediately after the germination of the haploid ascospores. Pairs of germinating ascospores, or the first vegetative cells produced from them, fuse to form diploid vegetative cells. Diploidy is then maintained throughout the entire subsequent period of vegetative development, and meiosis occurs immediately prior to the formation of ascospores. Thus, *S. cerevisiae* exists predominantly in the diplophase. Other ascomycetous yeasts do not share this pattern of behavior but form zygotes by fusion between vegetative cells immediately before ascospore formation. The germinating ascospores then give rise to haploid vegetative progeny.

Although budding is the predominant mode of multiplication in yeasts, there are a few that multiply by binary fission, much like bacteria; these are placed in a special genus, *Schizosaccharomyces*.

In ascomycetous yeasts, the vegetative cell or zygote becomes entirely transformed into an ascus at the time of ascospore formation. Yeasts of the genus *Sporobolomyces* form basidiospores, and in this case the entire vegetative cell becomes transformed into a basidium. Just as in the mushrooms, basidiospore discharge in *Sporobolomyces* is a violent process, and the colonies of this yeast are readily detectable on plates that have been incubated in an inverted position because the portion of the glass cover underlying a *Sporobolomyces* colony becomes covered with a deposit of discharged spores that form a mirror image of the colony above (Figure 13.20).

## the slime molds

We conclude this survey of the protists by discussing the *slime molds*, which are not classified as true fungi, although they possess certain characteristics that resemble those of the fungi. The best-known representatives of the slime molds are the Myxomycetes, organisms that are found most commonly growing on decaying logs and stumps in damp woods. The vegetative structure, known as a *plasmodium*, is a multinucleate mass of cytoplasm unbounded by rigid walls, which flows in ameboid fashion over the surface of the substrate, ingesting smaller microorganisms and fragments of decaying plant material. An actively moving plasmodium is charac-

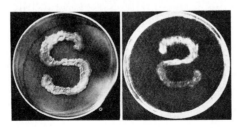

**Figure 13.20**
The formation of a mirror image of a colony of *Sporobolomyces* by basidiospore discharge in a petri dish incubated in the inverted position: (top) the colony on the agar surface, streaked in the form of an S; (bottom) the deposit of basidiospores formed on the lid of the petri dish as a result of spore discharge from the colony. From A. H. R. Buller, *Researches on Fungi*, Vol. 5, p. 175. New York: Longmans Green, 1933.

**Figure 13.21**
The plasmodium of a myxomycete, *Didymium*, growing at the expense of bacteria on the surface of an agar plate. Courtesy of Dr. K. B. Raper.

teristically fan-shaped, with thickened ridges of cytoplasm running back from the edge of the fan; it resembles a spreading layer of thin, colored slime (Figure 13.21). As long as conditions are favorable for vegetative development, the plasmodium continues to increase in bulk with accompanying repeated nuclear divisions. Eventually, the organism may become a mass of cytoplasm containing thousands of nuclei and weighing several hundred grams. Fruiting occurs when a plasmodium migrates to a relatively dry region of the substrate. Out of the undifferentiated plasmodium there is then produced a fruiting structure that is often of remarkable complexity and beauty [as illustrated by the case of *Ceratiomyxa* (Figure 13.22)]. As this fruiting body develops, small, uninucleate sections of the plasmodium become surrounded by walls to form large numbers of uninucleate spores, borne on the fruiting structure. After liberation, the spores germinate to produce uniflagellate ameboid gametes, which fuse in pairs to form biflagellate zygotes. After some time, a zygote loses its flagella and develops into a new plasmodium. The vegetative nuclei in a growing plasmodium are diploid, meiosis taking place just prior to the formation of spores in the fruiting body.

It is, of course, the fruiting stage of a myxomycete that at once reminds one of a true fungus: at first sight, the amorphous. plasmodial vegetative stage appears to resemble little, if at all, the branched, mycelial vegetative stage of the fungi, but suggests, rather, a relationship to the ameboid protozoa. In fact, the plasmodium and the mycelium are basically similar structures. Both are coenocytic, and in both the cytoplasm can flow, although in the mycelium cytoplasmic streaming is confined within the walls of branched tubes. The superficial difference between a plasmodium and a mycelium is essentially caused by the fact that in a plasmodium the cytoplasm is not bounded by rigid walls and is thus free to flow in any direction; in a mycelium it flows only inside the tubular structure.

The slime molds also include a small group, the Acrasieae (Figure 13.23), which show far greater resemblances to the unicellular ameboid protozoa than do the true Myxomycetes. The vegetative stage of an acrasian consists of small, uninucleate amebas, which multiply by binary fission and can in no way be distinguished, at this stage of their life history, from other small ameboid protozoa. Nevertheless, when conditions are favorable, thousands of these isolated amebas are capable of aggregating and cooperating, without ever losing their cellular distinctness, in the construction of an elaborate fruiting body. The first sign of approaching fructification is the aggregation of the vegetative cells to form a macroscopically visible heap. This heap of cells gradually differentiates into a tall stalked structure that bears a rounded head of asexual spores. At all stages in the formation of this fruiting body, the cells remain separate; some individuals form the stalk, which is surrounded and given rigidity by a cellulose sheath, while others are carried up the outside of the rising stalk to form the spore head. As this matures, each ameba in it rounds up and becomes surrounded by a wall.

**Figure 13.22**
Fruiting bodies of a myxomycete, *Ceratiomyxa*, on a piece of wood. From C. M. Wilson and I. K. Ross, "Meiosis in the Myxomycetes," *Am. J. Botany* **42**, 743 (1955).

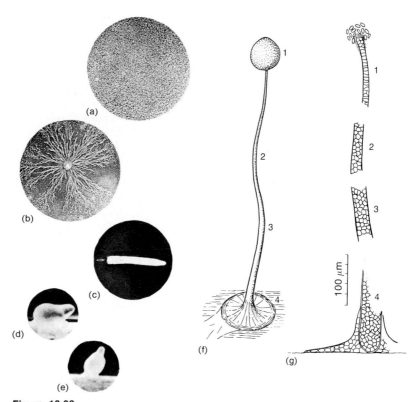

**Figure 13.23**

The life cycle of *Dictyostelium*, a representative of the Acrasieae. (a) A uniform mass of vegetative amebae. (b) Aggregation of the amebae to a fruiting center. (c) Motile mass of aggregated cells. (d), (e) Early stages in formation of the fruiting body. (f) A mature fruiting body. (g) Magnified sections through various regions of the fruiting body. From K. B. Raper, "Isolation Cultivation, and Conservation of Simple Slime Molds," *Quart. Rev. Biol.* **26,** 169 (1951).

These spores, following their release, germinate and give rise to individual ameboid vegetative cells once more. This remarkable kind of life cycle, where a communal process of fructification is imposed on a unicellular phase of vegetative development, occurs in one procaryotic group, the myxobacteria (described in Chapter 11).

# MICROORGANISMS AS GEOCHEMICAL AGENTS

The current chemical state of the elements on the outer surface of the earth is, to a considerable extent, a consequence of the chemical activities of living organisms. This fact is dramatically illustrated by the changes that have occurred in the earth's atmosphere. Before life evolved, the gases of the atmosphere were highly reduced: nitrogen was present in the form of ammonia ($NH_3$), oxygen as water ($H_2O$), and carbon as methane ($CH_4$). Now, these gases exist in oxidized form: nitrogen and oxygen as elemental gases ($N_2$ and $O_2$) and carbon as carbon dioxide ($CO_2$). The quantity of these and many other compounds found on the earth's surface represents the net balance between their rates of formation and their utilization in biological and geological processes. Such transformation occurs in all regions of the earth that contain living organisms, regions collectively known as the *biosphere*. The oceans, the freshwater and land surfaces of the continents, and the lower portion of the atmosphere comprise the biosphere. This thin film of life on the earth's surface exists in a more or less steady state, maintained by a cyclic turnover of the elements necessary for life, and powered by a continuous input of energy from the sun.

The various steps in the cyclic turnover of elements are brought about by different types of organisms. Thus, the continued existence of any particular group of organisms depends on the chemical transformation carried out by others. A break in the cycle at any point would eventually preclude all life. All the major bioelements (carbon, oxygen, nitrogen, sulfur, and phosphorus) are transformed cyclically.

The cyclic nature of transformations in the biosphere can be summarized as follows. Through solar-energy conversion by photosynthesis, $CO_2$ and other inorganic compounds are withdrawn from the environment and are accumulated in the organic constituents of living organisms. The major producers of organic matter through photosynthesis are unicellular algae (principally diatoms and dinoflagellates) in the ocean and seed plants on land. Organic material thus accumulated provides, either directly or indirectly, the energy sources for all other forms of life.

Insofar as photosynthetic organisms serve as food sources for animals or microorganisms, the elements of major biological importance remain, at least in part, in the organic state, during the transformations that lead to their incorporation into the cells and tissues of the primary consumers. The primary consumers may themselves provide food sources for other organisms, so that these elements may persist in organic *food chains* made up of many types of nonphotosynthetic organisms.

An example of a food chain in a freshwater pond is shown in Figure 14.1. Typically, large losses of mass of organic matter occur at each step in the food chain, because most organic matter ingested as food is respired; only a minor fraction is incorporated into the cells and tissues of the consumer. In the example cited in Figure 14.1, less than 0.5 percent of the mass of phytoplankton becomes incorporated into the tissues of bass.

Before the mineral elements contained in nonphotosynthetic organisms can be again utilized by photosynthetic organisms, they usually must be converted once more to inorganic form. This conversion, known as *mineralization*, is brought about largely by the decomposition of plant and animal remains and excretory products by microorganisms, principally fungi and bacteria. It is estimated that 90 percent of the mineralization of organic carbon atoms (i.e., their conversion to $CO_2$) is the result of the metabolic activities of these two groups of microorganisms. The remaining 10 percent results from the metabolism of all other organisms, as well as the combustion of fuels and other materials. The overwhelming contribution of microorganisms to this process reflects their ubiquity, their significant contribution to the total bulk of living material (their biomass), their high rates of growth and metabolism, and their collective ability to degrade a vast variety of naturally occurring organic materials.

**Figure 14.1**

Schematized food chain in a fresh water pond (without dotted lines) used for sports fishing. Numbers in parentheses represent input of biomass in terms of kilocalories per square meter per year. After E. P. Odum, *Fundamentals of Ecology,* Philadelphia: W. B. Saunders Co., 1971. Data from H. Welch, "Energy Flow through the Major Macroscopic Components of an Aquatic Ecosystem," Ph.D. dissertation, University of Georgia, Athens, 1967.

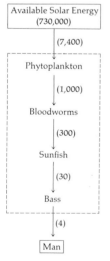

## the fitness of microorganisms as agents of geochemical change

### the distribution of microorganisms in space and time

The omnipresence of microorganisms throughout the biosphere is a consequence of their ready dissemination by wind and water. Surface waters, the floors of oceans over the continental shelves, and the top few inches of soil are teeming with microorganisms that are ready to decompose organic matter that may become available to them. It has been estimated that the top 6 in. of fertile soil may contain more than two tons of fungi and bacteria per acre. Any handful of soil contains many different kinds of microbes, presenting at different times microscopic ecological niches for different types of microorganisms to develop. Even on a single soil particle, the conditions may change from hour to hour and from facet to facet.

Let us consider what happens upon the death of a microscopic root hair or a worm in the soil. The organic compounds of the dead tissue are attacked by microorganisms that are capable of digesting and oxidizing these compounds. As oxygen is consumed, conditions may become anaerobic in the immediate proximity of the dead tissue, and fermentative organisms develop. The products of fermentation then diffuse to regions in which oxygen is still present, or they may be oxidized anaerobically by organisms capable of reducing nitrate, sulfate, or carbonate. Ultimately, the organic compounds will be completely converted to $CO_2$ or assimilated; the condition will again become fully aerobic; and autotrophs will develop at the expense of such reduced inorganic products as ammonia, sulfide, and hydrogen. Thus, the inorganic products of the decomposition of the plants or animals are eventually completely oxidized. This sequence of events, which occurs on a microscopic scale on a particle of soil, can be observed on a macroscopic scale in nature. When a tree falls into a swamp or a whale decomposes on a beach, the eventual chemical results are essentially the same. Seasonal and climatic conditions may retard or accelerate the cyclic turnover of matter. In cold climates, decomposition is most rapid in the early spring; in semiarid areas, it is largely restricted to the rainy season.

In nature, only those microorganisms that are favored by the local and temporary environment reproduce, and their growth may cease when they have changed their environment. Most of them are eventually consumed by such ever-present predators as the protozoa, but a few cells of each type persist to initiate a new burst of growth when conditions again become favorable for their development.

### the metabolic potential
### of microorganisms

The relatively enormous catalytic power of microorganisms contributes to the major role they play in the chemical transformations occurring on the earth's surface. Because of their small size, bacteria and fungi possess a large surface-volume ratio compared with higher animals and plants. This permits a rapid exchange of substrates and waste products between the cells and their environment.

Per gram of body weight, the respiratory rates of some aerobic bacteria are hundreds of times greater than that of man. On the basis of the known metabolic rates of microorganisms, one can estimate that the metabolic potential of the microorganisms in the top 6 in. of an acre of well-fertilized soil at any given instant is equivalent to the metabolic potential of some tens of thousands of human beings.

An even more important factor influencing the chemical role that microorganisms play in nature is their high rate of reproduction in favorable environments.

### the metabolic versatility
### of microorganisms

The remarkable ability of microorganisms to degrade a vast variety of organic compounds has led to a widely held conviction that has been termed the "principle of microbial infallibility," a principle that was clearly stated by E. F. Gale in 1952:* "It is probably not unscientific to suggest that somewhere or other some organism exists which can, under suitable conditions, oxidise any substance which is theoretically capable of being oxidised." With the increasing production of plastics as well as synthetic insecticides, herbicides, and detergents, it has become clear that some substances are remarkably resistant to microbial attack, because they persist and accumulate in nature. Even certain naturally occurring organic compounds are somewhat resistant; they accumulate and constitute the organic fraction of soil known as *humus,* which confers the deep brown or black color to fertile soils. Because of the importance of humus to agriculture, this complex mixture of persistent organic compounds has been studied extensively. In large degree, it appears to consist of degradation products of a particularly stable component of woody plants known as *lignin.* The remarkable stability of humus has been demonstrated by radiocarbon dating; humus from certain soils is thousands of years old.

These exceptions aside, most organic compounds that are no longer a part of a living organism are rapidly mineralized by microorganisms in the biosphere.

Although some nonphotosynthetic microorganisms (e.g., the Pseudomonas group) can attack many different organic compounds, the metabolic versatility of the microbial world *as a whole* is not primarily a reflection of the metabolic versatility of its individual members. Any single bacterial species is only a limited agent of mineralization. Highly specialized physiological groups of microorganisms play important roles in the mineralization of specific classes of organic compounds. For example, the decomposition of cellulose, which is one of the most abundant constituents of plant tissues, is mainly brought about by organisms that are highly specialized nutritionally. Among the aerobic bacteria capable of decomposing cellulose, the gliding bacteria that belong to the *Cytophaga group* are perhaps the most important. The cytophagas can rapidly degrade this insoluble compound, but cellulose is the only substance they can use as a carbon source.

It will be recalled that the autotrophic bacteria, responsible for the oxidation of reduced inorganic compounds in nature, are

*E. F. Gale, *The Chemical Activities of Bacteria,* p. 5. New York: Academic Press, 1952.

also highly specific. Each type of autotroph is capable of oxidizing only one class of inorganic compounds and, in some cases (the nitrifying bacteria), only one compound.

# the cycles of matter

The turnover of the elements that compose living organisms constitute the *cycles of matter*. All organisms participate in various steps of these cyclic conversions, but the contribution of microorganisms is particularly important, both quantitatively (as discussed previously) and qualitatively. For example, certain steps in the nitrogen cycle are exclusively brought about by procaryotes.

### the phosphorus cycle

Considered from a chemical point of view, the phosphorus cycle is simple, because phosphorus occurs in living organisms only in the $+5$ valence state, either as free phosphate ions ($PO_4^{3-}$) or as organic phosphate constituents of the cells. Most organic phosphate compounds cannot be taken into the living cell; instead, phosphorus requirements are met by the uptake of phosphate ions. Organic phosphate compounds are then synthesized within the cell, and upon the death of the organism, phosphate ions are rapidly released by hydrolysis.

In spite of the rapid functioning of the biological components of the phosphorus cycle and the relative abundance of phosphates in soils and rocks, phosphate is a limiting factor for the growth of many organisms because much of the earth's supply of phosphates occurs as insoluble calcium, iron, or aluminum salts. Fresh water often contains phosphate ions in mere trace amounts, which are available to animals only after they have been concentrated by the phytoplankton.

Soluble phosphates are constantly being transferred from terrestrial environments to the sea as a consequence of leaching, a transfer that is largely unidirectional. Only small quantities are returned to the land, principally by the deposits of guano by marine birds. Thus, the availability of phosphate for terrestrial forms of life depends on the continued solubilization of insoluble phosphate deposits, a process in which microorganisms play an important role. Their acidic metabolic products (organic, nitric, and sulfuric acids) solubilize the phosphate of calcium phosphate, and $H_2S$, which they produce, dissolves ferric phosphates.

### the cycles of carbon and oxygen

The cyclic conversions of carbon and oxygen are brought about primarily by two processes—*oxygenic photosynthesis* on the one hand, and *respiration* and *combustion* on the other. It is mainly through the process of oxygenic photosynthesis that the oxidized form of carbon ($CO_2$) is converted to the reduced state in which it occurs in organic compounds, and that the reduced form of oxygen ($H_2O$) is oxidized to molecular oxygen ($O_2$). Although other autotrophs can reduce $CO_2$ to organic material while oxidizing compounds other than water ($NO_2^-$, $NH_4^+$, $H_2$, $Fe^{2+}$, and reduced forms of sulfur), the contribution of these processes to the total fixation of $CO_2$ is relatively minor.

Heterotrophic metabolism coupled directly or indirectly with the reduction of molecular $O_2$ completes the cycle by regenerating the major nutrients of oxygenic photosynthesis: $CO_2$ and water. Algae and plants, as well as the animals that feed on them, contribute their share to this process through their respiratory activities. The bacteria and fungi, however, oxidize the bulk of organic material. Thus, through oxygenic photosynthesis on the one hand, and aerobic respiration on the other, the cyclic transformations of carbon and oxygen are obligately linked (Figure 14.2).

Air contains approximately 0.03 percent $CO_2$ by volume, this concentration being maintained relatively constant as a result of the dynamic balance between photosynthesis and mineralization. When dissolved in slightly alkaline water, bicarbonate ($HCO_3^-$) and carbonate ($CO_3^{2-}$) ions are formed:

$$CO_2 + OH^- \rightleftharpoons HCO_3^-$$

$$HCO_3^- + OH^- \rightleftharpoons H_2O + CO_3^{2-}$$

Therefore, bicarbonate serves as the reservoir of carbon for photosynthesis in aquatic environments. The bicarbonate concentration of ocean waters ($\sim$0.002 M) acts as a reservoir for $CO_2$ for the atmosphere; the oceans trap a large fraction of the $CO_2$ produced on land, keeping its concentration at a relatively low and constant level.

**Figure 14.2**
The carbon and oxygen cycles: oxidations of carbon and oxygen are shown as solid arrows, reductions as broken arrows, and reactions with no valence change as dotted arrows.

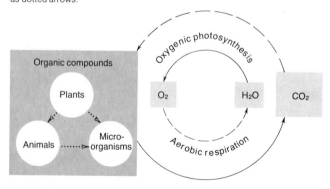

The importance of the carbon cycle can best be emphasized by the estimate that the total $CO_2$ contained in the atmosphere, if it were not replenished, would be completely exhausted in less than 20 years at the present rate of photosynthesis. This estimate does not appear too radical when it is realized that the carbon contained in a single giant redwood tree is equivalent to that present in the atmosphere over an area of approximately 40 acres. On land, seed plants are the principal agents of photosynthetic activity. A minor contribution is made by the algae. In the oceans, however, it is the unicellular photosynthetic organisms that play the most important role. The large plantlike algae (seaweeds) are confined in their development to a relatively narrow coastal strip. Since the light of photosynthetically effective wavelengths is largely filtered out at a depth of about 50 ft, these sessile algae cannot grow in deeper waters. Because they are free-floating, the microscopic algae of the ocean (known as the *phytoplankton*) are capable of developing in the surface layers wherever the environment is favorable. Their growth is largely limited by the relative scarcity of two elements: phosphorus and nitrogen. Where these elements are made available as phosphates and nitrates by the runoff of rain water from continents and subsequent distribution by ocean currents, profuse development of phytoplankton occurs. According to one estimate, the total annual fixation of carbon in the oceans amounts to approximately $1.2 \times 10^{10}$ tons, whereas that on the land is about $1.6 \times 10^{10}$ tons.

THE MINERALIZATION PROCESS: CARBON DIOXIDE FORMATION AND THE REDUCTION OF OXYGEN. The biological conversion of organic carbon to $CO_2$ with the concomitant reduction of molecular oxygen involves the combined metabolic activity of many different kinds of microorganisms. The complex constituents of dead cells must be digested, and the products of digestion must be oxidized by specialized organisms that can use them as nutrients. Many aerobic bacteria (e.g., pseudomonads, bacilli, actinomycetes), as well as fungi, carry out complete oxidations of organic substances derived from dead cells. However, it should be remembered that even those organisms that produce $CO_2$ as the only waste product of the respiratory decomposition of organic compounds usually use a large fraction of the substrate for the synthesis of their own cell material. In anaerobic environments, organic compounds are decomposed initially by fermentation, and the organic end products of fermentation are then further oxidized by anaerobic respiration, provided that suitable inorganic hydrogen acceptors (nitrate, sulfate, or $CO_2$) are present.

THE SEQUESTRATION OF CARBON: INORGANIC DEPOSITS. The carbonate ions in the oceans combine with dissolved calcium ions and become precipitated as calcium carbonate. Calcium carbonate is also deposited biologically in the shells of protozoa, corals, and mollusks. This is the geological origin of the calcareous rock (limestone) that is an important constituent of the surface of continents. Calcareous rock is not directly available as a source of carbon for photosynthetic organisms, and hence its formation causes a depletion of the total carbon supply available for life. Nevertheless, much of this carbon eventually reenters the cycle through weathering. The formation and solubilization of calcium carbonate are brought about primarily by changes in hydrogen ion concentration, and microorganisms contribute indirectly to both of these processes as a consequence of pH changes that they produce in natural environments. For example, such microbial processes as sulfate reduction and denitrification cause an increase in the alkalinity of the environment, which favors the deposition of calcium carbonate in the ocean and other bodies of water. Microorganisms also play an important role in solubilizing calcareous deposits on land, similar to the role they play in solubilizing phosphates, by production of acid during nitrification, sulfur oxidation, and fermentation.

THE SEQUESTRATION OF CARBON: ORGANIC DEPOSITS.   A high moisture content, causing oxygen depletion and the accumulation of acidic substances, is sometimes particularly favorable for the accumulation of partly decomposed organic material of plant origin: this complex mixture of materials, which influences the physical properties of soil (e.g., moisture retention) is termed *humus* (see p. 352). Such accumulation is most pronounced in *bogs,* where, in the course of time, deposits of undecomposed organic matter known as *peat* accumulate. These deposits may extend for hundreds of feet below the surface of the bog. In the course of geological time the compression of peat deposits, probably aided by other physical and chemical factors, has resulted in the formation of coal. Much carbon has thus been sequestered from the biosphere in the form of peat and coal deposits. A second kind of sequestration of carbon in organic form has occurred in deposits of petroleum.

Since the Industrial Revolution, man's exploitation of the stored deposits of organic carbon in the earth's crust has resulted in their very rapid mineralization. Although substantial deposits still remain to be exploited, it is estimated that at current rates of consumption most of the petroleum and natural gas will be used up within a few decades. The rapid rate of utilization of fossil fuels is producing a small but significant increase in atmospheric $CO_2$; its concentration increased by approximately 2 percent during the decade 1958–1968. Possibly this change will affect climate and rate of photosynthesis.

### the nitrogen cycle

Although nitrogen gas ($N_2$) is abundant, constituting about 80 percent of the earth's atmosphere, it is chemically inert and therefore not a suitable source of the element for most living forms. All plants and animals, as well as most microorganisms, depend on a source of combined, or fixed, nitrogen in their nutrition. Combined nitrogen in the form of ammonia, nitrate, and organic compounds is relatively scarce in soil and water, often constituting the limiting factor for the development of living organisms. For this reason, the cyclic transformation of nitrogenous compounds is of paramount importance in supplying required forms of nitrogen to the various nutritional classes of organisms in the biosphere. The main features of the nitrogen cycle are illustrated schematically in Figure 14.3.

NITROGEN FIXATION. The turnover of nitrogen through its cycle is estimated to be between $10^8$ and $10^9$ tons per year. The vast supply of nitrogen gas in the atmosphere and the relative scarcity of combined nitrogen on the earth's surface suggest that the process of nitrogen fixation is the rate-limiting step. This process is largely a biological one, and bacteria are the only organisms capable of causing it (see Chapter 4). Some nitrogen is fixed by lightning, ultraviolet light, electrical equipment, and the internal combustion engine, but these nonbiological processes are quantitatively minor, together accounting for about 10 percent of nitrogen fixation. Even industrial manufacture of fertilizer by the Haber process contributes only about 20 percent. Thus, over

**Figure 14.3**
The nitrogen cycle: the oxidations of nitrogen are shown as solid arrows, the reductions as broken arrows, and reactions with no valence change as dotted arrows.

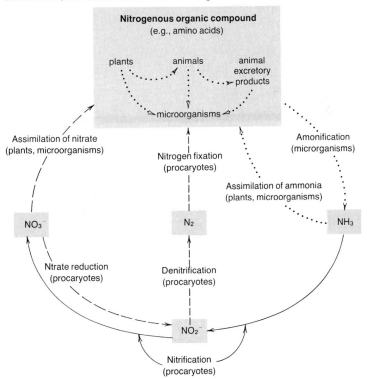

60 percent of all nitrogen fixation is brought about by bacteria.

Biological nitrogen fixation in nature is mediated in part by free-living bacteria (*nonsymbiotic nitrogen fixation*) and in part by bacteria (p. 375) that exist in mutualistic partnership with certain plants (*symbiotic nitrogen fixation*).

The most important agents of symbiotic nitrogen fixation are the bacteria of the genus *Rhizobium*, which invade the root hairs of leguminous plants and develop in nodules produced on the roots, where nitrogen fixation occurs. The enzymatic machinery for the fixation of nitrogen is synthesized by the bacterium; the host plant provides an environment that favors the expression of this property.

The most important agents of nonsymbiotic nitrogen fixation are heterocyst-forming blue-green bacteria such as *Anabaena* and *Nostoc*. A wide variety of other bacteria are also capable of fixing nitrogen; these include aerobic bacteria (e.g., *Azotobacter* and *Beijerinckia*), anaerobic bacteria (e.g., photosynthetic bacteria and *Clostridium* spp.) and some facultatively anaerobic bacteria when grown under anaerobic conditions (e.g., *Bacillus polymyxa* and some *Enterobacter* strains).

Evaluated as nitrogen fixed per acre of soil per year (Table 14.1), the contribution of the legume-*Rhizobium* symbiosis is far greater than that made by nonsymbiotic nitrogen fixers. Among the latter organisms, blue-green bacteria probably make the largest contribution to nitrogen fixation in nature. However, recent evidence suggests that nitrogen fixation by *Azotobacter* and *Spirillum* is highly effective when these bacteria develop in close association with the roots of plants; their contribution to nitrogen fixation may therefore have been hitherto underestimated.

Because of the critical agronomic importance of fixed nitrogen, the current world food crisis, and the fact that manufacture of nitrogen fertilizers by the Haber process requires large expenditures of en-

**Table 14.1**
Efficiencies of Some Nitrogen-Fixing Systems

| ORGANISM OR SYSTEM | POUNDS OF N FIXED/ACRE/YEAR[a] |
|---|---|
| SYMBIOTIC | |
| Alfalfa: *Rhizobium* | >264 |
| Clover: *Rhizobium* | 220 |
| NONSYMBIOTIC | |
| Blue-green bacteria | 22 |
| *Azotobacter* spp. | 0.26 |
| *Clostridium pasteurianum* | 0.22 |

[a] Recalculated from E. N. Mishustin and V. K. Shilnikova, *Biological Fixation of Atmospheric Nitrogen*. (London: Macmillan & Co., 1971).

ergy, biological nitrogen fixation has become an intensive subject of investigation. Among the important goals of these investigations are the development of new plants capable of harboring nitrogen-fixing symbionts. Although the current range of such plants is wide, it does not include the world's major food crops, wheat and rice, nor the major forage crop, grass.

THE UTILIZATION OF FIXED NITROGEN BY PHOTOSYNTHETIC ORGANISMS. Algae and plants assimilate nitrogen either as nitrate or as ammonia. If nitrate is the form in which nitrogen is assimilated, it must be reduced in the cell to ammonia. This process of nitrate reduction proceeds only to the extent to which nitrogen is required for growth; ammonia is not excreted. It is this feature in particular that distinguishes *nitrate assimilation* by plants (and also by microorganisms) from *nitrate reduction,* a process of anaerobic respiration that is limited to procaryotes (Figure 14.3).

THE TRANSFORMATIONS OF ORGANIC NITROGEN AND THE FORMATION OF AMMONIA. The organic nitrogenous compounds synthesized by algae and plants serve as the nitrogen source for the animal kingdom. During their assimilation by animals, these nitrogenous compounds of plants are hydrolyzed to a greater or lesser extent, but the nitrogen remains largely in reduced organic form. Unlike plants, however, animals do excrete a significant quantity of nitrogenous compounds in the course of their metabolism. The form in which this nitrogen is excreted varies from one group of animals to another. Invertebrates predominantly excrete ammonia; but among vertebrates, organic nitrogenous excretion products make their appearance as well. In reptiles and birds, uric acid is the major form in which nitrogen is excreted; in mammals, urea is the principal form. Urea and uric acid are rapidly mineralized by special groups of microorganisms, with the formation of $CO_2$ and ammonia.

Only part of the nitrogen stored in organic compounds through plant growth is converted to ammonia by animal metabolism and by the microbial decomposition of urea and uric acid. Much of the nitrogen remains in plant and animal tissues and is liberated only on the death of these organisms. Whenever a plant or animal dies, its body constituents are immediately attacked by microorganisms, and the nitrogenous compounds are decomposed with the liberation of ammonia. Part of the nitrogen is assimilated by the microorganisms themselves, and thus converted into microbial cell constituents. Ultimately, these constituents are converted to ammonia following the death of the microbes.

The first step in this process of *ammonification* is the hydrolysis of the proteins and nucleic acids, with the liberation of amino acids and organic nitrogenous bases, respectively. These simpler compounds are then attacked by respiration or fermentation.

Protein decomposition under anaerobic conditions (*putrefaction*) usually does not lead to an immediate liberation of all the amino nitrogen as ammonia. Instead, some of the amino acids are converted to amines. The putrefactive decomposition is characteristically brought about by anaerobic spore-forming bacteria (genus *Clostridium*). In the presence of air, the amines are oxidized by other bacteria, with the liberation of ammonia.

NITRIFICATION. Through all the transformations that nitrogen undergoes, from the time of its reductive assimilation by plants until its liberation as ammonia, the nitrogen atom remains in the reduced form. The conversion of ammonia to nitrate (*nitrification*) is brought about in nature by two highly specialized groups of obligately aerobic chemoautotrophic bacteria, the ammonia oxidizers and the nitrite oxidizers (see Chapter 10). As a result of the combined activities of these bacteria, the ammonia liberated during the mineralization of organic matter is rapidly oxidized to nitrate. Thus nitrate is the principal nitrogenous material available in soil for the growth of plants. The practice of soil fertilization with manure depends on the microbial mineralization of organic matter and results in the conversion of organic nitrogen to nitrate through ammonification and nitrification. Irrigation with dilute solutions of ammonia, which is one of the modern methods used for fertilization, is an even more direct means by which the nitrate content of soil is increased. Ammonia, which can be synthesized chemically from molecular nitrogen, is the most concentrated form of combined nitrogen available because it contains about 82 percent nitrogen by weight. Nitrates are very soluble compounds and are therefore easily leached from the soil and transported by water; hence, a certain amount of combined nitrogen is constantly removed from the continents and carried down to the oceans. In some special localities, notably in the semiarid regions of Chile, deposits of nitrate have accumulated in the soil as a result of the runoff and evaporation of surface water. Such deposits are a valuable source of fertilizer, although their importance has diminished greatly in the course of the last 50 years as a result of the development of chemical methods for fixing atmospheric nitrogen.

Nitrates have played an important role not only in the development of agriculture but also in the destructive activities of man. Gunpowder, which was the only explosive used for war before the invention of nitroglycerine (dynamite), is a mixture of sulfur, carbon, and saltpeter ($KNO_3$). During the Napoleonic wars, largely as a result of the British blockade, a shortage of nitrate for gunpowder production occurred in France. This led to the development of "nitrate gardens," in which nitrate was obtained by the mineralization of organic matter. A mixture of manure and soil was spread on the surface of the ground and was frequently turned to permit aeration. After the manure had decomposed, nitrate was extracted from the residue.

DENTRIFICATION. Certain representatives of almost every major group of procaryotes are able under anaerobic conditions to generate ATP by anaerobic respiration utilizing nitrate as a terminal electron acceptor.

Thus, whenever organic matter is decomposed in soil or water and oxygen is exhausted as a result of aerobic microbial respiration, these microorganisms will continue to respire the organic matter if nitrate is present. As a consequence, nitrate is reduced. Some bacteria (e.g., *Escherichia coli*) are able to reduce nitrate only to the level of nitrite; others (e.g., *Pseudomonas aeruginosa*) are able to reduce it to nitrogen gas. By this latter process, called *denitrification,* combined nitrogen is removed from soil and water, releasing $N_2$ gas to the atmosphere.

Denitrification is a process of major ecological importance. It depletes the soil of an essential nutrient for plants, thereby decreasing agricultural productivity. Such losses are particularly important from fertilized soils. Although precise values are not available, under certain conditions, the amount of fixed nitrogen fertilizer lost through denitrification may approach 50 percent.

Nevertheless, not all the consequences of denitrification are detrimental. Denitrification is vital to the continued availability of combined nitrogen on the land masses of the earth. The highly soluble nitrate ion is constantly leached from the soil, and it is eventually carried to the oceans. Without denitrification, the earth's supply of nitrogen, including $N_2$ of the atmosphere, would eventually accumulate in the oceans, precluding life on the land masses except for a fringe near the oceans. Denitrification also maintains the potability of fresh waters, because high concentrations of nitrate ions may be toxic.

### the sulfur cycle

Sulfur, an essential constituent of living matter, is abundant in the earth's crust. It is available to living organisms principally in the form of soluble sulfates or reduced organic sulfur compounds. Reduced sulfur in the form of $H_2S$ also occurs in the biosphere as a result of microbial metabolism and, to a limited extent, of volcanic activity. Except under anaerobic conditions, however, its concentration is low because it is oxidized rapidly in the presence of oxygen, either spontaneously or by bacteria.

The turnover of sulfur compounds is referred to as the *sulfur cycle.* The biological aspects of this cycle are shown in Figure 14.4. In many respects, it shows a striking resemblance to the nitrogen cycle.

In addition to the biological sulfur cycle, important nonbiological transformations of gaseous forms of sulfur occur in the earth's atmosphere. It is estimated that some 90 million tons of sulfur in the form of biologically generated $H_2S$ are released to the atmosphere annually; an additional 50 million tons are contributed in the form of $SO_2$ by the burning of fossil fuels; and about 0.7 million tons in the form of $H_2S$ and $SO_2$ come from the earth's volcanic activity. In the atmosphere, $H_2S$ is rapidly oxidized to $SO_2$, which can dissolve in water to form sulfurous acid ($H_2SO_3$); or be oxidized by a second and slower series of reactions (requiring hours or days)

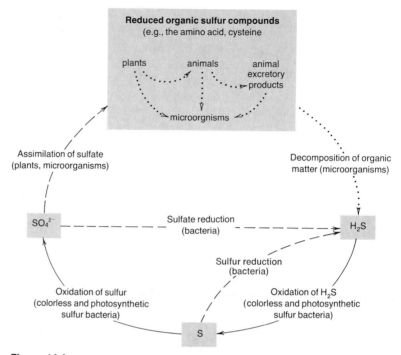

**Figure 14.4**

The sulfur cycle: the oxidations of the sulfur atom are shown as solid arrows, the reductions as broken arrows, and reactions with no valence change as dotted arrows.

to $SO_3$. When dissolved in water, $SO_3$ becomes sulfuric acid ($H_2SO_4$). Some sulfuric acid is neutralized by the small quantities of ammonia in the atmosphere, but much of it returns along with unoxidized $H_2SO_3$ to the earth's surface in acid form, where it causes considerable damage to stone structures and sculptures. The rate of generation of acidic sulfur compounds increases as more fossil fuels are being burned. The problem is particularly acute in areas of high population density, and even now it is causing the rapid destruction of much stone sculpture. In some industrialized areas (e.g., the northeastern United States), this *acid rain* has also caused a significant drop in the pH of lakes and streams, with adverse effects on the flora and fauna.

THE ASSIMILATION OF SULFATE. Sulfate is almost universally used as a nutrient by plants and microorganisms. The assimilation of sulfate resembles the assimilation of nitrate in two respects. First, like the nitrogen atom of nitrate, the sulfur atom of sulfate must become reduced in order to be incorporated into organic compounds, because in living organisms, sulfur occurs almost exclusively in reduced form as —SH or —S—S— groups. Second, in both cases, only enough of the nutrient is assimilated to provide for the growth of the organism, no reduced products being excreted into the environment.

THE TRANSFORMATION OF ORGANIC SULFUR COMPOUNDS AND FORMATION OF $H_2S$. When sulfur-containing organic compounds are mineralized, sulfur is liberated in the reduced inorganic form as $H_2S$. This process resembles ammonification, in which nitrogen is liberated from organic matter in its reduced inorganic form as ammonia.

THE DIRECT FORMATION OF $H_2S$ FROM SULFATE. The utilization of sulfate for the synthesis of sulfur-containing cell constituents and the subsequent decomposition of these compounds results in an overall reduction of sulfate to $H_2S$. $H_2S$ is also formed more directly from sulfate through the activity of the sulfate-reducing bacteria. These obligately anaerobic bacteria oxidize organic compounds and molecular hydrogen by using sulfate as an oxidizing agent. Their role in the sulfur cycle may therefore be compared to the role of the nitrate-reducing bacteria in the nitrogen cycle. The activity of the sulfate-reducing bacteria is particularly apparent in the mud at the bottom of ponds and streams, in bogs, and along the seashore. Since seawater contains a relatively high concentration of sulfate, sulfate reduction is an important factor in the mineralization of organic matter on the shallow ocean floors. Signs of the process are the odor of $H_2S$ and the pitch-black color of the mud in which it occurs. The color of black mud is due to the accumulation of ferrous sulfide. Some coastal areas, where an accumulation of organic matter leads to a particularly massive reduction of sulfate, are practically uninhabitable because of the odor and the toxic effects of $H_2S$.

THE OXIDATION OF $H_2S$ AND SULFUR. The $H_2S$ that is produced in the biosphere as a result of the decomposition of sulfur-containing compounds, of sulfur reduction, and of volcanic activity is largely converted to sulfate. Only a small part of it becomes sequestered in the form of insoluble sulfides or, after spontaneous oxidation with oxygen, as elemental sulfur.

The biological oxidation of $H_2S$ and of elemental sulfur is brought about by photosynthetic and chemoautotrophic bacteria. It can be effected either aerobically by the colorless sulfur bacteria or anaerobically by the photosynthetic purple and green sulfur bacteria. Since these oxidations result in the production of hydrogen ions, they result in the local acidification of soils. Sulfur is commonly added to alkaline soil to increase its acidity.

# the cycle of matter
# in anaerobic environments

Regions of the biosphere that are not in direct contact with the atmosphere can remain oxygen-free for long periods of time. Such anaerobic environments harbor distinctive microorganisms; provided that light can penetrate, these microorganisms are able to bring about an almost completely closed anaerobic cycle of matter.

The primary synthesis of organic material under these conditions is mediated by photosynthetic bacteria. The purple and green sulfur bacteria convert $CO_2$ to cell material, using $H_2S$ as a reductant; the purple nonsulfur bacteria perform an almost complete assimilation of acetate and other simple organic compounds. Upon the death of the photosynthetic bacteria, their organic cell constituents are decomposed by clostridia and other fermentative anaerobes, with the formation of $CO_2$, $H_2$, $NH_3$, organic acids, and alcohols. The hydrogen and some of the organic fermentation products are anaerobically oxidized by sulfur-reducing and methane-producing bacteria. The anaerobic oxidations performed by the sulfur reducers result in the formation of $H_2S$ and acetate, both utilizable in turn by photosynthetic bacteria. The metabolism of the methane bacteria results in the conversion of $CO_2$ and of some organic carbon (e.g., the methyl group of acetic acid) to methane. Present information suggests that methane cannot be further metabolized under anaerobic conditions. Much of it escapes, however, to aerobic regions, where it is oxidized by *Methanomonas* and other aerobic methane oxidizers. The loss of methane constitutes the only significant leak from this anaerobic cycle of matter. The anaerobic sulfur cycle is almost completely closed, since sulfate and $H_2S$ are interconverted by the combined activities of sulfate-reducing and photosynthetic bacteria; thus only a small amount of gaseous $H_2S$ escapes. The anaerobic nitrogen cycle is also closed and is chemically very simple, relative to the cycle in aerobic environments; the nitrogen atom does not undergo valence changes, alternating between ammonia and the amino groups ($R—NH_2$) in nitrogenous cell materials.

Since the participants in this anaerobic cycle of matter are all microorganisms, the cycle does not require much space. In fact, such a cycle can be established in the laboratory, in a closed bottle that contains the appropriate nutrients and that has been inoculated with water and mud from an anaerobic pool. Provided that the bottle is illuminated, microbial development will continue in it over a period of years.

# the influence of man on the cycle of matter

The emergence of man as a member of the biological community did not at first significantly affect the cycle of matter on earth. However, the rapid increases in the total size and local density of human populations that have occurred since the Industrial Revolution, coupled with the ever-increasing power of the human species to modify its environment, have begun to change the picture. Within the past century these factors have led to local environmental changes comparable in scale to those produced by major geological upheavals in the past history of the earth. The spread of agriculture, the denudation of forests, the mining and burning of fossil fuels, and the pollution of the environment with human and industrial wastes have profoundly affected the distribution and growth of other forms of life.

### sewage disposal and treatment

As a result of the concentration of human populations, which has occurred at an ever-increasing pace over the past 150 years, the disposal of waste, both domestic and industrial, has become a major public health and ecological problem.

The public health problems associated with sewage disposal are clearly defined and mostly solvable by present technology. Due to the numbers of people in any population who are ill or who carry pathogens (without exhibiting symptoms), all sewage contains pathogens. Therefore, sewage disposal must be accomplished in such a manner as to prevent dispersion of pathogens. A number of highly effective procedures have been developed, including the disposal of sewage in a manner that minimizes contamination of supplies of water and food, and the sterilization of sewage effluents with chlorine.

Ecological problems associated with sewage disposal are considerably more complex and varied. Chief among these are: (1) the large oxygen requirement, termed *BOD* (biological oxygen demand), associated with the mineralization of the organic constituents of sewage, and (2) the tendency of processed (i.e., mineralized) sewage to support prolific growth of algae and blue-green bacteria in the receiving lake, stream, or estuary. Such overfeeding of a body of water with mineralized nutrients is termed *eutrophication*. The BOD contributed to a receiving body of water can exceed the rate at which oxygen enters that body of water from the atmosphere. The result is anaerobiosis, which kills aerobic forms of life, including fish. Furthermore, conditions need not become fully anaerobic in order to produce a serious ecological disturbance. Even slight decreases in availability of oxygen can cause a rapid loss of certain fish (trout, salmon, pike, and bass) from the affected waters. Depletion of oxygen also allows the growth of anaerobic bacteria, the metabolic products of which include foul-smelling and toxic compounds (e.g., short-chain fatty acids, amines, and $H_2S$).

The prolific growth of photosynthetic organisms resulting from eutrophication effectively reverses the mineralization step of sewage treatment. When these photosynthetic organisms die and are mineralized, anaerobiosis can result with consequences similar to that caused by untreated sewage. Eutrophication of lakes and streams presents a local problem, whereas eutrophication of estuaries presents a potential global problem because these coastal bodies of water are the primary regions for production of marine life upon which our increasing world population is dependent for a major portion of its supply of protein. The primary nutrient in mineralized sewage that causes eutrophication appears to be the phosphate ions. Although the growth of many species of photoplankton is limited by the concentration of fixed nitrogen, phosphate limits growth of the blue-green bacteria that are capable of nitrogen fixation; enrichment with phosphate thereby indirectly causes an enrichment with fixed nitrogen.

Enrichment of inland waters with phosphate comes largely from sewage. For example, Lake Erie, which has suffered serious eutrophication, is estimated to receive 72 percent of its phosphate from the sewage it receives. And the major source of phosphate in sewage is from phosphate-based detergents.

SEWAGE TREATMENT SYSTEMS. Despite the damage associated with eutrophication caused by effluents of sewage treatment plants, any treatment markedly decreases the potential damage of sewage to receiving bodies of water. The overall process of sewage treatment can be divided into three stages: (1) *primary treatment*, consisting of mechanical separation of solids; (2) *secondary treatment*, the aim of which is the complete mineralization of sewage; and (3) *tertiary treatment*, designed to remove mineralized products.

If a sewage plant consists only of primary treatment, the separated solids (termed *sludge*) are buried, burned, or otherwise disposed of. However, in addition to this process, secondary sewage treatment is now widespread in the United States.

If secondary treatment is employed, the sludge is subjected to *anaerobic digestion* in deep tanks in which considerable breakdown and solubilization of high molecular weight materials occur along with the formation of methane by anaerobic reduction of $CO_2$. Large sewage-treatment installations find it economical to trap this methane and use it as a source of energy for powering the operations of the plant. Mineralization of the liquid effluent produced by anaerobic digestion (the solid material, termed *digested sludge,* is disposed of on land or dumped at sea), along with the effluent from primary treatment, proceeds aerobically, usually in an installation called a *trickling filter,* in which the sewage is sprayed repetitively over stones in an open concrete enclosure. The stones become covered with a heavy layer of slime in which large numbers of aerobic bacteria are embedded. In operation, the trickling filter is constantly covered with a thin layer of liquid; the high concentration of organisms on the stones along with the free access of oxygen brings about rapid mineralization.

In addition to trickling filters, many other devices are employed in the secondary treatment of liquid sewage. These include shallow ponds (*lagoons*) in which the sewage is held, and tanks that are sparged with air. The latter is termed the *activated sludge process*. Regardless of the design of the installation, the purpose is the same: to obtain rapid oxidation of dissolved organic matter by exposing sewage to a dense population of microorganisms under aerobic conditions.

The biomass that results from these treatment systems (a biomass consisting of microorganisms and the slime in which they are embedded) is constantly released into the effluent of the secondary treatment system. It is allowed to settle, then it is separated and subjected to anaerobic digestion along with the sludge from the primary treatment facility.

Tertiary treatment procedures are only rarely employed, but their use will have to become common if the bodies of water receiving effluent from sewage treatment plants are to be protected from eutrophication. Tertiary treatment procedures can be either chemical or biological. Phosphates can be removed chemically by precipitation with lime, iron, or aluminum salts. Nitrates can be removed biologically by establishing conditions that favor *anaerobic respiration* and thereby the conversion of nitrate to nitrogen gas.

### the dissemination of synthetic organic chemicals

In recent decades the chemical industry has produced an enormous variety of synthetic organic chemicals that are being used on an ever-increasing scale as textiles, plastics, detergents, insecticides, herbicides, and fungicides. Some of these synthetics, such as textiles and plastics, are virtually completely resistant to microbial decomposition; in nature they tend to remain as permanent unsightly litter.

The use of insecticides, herbicides, and fungicides involves their application over considerable areas. Some of these products are also remarkably resistant to microbial decomposition. Representative examples of the persistence of pesticides in soil are shown in Table 14.2. In other environments, such as more anaerobic ones (for example, the bottoms of lakes), they persist for even longer periods. Many of these compounds are toxic for forms of life other than those they are designed to control, and the long-term ecological effects of their dissemination are difficult, if not impossible, to predict, but it is already clear that their accumulation in nature presents very real hazards to many species.

DDT: AN EXAMPLE OF ECOLOGICAL DAMAGE. The persistence of one pesticide, DDT, has resulted in considerable ecological damage that was wholly unanticipated. Tests had shown that concentrations that might reasonably be expected to accumulate in the environment were without toxicity for organisms other than insects, the intended target. However, two factors were overlooked. First, DDT is concentrated as it passes through a food chain (Table 14.3), because it dissolves in fatty tissues and is thereby immobilized within the organism. At the bottom of the food chain shown in Table 14.3, phyto-

**Table 14.2**
Persistence of Pesticides in Soil

| COMMON NAME | CHEMICAL FORMULA | PERIOD OF PERSISTENCE |
|---|---|---|
| INSECTICIDES | | |
| Aldrin | 1,2,3,4,10,10-hexachloro-1,4,4a,5,8,8a-hexahydro-endo-1,4-exo-5,8-dime-thanonapthalene | >9 years |
| Chlordane | 1,2,4,5,6,7,8,8a-octachloro-2,3,3a,4,7,7a-hexahydro-4,7-methanoindene | >12 years |
| DDT | 2,2-*bis*(*p*-chlorophenyl)-1,1,1-trichlorethane | 10 years |
| HCH | 1,2,3,4,5,6-hexachloro-cyclohexane | >11 years |
| HERBICIDES | | |
| Monuron | 3-(*p*-chlorophenyl)-1,1-dimethylurea | 3 years |
| Simazene | 2-chloro-4,6-*bis*(ethyl-amino)-s-triazine | 2 years |
| FUNGICIDES | | |
| PCP | Pentachlorophenol | >5 years |
| Zineb | Zinc ethylene-1,2,-*bis*-dithiocarbamate | >75 years |

plankton contains only low concentrations of DDT. Organisms that feed on phytoplankton contain enriched concentrations of DDT. At the top of the chain, DDT has been concentrated over 500,000-fold.

Second, DDT exhibits toxicity only at specific stages of the life cycle of certain vertebrates. For example, it inhibits secretion of $Ca^{2+}$, which is an essential constituent of the shell of birds' eggs. As a result, birds exposed to DDT (principally by feeding on contaminated prey)

**Table 14.3**
An Example of Concentration of DDT in a Food Chain

| | PPM* DDT RESIDUE |
|---|---|
| Water | 0.00005 |
| Plankton | 0.04 |
| Minnow | 0.94 |
| Pickerel (a predatory fish) | 1.33 |
| Herring gull (a scavenger) | 6.00 |
| Cormorant (feeds on larger fish) | 26.4 |

*Source:* From E. P. Odum, *Fundamentals of Ecology* 3rd ed. (Philadelphia: W. B. Saunders Company, 1971).
*Parts per million; i.e., μg/g.

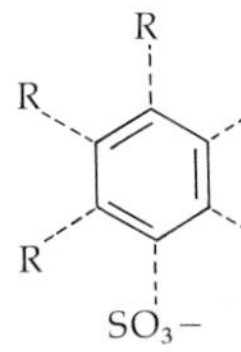

**Figure 14.5**
Generalized structure of alkylbenzene sulfonates. The alkyl R group
(either branched or linear) can be at
any of the positions indicated on the
dotted bonds.

lay eggs with very thin shells. The hatchability of such eggs is low. Thus, although DDT is not toxic to the adult form, it can eliminate the species. DDT exhibits a similar selective toxicity toward fish. It is without apparent effect on the adult, but often kills larvae at the time of yolk absorption.

BIODEGRADABILITY. It is now recognized as desirable that any synthetic organic compound widely disseminated in the natural environment should be susceptible to microbial decomposition. During the 1950s, alkylbenzene sulfonates (Figure 14.5) became major ingredients of household detergents. The side chains (R) of the compounds were branched aliphatic residues, rendering the entire molecule remarkably refractory to microbial decomposition. Accordingly, these products passed through sewage treatment plants largely unaltered, and subsequently contaminated supplies of potable water causing it to foam. During the early 1960s, the manufacturing process was altered in order to synthesize benzene sulfates with linear aliphatic side chains (linear alkylbenzene sulfonates); since these are highly susceptible to microbial decomposition, the problem was largely solved.

# 15

# SYMBIOSIS

Each group of organisms has had to adapt itself during evolution to its own nonliving environment, and to the other organisms by which it is surrounded. Adaptation to the environment sometimes involves the acquisition of special metabolic capacities that endow their possessor with the unique ability to occupy a particular physicochemical niche. The nitrifying bacteria, for example, can grow in a strictly inorganic environment with ammonia or nitrite as the oxidizable energy source; in the absence of light, no other living organisms are capable of developing in this particular environment, and the nitrifying bacteria are thus freed from biological competition. Withdrawl into a unique physicochemical niche is one means, and a highly effective one, of meeting the challenge of biological competition. A second method that has been adopted by large numbers of microorganisms has been to meet the challenge *by adapting to existence in continued close association with some other form of life.* This is the biological phenomenon known as *symbiosis.*

## types of symbioses

The symbiotic associations that microorganisms form with plants and animals, as well as with other microorganisms, vary widely in their degree of intimacy. In terms of the closeness of the association, symbioses may be roughly divided into two categories: *ectosymbioses* and *endosymbioses*. In ectosymbioses the microorganism remains external to the cells of its host;* in endosymbioses the microorganism grows within the cells

*The term *host* refers to the larger of two symbionts.

371

of its host. The distinction, however, is not always clear-cut; in lichens, for example, the fungal partner forms a projection that penetrates the cell wall, but not the cell membrane, of its algal partner.

Symbioses also differ with respect to the relative advantage accruing to each partner. In *mutualistic symbioses* both partners benefit from the association; in *parasitic symbioses* one partner benefits, but the second gains nothing and often suffers damage. It is sometimes difficult to determine whether a given symbiosis is mutualistic or parasitic. The degree to which each partner is benefitted or harmed can only be evaluated by comparing the fitness of the two members when living independently with their fitness when living in association. Furthermore, the nature of a particular symbiosis can shift under changing environmental conditions, so that a relationship that begins as mutualistic may become parasitic, or vice versa.

The fact that two organisms have evolved a symbiosis implies that at least one partner derives some advantage from the relationship. The extent to which one partner depends on symbiosis for its existence, however, varies considerably. At one extreme are the microorganisms that populate the *rhizosphere*—the region that includes the surface of the roots of plants together with the immediately surrounding soil. These microorganisms live successfully in other regions of the soil, but they attain higher cell densities in the rhizosphere, where they derive advantages from their proximity to the root hairs. At the other extreme are the obligate parasites, which cannot live outside their specific hosts.

Thus, symbioses vary with respect to the degree of intimacy (*ectosymbiosis* vs. *endosymbiosis*), the balance of advantage (*mutualism* vs. *parasitism*), and the extent of dependence (*facultative* vs. *obligate*). In the next few pages we describe some representative mutualistic symbioses of particular importance to man. Then, in the following two chapters (16 and 17), we will describe infectious disease, which constitutes the major form of parasitic symbiosis.

### the rhizosphere

The rhizosphere can be defined as the region extending a few millimeters from the surface of each root, in which the microbial population of the soil is influenced by the chemical activities of the plant. The major influence is a quantitative one: the numbers of bacteria in the rhizosphere usually exceed the numbers in the neighboring soil by a factor of 10 and often by a factor of several hundred.

There is also a qualitative influence. Short Gram-negative rods predominate in the rhizosphere, while Gram-positive rods and coccoid forms are less numerous in the rhizosphere than elsewhere in the soil. No specific association of a particular bacterial species with a particular plant has, however, been established.

The reason for the relative abundance of bacteria in the rhizosphere is, in most cases, the excretion by plant roots of organic nutrients, which favor certain nutritional types of bacteria. However, no clear-cut nutritional relationships of this type have been discovered, although many organic products excreted by plant roots have been identified. Our state of knowledge concerning the effects of the microbial population in the rhizosphere on the plant is even less satisfactory; despite numerous claims, it remains to be established that the plant benefits from the association. Many free-living soil bacteria, however, clearly perform functions essential for plants, such as nitrogen fixation and the mineralization of organic compounds. Recent evidence indicates that the bulk of nonsymbiotic nitrogen fixation occurs in the rhizosphere. It thus seems reasonable to assume that some plants do profit from the proximity of some microorganisms.

In one case a biochemical rationale for the symbiosis is emerging, as follows: rice is grown in flooded paddies, the soil of which is waterlogged and hence anaerobic immediately below the surface. As a consequence, large amounts of sulfide may be produced by sulfate reducers (see p. 261). Relatively large numbers of the filamentous sulfide-oxidizing organism *Beggiatoa* are found in the rice rhizosphere. Their presence insures that the sulfide concentrations, to which the rice seedlings are sensitive, will be low in the immediate vicinity of the plant. *Beggiatoa*, in turn, benefits from the excretion by the rice of a catalaselike activity (presumably an enzyme), which detoxifies $H_2O_2$. *Beggiatoa* is unusual among aerobic respiratory organisms in not possessing its own catalase. It is consequently very sensitive to $O_2$ tension, and in culture its growth can be markedly enhanced by the addition of purified catalase to the medium. Thus, this loose ectosymbiosis is one in which both partners are protected from specific toxic chemicals in their environment.

### mycorrhizas

The roots of most higher plants are infected by fungi. As in so many symbioses, a dynamic condition of mutual exploitation results, both partners benefitting as long as a balance between invasive and defensive forces is maintained. As a result of the infection, the plant root is structurally modified in a characteristic way. The composite root-fungus structure is called a *mycorrhiza*.

The formation of a mycorrhiza begins with the invasion of the plant root by a soil fungus; growth of the fungus toward the root is stimulated by the excretion into the soil of certain organic compounds by the plant. The fungal mycelium penetrates the root cells by means of projections called *haustoria*, and develops intracellularly. In some mycorrhizas, the fungus forms intracellular branching structures called *arbuscules* (Figure 15.1); in others it forms characteristic coils.

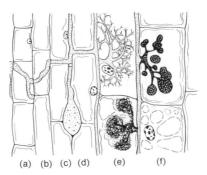

(a) (b) (c) (d)      (e)        (f)

**Figure 15.1**

Drawing showing the penetration of the root of *Allium* by a mycorrhizal fungus. In the first two cell layers (a and b) the fungal mycelium is intracellular. In the third and fourth layers (c and d) it is intercellular; a vesicular storage organ is shown between these layers. In the fifth and sixth layers (e and f) the fungus has formed intracellular branching structures (arbuscules). In (f) the arbuscules are undergoing digestion by the host cells. From F. H. Meyer, "Mycorrhiza and Other Plant Symbioses," in *Symbiosis*, Vol. I (S. M. Henry, editor). New York: Academic Press, 1966.

With few exceptions, mycorrhizas are not species-specific. A given fungus may be associated with any of several plant hosts, and in most cases a given plant may form mycorrhizas with any of a number of soil fungi. One species of pine tree, for example, has been found to associate with any of 40 different fungi. A great many free-living soil fungi are capable of forming mycorrhizas. In an experiment performed with pure cultures of free-living fungi and sterile plant roots, over 70 fungal species were found to form mycorrhizas; many times that number are undoubtedly capable of doing so in nature.

A typical mycorrhiza is shown in Figure 15.2. The stocky, club-shaped appearance results from several effects of the fungus on the root: cell volumes increase but root elongation is inhibited, and lateral root formation is stimulated by *auxins* (plant growth hormones) produced by the fungi.

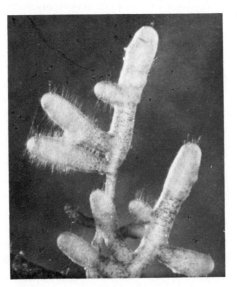

**Figure 15.2**

Mycorrhiza of *Fagus sylvatica*, showing the club-shaped apices of roots and hyphae radiating from the surface. From F. N. Meyer, "Mycorrhiza and Other Plant Symbioses," in *Symbiosis*, Vol. I, S. M. Henry (editor). New York: Academic Press, 1966.

The mutualistic nature of the mycorrhiza symbiosis can be readily demonstrated in many cases. The fungi that participate are characteristically those that are unable to use the complex polysaccharides that are the principal carbon sources for microorganisms in forest soils and humus. By invasion of plant roots, these fungi avail themselves of simple carbohydrates such as glucose. Moreover, the auxins excreted by the fungi induce an increased flow of carbohydrate from the leaves to the roots of the host plant, thereby increasing the nutrients available to the fungus.

The plant also benefits from the association. Many forest trees become stunted and die when deprived of mycorrhizal fungi. Stunted trees can be restored to health by the introduction of these fungi into the soil. The fungus seems to facilitate the absorption of water and minerals from the soil; the absorbing surface of the plant's root system is increased manyfold by the fungal hyphae. The function of a mycorrhiza as an absorbing organ has been confirmed by comparing the uptake of minerals from the soil by plants with and without mycorrhizas. Pines, for example, absorb two to three times more phosphorus, nitrogen, and potassium when mycorrhizas are present than when they are absent.

### root nodule bacteria and leguminous plants

It has long been known that the fertility of agricultural land can be maintained by a "rotation of crops." If a given plot is sown year after year with a grass, such as wheat or barley, its productivity declines; productivity can be restored by interrupting this annual cycle with a crop of some leguminous plant such as clover or alfalfa. Roman writers recognized that leguminous plants possess this ability to restore or maintain soil fertility, a property that is not exhibited by other types of plants. It was also known that the leguminous plants have peculiar nodular structures on their roots (Figure 15.3).

About the middle of the nineteenth century, total nitrogen analyses showed that when leguminous plants are grown on nitrogen-poor soil, there is a net increase in the amount of fixed nitrogen in the soil. Since the only possible source of this extra nitrogen is the atmosphere, such experiments suggested that leguminous plants, unlike other higher plants can fix atmospheric nitrogen. Hence, the growth of a crop of legumes on a nitrogen-poor soil increases the total fixed nitrogen content of the soil, particularly if the crop is plowed under. This is the chemical basis for the long-established practice of crop rotation.

Once these facts were established, the question naturally arose as to whether the peculiar nodulations on the roots of leguminous plants had any connection with the plants ability to fix nitrogen. Occasionally, leguminous plants fail to form nodules, and analyses showed that these plants do not fix nitrogen. When the contents of nodules were examined

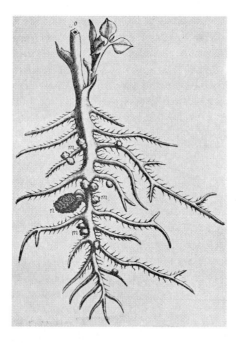

**Figure 15.3**
A seventeenth-century drawing by Malpighi of the root of a leguminous plant, showing the root nodules (m). The large dark object (n) is the coat of the seed from which the plant has developed.

microscopically, they were found to contain large numbers of "bacteroids": small, rod-shaped, or branched bodies similar in size and shape to bacteria (Figure 15.4). These facts suggested that the nitrogen-fixing ability of leguminous plants is a property of the bacteria that infect their roots, such infection leading to the formation of nodules. About 1885 the correctness of this hypothesis was established by showing that if the surface of seeds is sterilized chemically, the resulting plants will never form nodules, and their growth is strictly limited by the supply of combined nitrogen in the soil. Nodulation can be induced by adding crushed nodules from plants of the same species to the

**Figure 15.4**
A stained smear of the contents of a root nodule, showing bacteroids (×1,050). Courtesy of H. G. Thornton and the Rothamsted Experimental Station, United Kingdom.

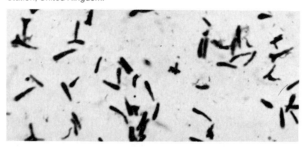

**Figure 15.5**
The effect of nodulation on plant growth. Two red clover plants grown in a medium deficient in combined nitrogen. The one at left, without nodules, shows very poor growth as a result of nitrogen deficiency. The plant at right, with nodules, shows normal growth. Courtesy of H. G. Thornton and the Rothamsted Experimental Station, United Kingdom.

soil. Once nodulation has occurred, the growth of the plants becomes independent of the supply of combined nitrogen (Figure 15.5). The final proof came in 1888, when M. W. Beijerinck succeeded in isolating and cultivating the bacteria present in the nodules and demonstrated that sterile seeds produced the characteristic nodules once more, when treated with pure cultures of these bacteria.

The agricultural importance of nitrogen fixation led to extensive work on the nodule bacteria. These organisms are Gram-negative motile rods that are classified in the genus *Rhizobium*. It was soon found that the nodule bacteria isolated from the roots of the various kinds of leguminous plants resemble one another closely in their morphological and cultural properties. When inoculated back into plants, however, they show a considerable degree of host specificity. The nodule bacteria isolated from the roots of lupines cannot evoke nodule formation on peas, and vice versa. In contrast, the nodule bacteria from peas, lentils, and broad beans can evoke nodulation in every member of this group of legumes. Thus, nodule bacteria can be classified into a series of cross-inoculation groups. Strains of one group have the same host range, which differs from strains of the other groups.

The nodule bacteria are normally present in soil. Their numbers are variable, depending on the nature of the soil and on its previous agricultural use. Hence, it often happens that a leguminous crop will develop poorly in a given plot of soil as a consequence of the fact that the nodule bacteria specific for it are either absent or present in such small numbers that effective nodulation does not occur. Nodulation can be ensured by inoculating the seed with a pure strain of nodule bacteria belonging to the correct inoculation group. Bacterial cultures of proved effectiveness were first made commercially available at the beginning of the twentieth century, and seed inoculation is now a routine agricultural operation. This is by far the most important contribution that the science of bacteriology has made to agricultural practice.

Whereas the soil under a nonleguminous crop, such as wheat, may have fewer than 10 *Rhizobium* cells per gram, the same soil will contain between $10^5$ and $10^7$ *Rhizobium* cells per gram following the development of a flourishing legume crop. The ability of legume plants to stimulate the growth of *Rhizobium* in the soil extends as far as 10 to 20 mm from the roots. The effect is highly specific: bacteria other than *Rhizobium* show little or no stimulation, and growth of the species of *Rhizobium* able to infect that particular leguminous plant is stimulated more than the growth of other species of *Rhizobium*. The substances responsible for this stimulation have not been identified. It has been experimentally established that the high number of *Rhizobium* cells in the rhizosphere of legumes represents stimulation of free-living cells rather than their liberation from nodules, by showing that the increase occurs in the absence of active nodulation.

The number of nodules formed on the roots of

the legumes is directly proportional to the density of *Rhizobium* in the soil, up to about $10^4$ cells per gram. Above this number, no further increase takes place, and nodule formation may even decline. When the number of *Rhizobium* cells is limited, so that fewer nodules are formed, the size of the nodules is proportionately larger. The result is that the total *volume* of nitrogen-fixing tissue remains fairly constant per acre of leguminous plants.

The symbiosis begins with the penetration of a root hair by a group of rhizobial cells, and involves the invagination of the root hair membrane. Thus a tube is formed containing bacteria, and is lined with cellulose produced by the host cell. This tube, termed the *infection thread* (Figure 15.6), penetrates the cortex of the root, passing through the cortical cells rather than between them.

As the thread passes through a cell, it may branch to produce vesicles that contain bacteria; the walls of the thread and vesicles are continuous with the host cell membrane. The bacteria are finally liberated into the cytoplasm of the host cell; electron micrographs of thin sections of legume roots show that the bacteria are enclosed, either singly or in small groups, in a membranous envelope (Figure 15.7).

Development of the nodule itself is initiated

**Figure 15.6**

(a) A newly infected root hair. The bacterial infection thread can be seen passing up a root hair, which has curled at the tip as a result of infection. Courtesy of H. G. Thornton and the Rothamsted Experimental Station, United Kingdom. (b) Infection thread crossing a central tissue cell of a nodule aged 1 to 2 days. From D. J. Goodchild and F. J. Bergersen, "Electron Microscopy of the Infection and Subsequent Development of Soybean Nodule Cells," *J. Bacteriol.* **92,** 204 (1966).

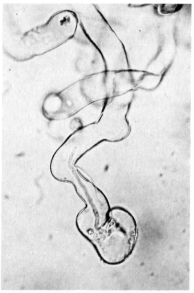

(a)

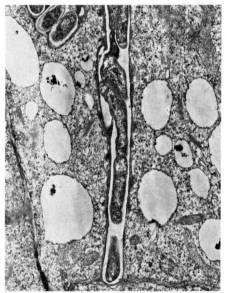

(b)

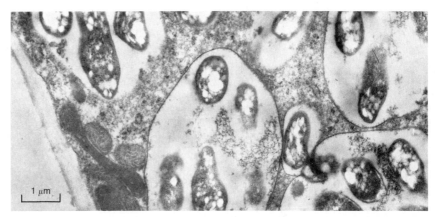

**Figure 15.7**
Mature nodule cell with large membrane envelopes containing four to six bacteroids. No further bacterial growth occurs. From D. J. Goodchild and F. J. Bergersen, "Electron Microscopy of the Infection and Subsequent Development of Soybean Nodule Cells," *J. Bacteriol.* **92,** 204 (1966).

when the infection thread reaches a tetraploid cell of the cortex. This cell, along with neighboring cells, is stimulated to divide repeatedly, forming the nodule (Figure 15.8). In young nodules, the bacteria occur mostly as rods but subsequently acquire irregular shapes, becoming branched, club-shaped, or spherical (the typical bacteroids). At the end of the period of plant growth, the bacteria have often disappeared completely from the nodules; they die, and their cell materials are absorbed by the host plant.

**Figure 15.8**
(a) Section of a root nodule. The dark cells are filled with bacteria. (b) Section of a nodule at high magnification, showing the individual bacteria in the infected cells. Courtesy of H. G. Thornton and the Rothamsted Experimental Station, United Kingdom.

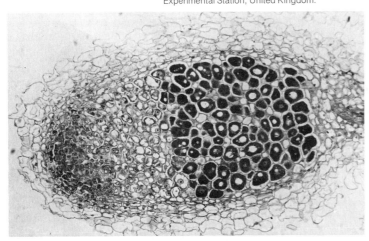

(a)

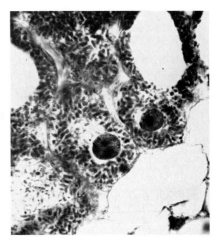

(b)

The definition of this symbiosis as mutualistic poses conceptual problems. The establishment of symbiosis bears a striking similarity to initial stages of infection in some plant diseases. In addition, the ultimate fate of most bacteroids is undoubtedly digestion by the host. Nevertheless, the nodulation of the plant stimulates the growth of free-living rhizobia, so that the rhizobial population as a whole increases. The plant benefits by receiving fixed nitrogen, although in some cases (if the infecting *Rhizobium* is of the wrong inoculation group), nodulation may occur without nitrogen fixation. In such cases the symbiosis would be clearly parasitic.

The adaptations of the two symbionts to each other are undoubtedly extensive. One of the most striking examples is the production (by the plant) of *leghemoglobin* in the nodules. This hemoglobin, produced in no other plant tissue, strongly binds $O_2$ and maintains a very low partial pressure of $O_2$ in the nodule. This protects the $O_2$-sensitive nitrogenase from inactivation while insuring an adequate supply of $O_2$ for the obligately aerobic rhizobia. Little or no $N_2$ fixation by *Rhizobium* occurs outside the nodule; in these bacteria nitrogenase is inactivated by $O_2$.

### the *Azolla-Anabaena* symbiosis

A symbiosis that is now being intensively investigated because it has the potential to increase nitrogen fixation in aquatic environments is that between *Azolla* (a small aquatic fern) and *Anabaena* (a heterocyst-forming blue-green bacterium). *Azolla*, which grows on the surfaces of still water in the tropical and temperate zones (Figure 15.9), has on the

**Figure 15.9**
*Azolla* frond from culture grown for several months on nutrient solution without combined $N_2$. From G. A. Peters and B. C. Mayne, "The Azolla, Anabaena Azollae Relationship. I. Initial Characterization of the Association," *Plant Physiol.* **53**, 813 (1974).

lower surfaces of its leaves mucilage-containing cavities which always contain the bacterial symbiont. During development, each new leaf becomes inoculated (Figure 15.10) most probably by hormogonia (short gliding trichomes) of the blue-green bacterium that migrate to the new leaf from cavities on older leaves. It is possible to separate the bacterium from the fern (Figure 15.11) and thereby demonstrate that the blue-green bacterium alone is responsible for the synthesis of nitrogenase. The association of *Azolla* and *Anabaena* appears to be mutualistic: the plant supplies the bacterium with a favorable exposure to light energy and probably with nutrients; the bacterium supplies the plant with fixed nitrogen.

Heavy surface growth of *Azolla-Anabaena* is capable of fixing $N_2$ equivalent in amount to that fixed by a legume-*Rhizobium* crop. This fixation has considerable agronomic importance as has been demonstrated by recent experiments and by the cultivation of *Azolla-Anabaena* for centuries as fodder and fertilizer in Southeast Asia, particularly in Viet Nam and in Thailand. A small rural industry has flourished there to provide live plants for the rice paddies. The expansion of these practices is probable, as the cost of nitrogenous fertilizers continues to increase.

### the intestinal flagellates of
### wood-eating termites and roaches

The woody tissue of trees, consisting mainly of cellulose and lignin, is unavailable as a source of food for most animals; in general, animals do not possess the enzymes necessary to degrade these polymers. Nevertheless, many species of insects obtain the bulk of their food from wood by virtue of an ectosymbiotic relationship with cellulose- and lignin-digesting microorganisms.

Both the termites and cockroaches, which have evolved from a common ancestral group, include some species that eat wood. All the wood-eating species of both groups harbor in their gut immense

**Figure 15.10**

*Azolla filiculoides* Lam. Longitudinal sections through the dorsal lobe of an immature (a) and a mature (b) leaf. Short trichomes of *Anabaena* are visible within the large cavity of both leaves. [(a) ×215; (b) ×80.] From Smith, *Cryptogamic Botany* **2,** 356 (1938).

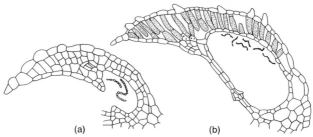

(a)                                     (b)

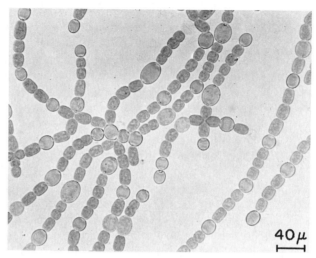

**Figure 15.11**
*Anabaena azollae* isolated from leaf cavity with micromanipulator. Note heterocysts. From G. A. Peters and B. C. Mayne, "The Azolla, Anabaena Azollae
Relationship. I. Initial Characterization of the Association," *Plant Physiol.* **53,**
813 (1974).

numbers of specialized flagellated protozoa. The flagellates are packed in a
solid mass within a saclike dilation of the hindgut; it has been reported that
they constitute over one-third of the body weight of the insect in some cases.
The flagellates are responsible for cellulose digestion, of which the insects
themselves are incapable. The flagellates, in turn, are themselves hosts to
intracellular bacteria, and it is possible that some— if not all— of the cellulases produced by the flagellates derive from their intracellular symbionts.
Nitrogen fixation also occurs in the termite gut, and it is assumed to reflect the
activity of nitrogen-fixing bacteria. Whether these occur free in the gut, or as
intracellular symbionts of the flagellates, is not known.

### the ruminant symbiosis

The ruminants are a group of herbivorous mam-
mals that includes cattle, sheep, goats, camels, and giraffes. Ruminants, like
other mammals, cannot make cellulases. They have evolved an ectosymbiosis
with microorganisms, however, that enables them to live on a diet in which
the major source of carbon is cellulose.

The digestive tract of a ruminant contains four
successive stomachs. The first two, known as the *rumen*, are essentially vast
incubation chambers teeming with bacteria and protozoa. In the cow the
rumen is a bag with a capacity of about 100 liters. The plant materials ingested
by the cow are mixed with a copious amount of saliva and then passed into the
rumen, where they are rapidly attacked by bacteria and protozoa. The total

microbial population of the rumen is of the same order as that of a heavy laboratory culture of bacteria ($10^{10}$ cells per milliliter). Many different microorganisms are present (Figure 15.12), and the full details of their biochemical activities are not yet understood. However, the net effect is clear: the cellulose and other complex carbohydrates present in the ingested fodder are broken down with the eventual formation of simple fatty acids (acetic, propionic, and butyric) and gases (carbon dioxide and methane). The fatty acids are absorbed through the wall of the rumen into the bloodstream, circulating in the blood to the various tissues of the body where they are respired. The cow gets rid of the gases formed in the rumen by belching at frequent intervals. The microbial population of the rumen grows rapidly, and the microbial cells pass out of the rumen with undigested plant material into the lower regions of the cow's digestive tract. The rumen itself produces no digestive enzymes, but the lower stomachs secrete proteases, and as the microbial cells from the rumen reach this region, they are destroyed and digested. The resulting nitrogenous compounds and vitamins are absorbed by the cow. For this reason, the nitro-

**Figure 15.12**

Some microorganisms from the rumen of the sheep. (a), (b), and (c) Bacteria (ultraviolet photomicrographs, ×732). From J. Smiles and M. J. Dobson, "Direct Ultraviolet and Ultraviolet Negative Phase-contrast Micrography of Bacteria from the Stomachs of the Sheep," *J. Roy. Micro. Soc.* **75**, 244 (1956). (d) Ciliate protozoa (×9). Courtesy of J. M. Eadie and A. E. Oxford.

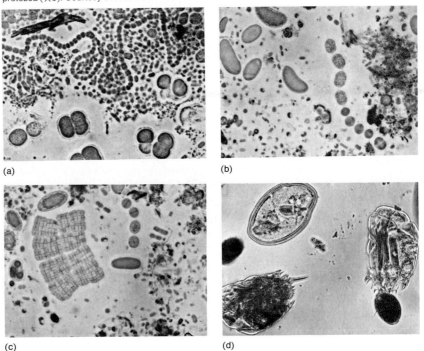

(a)

(b)

(c)

(d)

gen requirements of the cow and other ruminants are much simpler than those of other groups of mammals. Whereas man or the rat requires many amino acids preformed in the diet, the ruminant can grow on ammonia or urea. These simple nitrogenous compounds are converted to microbial proteins by the rumen population.

All the microbial processes that occur in the rumen are anaerobic ones. As the ruminant grazes, the rumen receives a steady flow of finely ground plant materials mixed with saliva. The plant materials consist chiefly of cellulose, pectin, and starch, together with some protein and lipid. The first stage in the process is the digestion of these biopolymers. A great deal of attention has been given to the identification of the microorganisms responsible for the digestion of cellulose, since this is the major digestive process in the rumen. Between 1 and 5 percent of the bacterial cells in the rumen are cellulolytic; they produce an extra-cellular cellulase that hydrolyzes cellulose to glucose and cellobiase.

The great bulk of the bacterial population, however, is noncellulolytic. These organisms rapidly utilize the glucose and cellobiose produced by the cellulolytic species; the great efficiency of the noncellulolytic bacteria in scavenging these molecules presumably accounts for their predominance over the cellulolytic forms. Furthermore, many of the rumen bacteria (including some of the cellulolytic species) are capable of digesting starch, pectin, proteins, and lipids. Indeed, only the lignin of the ingested plant material escapes digestion by the rumen flora.

The evolution of the rumen has involved both structural and functional modifications of the gastrointestinal tract. The principal structural modification is the development of a complex stomach, of which the largest compartments are essentially fermentation vats. The functional modifications that ruminants have undergone are even more profound. In the first place, the salivary glands do not secrete enzymes, the saliva being essentially a dilute salt solution (principally sodium bicarbonate and sodium phosphate) that provides a suitable mineral base for the microbes of the rumen. In the second place, the lower fatty acids have very largely replaced sugar as the primary energy-yielding substrate. This, in turn, has led to changes in the enzymatic makeup of nearly all the tissues in the body. Finally, the source of amino acids and vitamins has become very largely internalized (microorganisms instead of ingested food materials).

For the microorganisms that have taken up residence in the rumen, the situation also offers advantages: they are provided with an environment always rich in fermentable carbohydrates, well buffered by the saliva, and maintained at a constant favorable temperature. Individually, their ultimate fate is to fall prey to the proteolytic enzymes in the lower regions of the digestive tract; for the species, however, the rumen provides a safe and constant ecological niche.

The rumen association is in a delicately balanced equilibrium, easily disturbed by slight changes of the environment. The principal failure to which this symbiosis is liable is a mechanical one. The gas production in the rumen of a cow is some 60 to 80 liters per day, and since the total volume of the rumen is only 100 liters, steady belching is necessary to get rid of the accumulating gases. For reasons that are not fully understood, certain diets lead to foaming of the rumen contents, and when this happens, the belching mechanism of the cow fails to function properly. This causes a painful and, if untreated, eventually fatal affliction known as "bloat" (i.e., distention of the rumen by the trapped gases).

### the "normal flora" of the human body

Both the skin and the mucous membranes of the body are directly accessible to the external environment; soon after an infant is born these surfaces become populated by a characteristic flora. The skin becomes contaminated during passage through the birth canal; the mucous membranes may be sterile at birth but become contaminated within hours. Of the many microorganisms that reach these surfaces, only those that are particularly suited to growth in such environments become established; these constitute the "normal flora," which remains remarkably constant. The bacteria of the normal flora do not cause disease unless accidentally introduced into normally protected regions of the human body or as a result of physiological changes within the host. For example, a radical change in diet or infection with a virus may alter the conditions within the body in such a fashion that a member of the normal bacterial flora can become pathogenic.

Each region of the body provides a distinctive ecological niche that selects for the establishment of a characteristic flora. The bacteria that populate the skin, for example, include mainly corynebacteria, micrococci, streptococci, and mycobacteria. Moist regions of the skin harbor yeasts and other fungi. The skin flora is selected both by nutritional conditions and by antibacterial agents, such as the fatty acids secreted by the skin. The microorganisms of the skin flora are so firmly entrenched in the sweat glands, sebaceous glands, and hair follicles that no amount of bathing or scrubbing can totally remove them.

The throat and mouth support a variety of microorganisms, including representatives of most of the common eubacterial groups. Gram-positive cocci, including micrococci and streptococci, are common throat inhabitants. The Gram-negative cocci are represented by members of both the aerobic genus *Neisseria* and the anaerobic genus *Veillonella*. The mouth and throat also harbor large numbers of both Gram-positive and Gram-negative rods. The former include mainly lactobacilli and corynebacteria, while the latter include members of the genera *Bacteroides* and *Spirillum*. Spirochetes (*Treponema*), yeasts (*Candida*), and actinomycetes (*Acti-*

*nomyces*) are also common mouth inhabitants. All these organisms are normally harmless, but if the mucous membranes are injured, many can invade the body tissues and produce disease.

The same types of organisms inhabit the nasopharynx and are at least potentially able to reach the lungs. However, a series of protective mechanisms keep the trachea and bronchi relatively free of live bacteria. First, most of the bacteria adhere to the mucous lining of the nasopharynx and cannot readily move to the lungs because the ciliated epithelial cells that line the trachea constantly sweep the mucus upward. Second, the lungs are the site of extremely active phagocytosis, a mechanism whereby foreign particles are engulfed and destroyed by special amebalike cells.

The stomach and small intestine are unsuitable environments for most bacteria, but the large intestine harbors an extremely large resident flora. In the intestines the nature of the flora changes with diet and with age. Breast-fed infants, for example, have always a predominant intestinal population of *Bifidobacterium*, an organism not commonly found elsewhere. This organism disappears soon after weaning.

In adults the predominant bacteria of the intestinal tract are *Bacteroides* spp., *Escherichia coli*, and *Streptococcus faecalis*. Other genera, such as micrococci, are present in lesser numbers. The yeasts are represented by *Candida* and *Torulopsis*, and protozoa by several genera: *Balantidium*, a ciliate found only in man; *Entamoeba*; and flagellates of the genus *Trichomonas*. The proportions of the different microbial types are largely dependent on diet. Oral chemotherapy with antibiotics or sulfonamides can cause striking changes in the intestinal flora.

The principal bacteria of the vagina are the lactobacilli, which maintain a low pH as a result of their fermentative activity. This acidity is responsible for preventing the establishment of other forms. When the lactobacilli of the vagina are reduced in numbers during chemotherapy with antibacterial drugs, or during the use of oral contraceptives, vaginal infections by bacteria and by yeasts commonly occur.

In hospital practice today many serious infections are seen with bacteria that normally reside harmlessly in the host. Probably the most common source of such endogenous infections are the intestinal contents, which normally contain $10^{10}$ to $10^{11}$ bacteria per gram of feces. In persons with diseases that interfere with host defense mechanisms or in circumstances when drug treatment has the same effect, it is common for bacteria that reside normally in the bowel to invade the bloodstream. Examples of diseases that may interfere with host defense mechanisms include leukemia, which may produce a severe decrease in the number or kind of circulating leukocytes, and multiple myeloma, which is a disease of antibody-producing cells. Drugs like cortisone may also increase susceptibility to invasion by organisms of the normal intestinal flora.

The normal flora of the mammalian body thus provides another example of a symbiotic relationship that can shift from mutualism to parasitism and back again. In the absence of circumstances that would permit the organisms of the normal flora to invade the tissues of the host, such organisms benefit the host by preventing the establishment of virulent pathogens to which the host is often exposed.

### germ-free animals

By the use of complex equipment and elaborate techniques, it has been possible to deliver germ-free animals by caesarean section and to rear them in a germ-free environment. Adult animals in such an environment will mate and produce germ-free litters, so that colonies of germ-free animals can be maintained.

The chambers in which the animals are delivered and reared are equipped to prevent the entry of microorganisms, thus permitting experimentation on the role of the normal flora in the growth and development of the host, as well as on the resistance to infection.

The development of the germ-free animal is abnormal in several aspects. The cecum of the germ-free animal is greatly enlarged, the lymphatic system is poorly developed, and the germ-free animal makes much less immunoglobulin than does the normal animal.

Comparisons of normal and germ-free animals have not revealed any effects of the normal flora on the nutritional requirements of the host. However, striking effects have been observed on the *resistance or susceptibility of the host to infectious diseases*. For example, germ-free rats do not develop dental caries whereas control animals do. Cavities in the teeth appear, however, when the germ-free rats are infected with streptococci. A more complicated role of the normal flora in producing disease is revealed in the case of infection by the protozoan pathogen, *Entamoeba histolytica*, the agent of amebic dysentery. This phagotrophic organism cannot produce disease in germ-free guinea pigs, but it does so if the animals are first infected with bacteria (e.g., *Escherichia coli* or *Enterobacter aerogenes*) that serve as food source for the amebas. Thus, organisms typical of the normal flora potentiate the virulence of a pathogen, which, by itself, cannot survive in the intestine.

In contrast, the presence of the normal flora protects the host against some infectious diseases. For example, normal rats develop resistance to *Bacillus anthracis* spores at an early age, the number of spores in a dose required to kill 50 percent of the animals in a given experiment rising from $10^4$ to $10^9$ very soon after birth. Germ-free rats never develop this resistance, which can thus be attributed to the presence of the normal flora.

# MICROBIAL
# PATHOGENICITY

The term *pathogenicity* denotes the ability of a parasite to cause disease. Pathogenicity is a taxonomically significant attribute, being the property of a species; thus, the bacterial species *Corynebacterium diphtheriae* is said to be pathogenic for man. The individual strains of a bacterial species may, however, vary widely in their ability to harm the host species; this relative pathogenicity is termed *virulence*. Virulence is accordingly an attribute of a strain, not a species; one may speak of a highly virulent, a weakly virulent, or even an avirulent strain of C. *diphtheriae*.

In general, the virulence of a strain of a pathogenic species is determined by two factors: its *invasiveness*, or ability to proliferate in the body of the host, and its *toxigenicity*, or ability to produce chemical substances—*toxins*—that damage the tissues of the host. Certain pathogenic microorganisms, however, damage the host by a mechanism that is more indirect and does not come into play unless the host has previously experienced specific infection. This mechanism is known as *hypersensitivity*, or *allergy*, and involves an immune response to a cell component of the parasite.

The role played by invasiveness in damaging the host varies widely. Some pathogens are so toxigenic that an extremely localized infection may result in the production and diffusion through the host of sufficient toxin to cause death. The classical example of such a disease is diphtheria, in which *Corynebacterium diphtheriae* multiplies in the throat and produces a toxin that affects virtually all tissues of the body. At the other extreme are pathogens that must invade and multiply extensively to cause damage. The classical example of such a disease is anthrax, in which the pathogen, *Bacillus anthracis,* is present in the bloodstream in enormous numbers in the terminal stages of infection.

The transmission of microbial parasites from one host to another occurs by a number of routes: by contaminated food or drink, by airborne droplets, by direct contact with an infected individual, by animal bite, or by contamination of a wound. However, exposure to a pathogen is not necessarily harmful. The pathogen must become established in the host (a process called *infection*) for disease to result. The unbroken skin is a major barrier to infection. The mucous membranes, however, are readily penetrated by virulent microorganisms; thus, most microbial diseases (other than those transmitted by animal bites or wound contamination) begin as infections of the mucosal membranes of the respiratory, intestinal, or genitourinary tracts.

The ability of a pathogen to develop in its host depends on the balance between the growth-promoting and growth-inhibiting factors that the host environment provides. The growth-promoting factors are the nutrients required by the microorganism; the growth-inhibiting factors include a variety of cellular and chemical *host defense mechanisms*. Variations in these factors among animal species, and among the tissues of a given species, determine the host range and tissue specificities of each pathogenic microorganism.

There are three general modes of infection among the various pathogens: (1) some multiply only on the surface of the mucosal epithelium; (2) some penetrate and multiply in the epithelial cells; and (3) others pass through the epithelium, entering the deeper tissues of the body and the circulatory system.

## microbial toxins

Certain toxins cause highly specific types of tissue damage in the host. For example, when a wound is infected by *Clostridium tetani*, a toxin is elaborated in the localized region of growth of the pathogen at the site of infection; the toxin then diffuses through cells to the bloodstream which carries it to the central nervous system where it interferes with the normal transmission of nerve impulses. The action of tetanus toxin is so highly specific that none of the other tissues of the body are affected. However, other toxins shown no tissue specificity. For example, *Corynbacterium diptheria* toxin, which is also elaborated as a consequence of a localized infection, affects virtually every tissue of the body.

Another consequence of infection is the nonspecific *host* response, termed *inflammation*, which will be described later in this chapter. The nonspecific character of the inflammatory response makes it difficult to discover whether the microbial product that initiates the response is a true toxin or an *allergen*—that is, a product of the microorganism that induces the inflammatory allergic response.

## the nature of toxins

Most of our knowledge of microbial toxins has come from work on bacteria. The search for bacterial toxins began shortly after the discovery of the role of bacteria as etiological agents of human disease. By 1890 the toxins of two important human pathogens, *Corynebacterium diphtheriae* and *Clostridium tetani*, had been discovered. In each case, the discovery was made in the same manner: a sterile filtrate of the bacterial culture was observed to cause death when injected into experimental animals. Autopsies revealed that these animals showed the characteristic lesions associated with the natural infection. These toxic substances are now known to be proteins. Because they are present in the medium, and not associated with the bacterial cells, they are termed *exotoxins*.

A number of other pathogenic bacteria have been subsequently shown by comparable methods to produce exotoxins. However, filtrates prepared from cultures of many other pathogens have failed to show toxicity. This led to the discovery that the *cells of nearly all Gram-negative pathogenic bacteria are intrinsically toxic;* furthermore, heat-killed cells of many *nonpathogenic* all Gram-negative bacteria show similar toxic effects. The heat-stable toxins associated with the cells of Gram-negative bacteria came to be known as *endotoxins*. As we shall describe later, the endotoxins all produce similar symptoms when injected into experimental animals. Many years of intensive study have been required to reveal their nature; it is now known that *endotoxins are lipopolysaccharide-protein complexes, derived from the outer membrane of Gram-negative bacteria.* The lipopolysaccharide component of the complex is responsible for the toxicity of the endotoxins.

The names *exotoxin* and *endotoxin* can be misleading, since there is now good evidence to show that many exotoxins are associated with the bacterial cells during growth and are liberated only after lysis of the bacteria. Exotoxins can, however, be distinguished from endotoxins chemically. The former are simple proteins, whereas the latter are molecular complexes of protein, lipid, and polysaccharide. Nevertheless, these names are now so firmly entrenched that they are not likely to be abandoned.

In laboratory culture, many important bacterial pathogens, including the causative agents of anthrax and plague, do not produce any significant toxic product. The conditions in a laboratory culture are always different from the body of an infected animal, and the recognition of this obvious but previously overlooked fact led to *the search for bacterial toxins that are produced by pathogens only in the animal body.* Such work, conducted mainly by H. Smith and his collaborators, was centered originally on anthrax and led to the discovery of the specific exotoxin produced by *Bacillus anthracis.* Later, the toxin of *Yersinia pestis*, the agent of plague, was demonstrated for the first time by comparable experiments.

The toxins of both the anthrax and plague organisms were found to be complexes of two or more substances, each of which was nontoxic by itself but which acted together to produce a toxic effect. Such knowledge permitted the refinement of the assay systems for the toxins to such an extent that—in each case—it became possible to establish the production of toxin by cultures of bacteria in vitro.

The failure to discover the toxin of a virulent organism may often result from the lack of a suitable assay system, particularly in the case of organisms specifically pathogenic for man. For example, the toxin of *Vibrio cholera*, the agent of cholera, escaped detection for many years; neither culture filtrates nor cell extracts exhibited toxicity when injected into experimental animals. The toxin was discovered, however, when filtrates of cultures of *V. cholerae* were injected into artificially closed sections of rabbit intestine. Such filtrates produced the effects characteristic of the natural disease. Using this assay system, it was possible to purify the cholera toxin and characterize it as a heat-labile protein. The same assay system has permitted the detection and isolation of the toxins of virulent strains of *Escherichia coli*.

### bacterial exotoxins

Many virulent bacteria liberate toxic substances into the medium when grown in vitro. Some of these appear to be directly involved in pathogenesis; they are listed in Table 16.1. Others seem unrelated

**Table 16.1**
Some Ecologically Significant Exotoxins

| ORGANISM | EXOTOXIN | TARGET TISSUE | MODE OF ACTION |
|---|---|---|---|
| *Clostridium botulinum* | Neurotoxin | Myoneural junctions | Inhibits release of acetylcholine |
| *Clostridium tetani* | Neurotoxin (tetanospasmin) | Central nervous system | Suppresses synaptic inhibition |
| *Clostridium perfringens* | α-toxin | General (local wound tissue) | Lecithinase activity (cytolysis) |
| *Corynebacterium diphitheriae* | Diphtheria toxin | General (disseminated) | Inhibits protein synthesis |
| *Staphylococcus aureus* | α-toxin | General | Cytolytic |
| | "Enterotoxin" | Nerve cells | Unknown |
| *Shigella dysenteriae* | Enterotoxin | Intestinal epithelium | Interferes with regulation of electrolyte transport |
| *Vibrio cholerae* | Enterotoxin | Intestinal epithelium | Interferes with regulation of electrolyte transport |
| *Yersinia pestis* | "Guinea pig toxin" | General | Unknown |

to the disease produced by the pathogen. *Streptococcus*, *Staphylococcus*, and *Clostridium*, for example, produce numerous cytolytic toxins when grown in vitro. These toxins bring about the lysis of a variety of types of mammalian cells.* The extent, if any, to which the cytolytic toxins contribute to the disease process is unclear.

THE PRODUCTION AND LIBERATION OF EXOTOXINS. The toxins produced by Gram-positive bacteria are liberated into the medium during logarithmic growth; they are not found in the cytoplasm of the bacteria. The mechanism by which the toxins traverse the cell membrane is unknown. The extracellular toxins of the Gram-negative bacteria, however, are not actively excreted into the medium; they accumulate intracellularly and are liberated when the cells lyse.

The function of toxins is not apparent; many nontoxigenic strains develop just as well as their toxigenic counterparts in the host. Indeed, the genes that determine bacterial exotoxins have been found in many cases to be carried by plasmids or prophages. The diphtheria toxin, the erythrogenic toxin of *Streptococcus pyogenes*, enterotoxin and the α-toxin of *Staphylococcus aureus*, and the toxin of one group of *Clostridium botulinum* strains have all been found to be determined by bacteriophage genes, while several toxins produced by strains of *Escherichia coli* have been found to be determined by plasmid genes. In each of these cases, the loss of the prophage or plasmid renders the cell nontoxigenic, and toxin production is regained when the prophage or plasmid is reintroduced into the cell.

Sometimes toxins (for example that of *C. botulinum*) are produced in a nontoxic form; proteolytic cleavage by host digestive enzymes is required to produce the active form.

THE MODES OF ACTION OF EXOTOXINS. Some exotoxins have been shown to act as *hydrolytic enzymes*, degrading essential components of host cells or tissues. The toxin of *Clostridium perfringens*, for example, hydrolyzes important lipid constituents of cell membranes, resulting in lysis. This could be the primary cause of tissue damage in gas gangrene, although *Clostridium perfringens* produces a number of other exotoxins, including one that attacks collagen, the substance that cements animal cells together.

The diphtheria toxin is an example of a toxin that acts by interfering with a cell function rather than by destroying a cell component; it inhibits the chain elongation step in protein synthesis. It is another toxin that requires proteolytic cleavage for toxic activity.

A number of pathogens produce toxins that act specifically on the epithelial cells of the intestine; these are termed *enterotoxins*. They include *Vibrio cholerae*, *E. coli*, *C. perfringens*, and *Shigella dysenteriae*.

Enterotoxins act by binding to specific receptors

---

*Those that lyse red blood cells are termed *hemolysins*.

on the membranes of sensitive cells. The bound toxin stimulates a sharp rise in the intracellular concentration of cyclic AMP (see p. 117), which causes an abnormally high rate of electrolyte transport and $H_2O$ loss from tissues into the intestine. Death often occurs from dehydration, unless the fluid and electrolytes lost from the circulation are replaced.

### bacterial endotoxins

Endotoxins have been isolated from all pathogenic Gram-negative bacteria; the best known are those from the enteric bacterial genera *Salmonella, Shigella,* and *Escherichia.* These endotoxins exhibit two distinct effects: *pyrogenicity* (fever production) and *inflammation* (see p. 399). Both activities are due to the lipopolysaccharide component of the outer membrane. The isolated lipopolysaccharides are highly active; a microgram will produce fever in a horse.

The body temperature of mammals is controlled by certain centers in the brain. Endotoxins do not act directly on these centers; rather they cause the release of an *endogenous pyrogen* from polymorphonuclear leucocytes (a particular type of white blood cell).

The purified endotoxins of virulent, as well as of avirulent, enteric bacteria cause many of the symptoms of disease when injected into animals. The inflammatory and pyrogenic effects undoubtedly contribute to the general pathology of the infection; however, they are not responsible for the specific symptoms caused by the Gram-negative pathogens. It is important to emphasize that fever accompanies severe infections from Gram-positive organisms that do not contain endotoxins.

The endotoxins of enteric bacteria cause transient resistance to infection when injected into animals in minute doses. Figure 16.1 shows what happens when mice are injected with a purified lipopolysaccharide and then challenged at a later time by injection with a virulent strain of *E. coli*. Immediately after the administration of the lipopolysaccha-

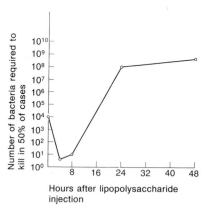

**Figure 16.1**
The effect of endotoxin lipopolysaccharide on resistance to infection (see text for explanation). Replotted from data of O. Westphal.

ride, the mice become quite susceptible to infection. Within 24 hours, however, their susceptibility to infection diminishes markedly. The state of increased resistance lasts for 3 to 5 days and then disappears. Purified lipopolysaccharide prepared from one species of Gram-negative bacterium confers resistance to infection by other species of Gram-negative bacteria. The mechanism of this effect is not known; however, it may play an important role in nature, because all mammals are constantly exposed to the endotoxins of enteric bacteria.

## infections of the mucosal epithelium

As stated earlier, almost all microbial infections begin at the mucosal membranes. These infections can be divided into two groups: those that progress to the subepithelial tissues of the body and those that remain localized at the surface. Even in the latter case, the toxins may diffuse throughout the body and produce lesions in other tissues.

### attachment without penetration

The epithelia of the respiratory tract, the intestinal tract, and the vagina carry a dense coat of microorganisms [the "normal flora" (see Chapter 15)]. The normal flora constitutes a barrier to the establishment of invading pathogens, by competing for attachment sites and nutrients and by producing inhibitory substances such as organic acids. The mucous membranes themselves secrete antimicrobial substances, including long-chain fatty acids and lysozyme. In the respiratory tract the epithelial cells beat their cilia constantly and rhythmically, sweeping the film of secreted mucus and loosely adhering microorganisms toward the outer portals of the body. All these defense mechanisms must be overcome in order for a pathogen to establish itself on the mucosal epithelium.

Some examples of pathogens that remain localized on the epithelial surface are *Bordetella pertussis* and *Corynebacterium diphtheriae*, which infect the respiratory tract of mammals, and *Vibrio cholerae*, which infects the intestinal tract. *B. pertussis* is the agent of whooping cough; it produces an exotoxin that acts locally to produce tissue damage and inflammation. The exotoxin of *C. diphtheriae* diffuses through the submucosa and is disseminated throughout the body by the bloodstream, causing damage to many tissues. The exotoxin of *V. cholerae* acts locally on the intestinal epithelial cells, as described earlier in this chapter.

### penetration of the epithelial cells

*Shigella dysenteriae* is an example of a pathogen that enters the epithelial cells, causing an erosion and ulceration of the mucosal epithelium of the intestinal tract. The mechanism of penetration appears to involve an extracellular bacterial product that induces a degeneration of the epithelial cells; the bacteria then enter the cells, ultimately killing them.

# infections of the subepithelial tissues

Once past the mucosal epithelium, the invading microorganism encounters several new and effective host defense mechanisms, as well as encountering a considerable variation from tissue to tissue in the biochemical milieu. The invasiveness of the microorganism depends on its ability to multiply in the environment provided by the host tissues and fluids, and on its ability to overcome the host defenses.

### the biochemical milieu

During their evolution, these invasive microorganisms have adapted in order to grow in specific tissues of the host. This high degree of tissue specificity exhibited by many of these organisms may be inferred to reflect differences between tissues. To date, however, only one such biochemical difference has been clearly documented. Erythritol* is the preferred carbon source of several species of *Brucella,* which infect placental tissue and hence cause abortion in ruminants; it is found in high concentrations in the placenta of these animals, but not in other tissues or in the placenta of various species (man, rat, rabbit, guinea pig) resistant to placental infection by these organisms.

In some infectious diseases, notably tuberculosis, the limiting factor for microbial growth is the *availability of free iron.* Both the vertebrate host and the microbial parasite depend on chelating agents, which they excrete, for the transport of iron into their cells. Mammals liberate iron-chelating proteins called *transferrins* into the bloodstream. Bacteria excrete a variety of iron chelators of low molecular weight; e.g., tubercle bacilli excrete a chelator called *mycobactin.*

The interactions between chelators and cells are highly specific. Thus, mammalian cells, but not tubercle bacteria, can utilize transferrin-bound iron, but the opposite is true for mycobactin-bound iron. The result is competition in which the outcome depends on the relative binding strengths and concentrations of the chelating agents excreted by host and parasite. Injection of iron into an infected host favors the parasite, while the infection of agents that depress iron levels protects the host against infection.

The effect of iron on the course of some microbial infections may be complicated by its influence on toxigenicity: high concentrations of available iron depress exotoxin production by such pathogens as *Corynebacterium diphtheriae* and *Clostridium tetani,* while favoring their invasiveness.

---

*Erythritol is a 5-carbon sugar alcohol.

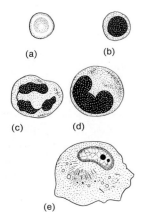

(a)    (b)

(c)    (d)

(e)

**Figure 16.2**
Some types of blood cells: (a) an erythrocyte; (b) a small lymphocyte; (c) a granulocyte; (d) a monocyte; (e) a macrophage.

# constitutive host defenses

Vertebrates have evolved numerous mechanisms that keep microbial parasites in check. Some of these mechanisms are *constitutive:* they are properties of the normal host. Others are *inducible:* they appear only in hosts that have undergone an induction process such as prior exposure to the infectious organism or its products.

The principal constitutive defenses encountered by microorganisms that penetrate beyond the epithelial layers are: the *phagocytic cells,* which engulf and digest foreign particulate matter; certain *antimicrobial substances* in the tissues and circulating fluids; and a complex of reactions known as *inflammation.*

## the phagocytic cells

Phagocytic cells are found both in the circulatory system and in the tissues. The white blood cells, or leucocytes, are described in Table 16.2 and Figure 16.2. They include two types that are actively phagocytic: *monocytes* and *granulocytes.* * Granulocytes owe their name to the presence of granules in their cytoplasm; they are further differentiated, on the basis of the staining properties of the granules, into *eosinophils* (granules stainable by the acid dye eosin), *basophils* (granules stainable by the basic dye methylene blue), and *neutrophils* (granules stainable by a mixture of acidic and basic dyes). All the granular leucocytes are capable of phagocytosis, but only the neutrophils are sufficiently active to play a significant role in host defense.

Infected tissue releases substances that attract both neutrophils and monocytes. During this process, which we shall describe in more detail in the section on inflammation, the monocytes enlarge and develop into cells called *macrophages,* which have enhanced phagocytic activity. Macrophages also occur as fixed cells in certain organs: the liver, spleen, bone marrow, and lymph nodes. Since they are for the most part

*Granulocytes are also known as *polymorphonuclear leucocytes,* because of the irregular shape of their nuclei.

**Table 16.2**
**The Leucocytes**

| CELL TYPE | LOCATION AND DEVELOPMENT | PRINCIPAL FUNCTION IN DEFENSE |
|---|---|---|
| Monocytes | Circulate in the bloodstream; move by chemotaxis into inflamed tissues; develop into tissue macrophages | Phagocytosis |
| Granulocytes | Circulate in the bloodstream; move by chemotaxis into inflamed tissues | Phagocytosis (principally in the neutrophilic granulocytes) |
| Fixed macrophages | Lining the sinuses of the liver, spleen, bone marrow, and lymph nodes (the reticuloendothelial system) | Phagocytosis |
| Lymphocytes | Circulate in the bloodstream; move by chemotaxis into inflamed tissues. Antigenic stimulation of one type (B cells) causes enlargement and development into plasma cells | Secretion of antibody (plasma cells); cellular immune functions (T cells) (see Table 16.3) |

reticular (forming a supporting network) or endothelial (lining the cavities) in these organs, the fixed macrophages are collectively referred to as the *reticulo-endothelial system*. This system constitutes a major defense in the animal body, actively filtering out microorganisms that have penetrated outer barriers.

The process of phagocytosis occurs in two steps. In the first step the microorganism becomes *attached to the membrane of the macrophage*. Attachment then triggers the ingestion process: pseudopods extend around the microorganism, fusing with one another so as to enclose it within a membrane-limited vacuole (Figure 16.3). Ingestion requires meta-

**Figure 16.3**

The ingestion of pneumococci by a phagocyte. (a) Two pneumococci are in contact with a pseudopodium of the phagocyte. (b) The same two pneumococci are inside the phagocyte, one on each side of a lobe of the nucleus; a group of four pneumococci are in the process of being ingested. (c) Six of the eight pneumococci have been engulfed. (d), (e), (f) Schematic diagram of the stages in phagocytosis, showing the formation of the phagocytic vacuole by an inversion of the cell membrane. (a), (b) and (c), from W. B. Wood, Jr., M. R. Smith, and B. Watson, "Studies on the Mechanism of Recovery in Pneumococcal Pneumonia: IV. The Mechanism of Phagocytosis in the Absence of Antibody," *J. Exptl. Med.* **84,** 402 (1946).

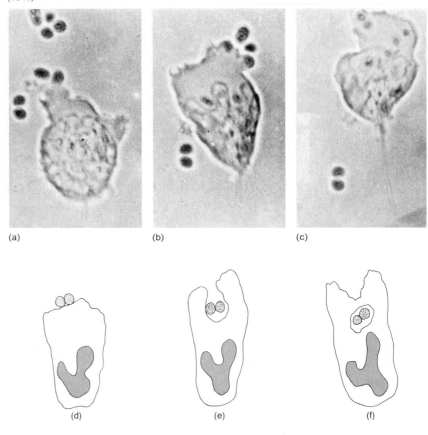

(a)            (b)            (c)

(d)            (e)            (f)

bolic energy, most of which is derived from glycolysis. The lactic acid produced diminishes the pH in the vacuole to 5.5 or below.

Phagocytic vacuoles fuse with lysosomes, and the enclosed bacteria are killed and digested through the action of a mixture of enzymes. A variety of lysosomal hydrolases have low pH optima; thus their activity is stimulated by the low pH of the phagocytic vacuole.

Many pathogens owe their invasiveness, and thus their virulence, to their ability to resist phagocytosis and intracellular destruction. In some cases, this resistance reflects the excretion by the microorganism of substances that block the phagocytic process. Some of these substances inhibit the chemotactic response of blood phagocytes; some (notably the capsule) block attachment and ingestion; and some inhibit the intracellular destruction of phagocytized cells. In addition, many pathogens excrete substances, called *leucocidins,* which kill phagocytes.

### antimicrobial substances in body tissues and fluids

Phagocytes contain a variety of antimicrobial substances. One of these, *lysozyme,* is also present in many other parts of the body: in tears, nasal secretions, saliva, mucus, and various organs including skin. These tissues have been found to contain other antimicrobial substances as well, including basic polypeptides and polyamines.

A different group of antimicrobial substances, called *beta lysins,* occur in the serum. Beta lysins are heat-stable substances of unknown composition, which are formed when the blood clots. They are active against a number of Gram-positive bacteria.

### inflammation

When a tissue of a higher animal is subjected to any of a variety of irritations, it becomes "inflamed." The characteristics of inflammation—reddening, swelling, heat, pain—are familiar. Although the causes of the heat and pain are not well understood, the reddening and swelling are. The reddening is due to dilation of the capillaries, which allows increased blood flow. As this dilation is accompanied by increased porosity of the capillary walls, fluid moves into the tissue, causing swelling.

The fact that inflammation is produced by such widely different irritants as heat, mechanical injury, and microbial infection suggests that the symptoms are caused by a substance derived from the host. Among the many compounds that have been extracted from cells or serum and shown to produce inflammation, the most thoroughly studied are the small molecules *histamine* and *serotonin.* Both compounds are released in response to a variety of stimuli.

As the inflammatory response develops, a striking change occurs in the behavior of the granulocytes. At first, they adhere to the inner walls of the capillaries. Next, they push their way between the cells of the capillary walls and enter the tissues, a process that can take as little as 2 minutes [Figure 16.4(a–d)]. If the inflammation has been initiated by a bacterial infection, the granulocytes move toward the focus of infection in response to substances liberated by the bacteria. As the inflammation progresses, the phagocytes release lysosomal enzymes that damage and eventually destroy neighboring tissue cells.

In the later stages of inflammation the granulocytes that have accumulated at the inflammatory site are replaced by monocytes, which respond chemotactically to the same substances as do the granulocytes that they replaced.

Lymphocytes, the antibody-forming cells, also leave the capillaries and accumulate at the site of injury. They apparently pass *through* rather than between the endothelial cells [Figure 16.4(f–i)].

Inflammation acts as a mechanism of defense, although it is itself a cause of tissue damage. First, the tissues at the site of infection become enriched with phagocytes. Second, the supply of plasma to the tissues is increased, raising the local concentration of antibacterial substances and (in immune animals) of antibodies. Third, progressive inflammation leads to the accumulation of dead host cells, from which additional antimicrobial substances are released. In the center of the necrotic area, oxygen tension is diminished and lactic acid accumulates; these conditions also inhibit the growth of many types of pathogenic bacteria. Finally, the higher temperatures characteristic of fever slow the multiplication of some viruses.

## inducible host defenses: the immune response

In both epithelial and subepithelial infections, prolonged contact of the vertebrate host with the microbial parasite leads to the induction of new and specific defense reactions collectively referred to as the *immune response*. There are two distinct classes of immune response: the production of circulating *antibodies*, and the production of specifically *sensitized cells*. The immune response usually reduces the toxigenicity and invasiveness of the parasite to the point where the constitutive host defenses can eliminate it. In some cases, however, the pathogen possesses factors that interfere with the immune response. In some infectious diseases the immune response itself seriously damages host tissues by the reaction known as *hypersensitivity;* being hypersensitive to a particular agent constitutes an *allergy.*

### antibody-mediated immunity

The term *antibody* refers to a group of related proteins that are capable of specific, noncovalent binding to the substances

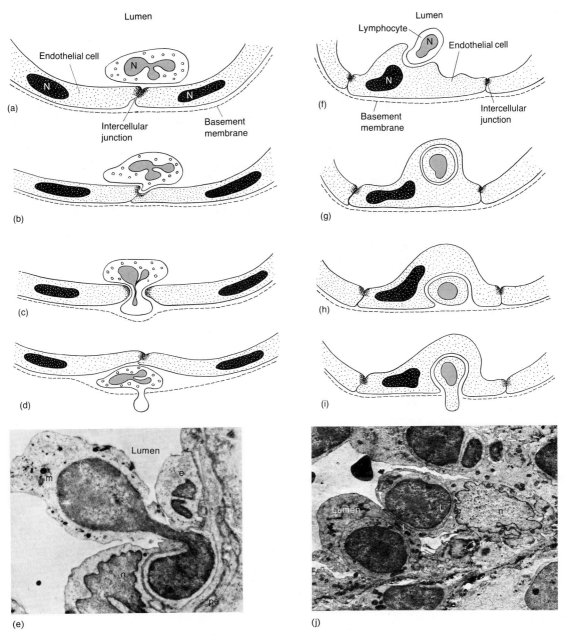

**Figure 16.4**

(a to d) Stages in the migration of a granulocyte through the venule wall. The cell penetrates an intercellular junction and remains extracellular at all times. (e) Part of an inflamed venule. Cell m is a monocyte, which is penetrating an intercellular junction by the same mechanism; e, endothelium; n, nucleus of an endothelial cell; pe, periendothelial sheath. (f to i) Stages in the migration of a small lymphocyte through an endothelial cell of a venule. The lymphocyte is totally intracellular at one stage in the passage. (j) Part of a venule from a normal lymph node; a lymphocyte (l) is completely enclosed by the cytoplasm of an endothelial cell; n, nucleus of the endothelial cell. From V. T. Marchesi and J. L. Gowans, "The Migration of Lymphocytes through the Endothelium of Venules in Lymph Nodes: An Electron Microscope Study," *Proc. Royal Soc. B* **159,** 283 (1964).

that induce their formation. Such substances are called *antigens;* practically all proteins are antigens, as are many polysaccharides. In addition, many compounds that are not intrinsically antigenic become so when attached to proteins; this class includes lipids, nucleic acids, and numerous small molecules.

There are several classes of antibodies, or *immunoglobulins,* but they all have certain common structural features. The basic unit of each antibody molecule is a complex of four polypeptide chains, two identical heavy chains and two identical light chains, folded in such a manner as to produce on its surface two identical regions, called *Fab,* which contain the *antigen-binding sites.* There is also a third region, called *Fc,* which gives the antibody additional specificity, such as the ability to bind to the membranes of certain cells (Figure 16.5).

The antigen-combining sites are shallow cavities in the molecule that fit, as a lock fits a key, complementary projections on the surfaces of antigens. Such a projection is called an *antigenic determinant.*

An individual vertebrate is capable of forming antibodies against many thousands of different antigens; each of these has a

**Figure 16.5**

Schematic diagram of an immunoglobulin. The molecule consists of two heavy polypeptide chains and two light polypeptide chains, held together by disulfide bridges as shown. The molecule can be hydrolysed with papain, which cleaves it at the site indicated by the arrows to produce three fragments: two identical fragments, called Fab, each of which contains one antigen-binding site, and one crystallizable fragment, called Fc, which contains sites that determine the binding of antibody to specific host cells.

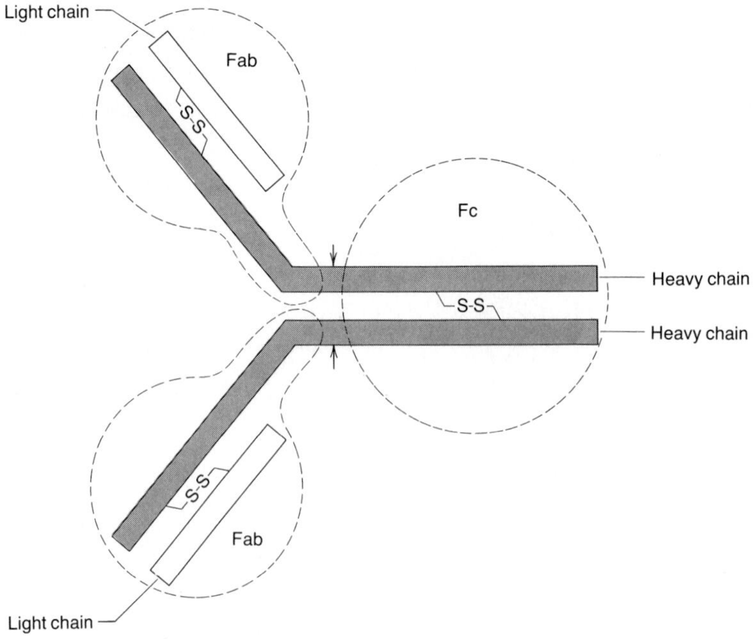

unique amino acid sequence. It is known from molecular genetics that differences in amino acid sequence reflect differences in the nucleotide sequence of DNA. The inescapable conclusion is that the genome of the vertebrate animal contains many thousands of genes coding for different immunoglobulin polypeptides, each determining a different antigen-binding specificity, or, alternatively, that the genetic differences arise by mutation or recombination during the development of the antibody-producing cells of the individual.

THE PRODUCTION OF ANTIBODY. Antibodies are produced by a class of white blood cells called *lymphocytes*. There are two types of lymphocytes, which play very different roles in the immune response: "B cells," which actively secrete antibody, and "T cells," which mediate cellular immunity. The formation of antibodies against some antigens requires interaction between B cells, T cells, and macrophages.

B cells are each capable of producing a single set of immunoglobulins against one antigen. Limited cell division (mostly prenatal) gives rise to clones of B cells capable of producing specific antibodies. Although these antibodies coat the cell surface, they are not actively produced prior to exposure to the antigen. Contact between an antibody and its cognate antigen on the surface of the B cells stimulates proliferation of the B cells capable of producing that antibody. This process, termed *clonal selection*, leads to a massively increased population of antibody-producing cells.

When infection subsides, antibody production diminishes. However, the organism still retains an elevated population of B cells able to produce antibody against the antigen that initiated the response. Thus if subsequently reexposed, the immune response is much more rapid and effective, leading frequently to immunity. This response (Figure 16.6) is termed the *anamnestic response*.

A parallel process of selection of T cells occurs upon exposure to antigens. The role of T cells is described later.

ANTIMICROBIAL CONSEQUENCES OF ANTIGEN-ANTIBODY BINDING. Both virions and microbial toxins must bind to receptors on host cell membranes if they are to damage the host. The combination of antibody with a virion or toxin molecule may be sufficient to block such binding, and thus to protect the host completely. This is termed *neutralization*.

The binding of antibody to the surface of microbial cells has little direct effect on microbial growth or toxin production. However, it makes bacterial cells much more *susceptible to phagocytosis*. Furthermore, antibody-binding makes Gram-negative bacteria susceptible to lysis by a set of blood proteins called *complement*. Finally, the combination of antibody with soluble microbial antigens triggers a greatly increased *inflammatory response* by the host.

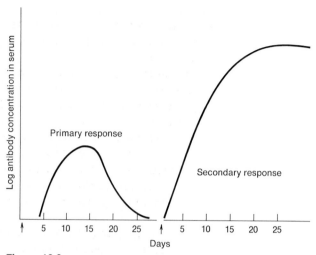

**Figure 16.6**
The anamnestic response. Antigen injections are indicated by the arrows.
The timing and kinetics of the primary response are quite variable.

THE LYSIS OF ANTIBODY-COATED CELLS BY COMPLEMENT. *Complement* is a group of proteins; in humans complement consists of 11 different proteins. Antibody-coated Gram-negative bacteria are generally susceptible to lysis by complement, whereas Gram-positive bacteria are totally resistant to this host defense mechanism. The basis of this difference is not known.

The reactions of complement with antibody-sensitized cells (termed *complement fixation,* since these reactions result in the removal of complement components from the serum) have been studied using red blood cells as an antigen. The process is actually a series of sequential reactions among the proteins of complement, which result in the binding of a number of the proteins to the red blood cell membrane, leading to lysis.

ANTIBODY-FACILITATED PHAGOCYTOSIS (OPSONIZATION). As already mentioned, microbial cells coated with antibody are more susceptible to phagocytosis. Their susceptibility is further enhanced by the binding of the complement. The process by which antibodies and complement molecules render cells susceptible to phagocytosis is called *opsonization* (Gr. *opsonein,* to prepare food for). Opsonizing antibodies act by combining with, and thus neutralizing, antigenic components of microbial cells that block phagocytosis.

*Agglutinating* antibodies also promote phagocytosis. Since antibodies are bivalent, and the microbial cell has many antigenic determinants on its surface, the combination of antibody with cells in appropriate proportions leads to the formation of clumps of cells, which are more readily phagocytized than single cells.

INDUCTION OF THE INFLAMMATORY RESPONSE. Antigen-antibody complexes that have bound complement release polypeptide fragments of certain complement proteins that trigger the release of histamine; the complexes also serve as chemotactic attractants for some phagocytes. Antigen-antibody complexes, through their activation of complement, thus increase the intensity of the inflammatory response over that induced by the infectious process alone.

HOST DAMAGE CAUSED BY ANTIBODY-MEDIATED HYPERSENSITIVITY. We have seen that a primary contact with antigen may produce the immune state, in which a greatly amplified population of specific small lymphocytes persists, ready to respond to subsequent antigen challenge. Under some circumstances, however, the primary contact with antigen may leave the host not only with lymphocytic memory cells, but also with *specific antibodies attached to the membranes of cells that produce histamine and serotonin.* If antigen molecules of the original type are reintroduced into the body at a later date, they combine with the cell-bound antibody and trigger the release of massive amounts of histamine and serotonin. These compounds mediate excessive inflammatory reactions, and sometimes trigger a fatal response called *anaphylactic shock.*

A different type of inflammation is produced when antigen-antibody aggregates accumulate in the blood and lymphatic vessels, fixing complement and inducing an inflammatory response so severe as to cause extensive host damage. The damage in such cases is mainly produced by lysosomal enzymes released from neutrophils, and is characterized by the destruction of small blood vessels. The lesions that are seen are called *Arthus reactions,* after their discoverer.

Two serious diseases occur as sequels to localized, superficial infections by streptococci: acute *glomerulonephritis,* in which kidney tissue is damaged; and *rheumatic fever,* in which heart valve tissue is damaged. In both cases, antibodies to streptococcal antigens cross-react with host tissue antigens, producing the specific damage, a phenomenon known as *autoimmunity.* Further damage is produced every time the individual is reexposed to streptococcal antigens.

The pathological results of immune sensitization are called *hypersensitivity,* or *allergy.* Sensitivity mediated by circulating antibody molecules is referred to as *immediate type,* since the inflammatory response to the secondary introduction of antigen is very rapid. Sensitivity that results from the accumulation of specifically sensitized T cells is referred to as *delayed type,* since the inflammatory response to the secondary introduction of antigen is much slower. The latter process is discussed below, in the section on cell-mediated immunity.

MICROBIAL RESISTANCE TO ANTIBODY-MEDIATED HOST DEFENSES. We have seen that the coating of bacteria by antibody and complement renders the bacteria susceptible to phagocytosis and—in the case of Gram-negative bacteria—to lysis by the further actions of complement. Some Gram-negative pathogens, however, produce surface components that block the processes of opsonization and lysis.

## cell-mediated immunity

A large number of immune responses have been found to be mediated not by antibodies secreted by B cells, but rather by the activities of immune (antigen-specific) T cells. This phenomenon is dramatically revealed in individuals who are genetically deficient in T cell production: they are highly susceptible to infections by protozoa, fungi, viruses, and intracellularly proliferating bacteria.

However, the activities of sensitive T cells in the immunized normal individual may, upon subsequent exposure to antigen, produce inflammatory responses of pathological severity. This is the condition of *cell-mediated hypersensitivity,* or *delayed-type hypersensitivity.*

PROTECTION AGAINST MICROBIAL INFECTION. When a T cell binds its cognate antigen, it acquires *cytotoxicity:* direct contact between the T cell and a target cell bearing the specific antigen leads to the death of the target cell. T cell cytotoxicity probably plays an important role in suppressing viral infections, by the killing of infected host cells which bear viral antigens on their surfaces (see Figure 7.3, p. 150).

Another modification of T cells that results from antigen binding, and is of major significance in host protection against bacterial infections, is the liberation of a substance that activates macrophages. The activated macrophages, termed *angry killers* by some authors, have greatly increased phagocytic activity and much higher levels of lysosomal enzymes. Their increased ability to kill ingested microorganisms is nonspecific. Thus, a host who has recovered from a microbial infection possesses a population of T cells, which, on later exposure to the antigens of the same microorganism, will activate macrophages to become angry killers not only of this microorganism but of others as well.

HOST DAMAGE CAUSED BY CELL-MEDIATED HYPERSENSITIVITY. Antigen-activated T cells not only activate macrophages, but also attract them and subsequently inhibit their movement. Since T cells themselves are attracted to loci of infection, the net result is the concentration of both activated T cells and macrophages at the site of an infection. The liberation of soluble toxins (lymphotoxins) by the T cells and of lysosomal enzymes by macrophages produces extensive tissue necrosis. Such necrosis is dramatically evident in the lesions that form around tubercle bacilli; the very reactions that serve to contain the bacteria (cell-mediated immunity) also produce the characteristic pathology of the disease (cell-mediated hypersensitivity).

# MICROBIAL
# DISEASES
# OF
# HUMANS

Man is host to a variety of pathogenic bacteria, protozoa, and viruses. In addition, certain members of the normal flora may become invasive and produce disease when the normal mechanisms of host immunity are suppressed.

The properties of the parasite that cause damage to the host vary among the major groups. As discussed in the previous chapter, toxins are responsible for the pathological consequences of many bacterial infections; they are absent or rare in infections by fungi and protozoa, most of which owe their pathogenicity to the induction of hypersensitivity reactions. Hypersensitivity also plays a role in many viral diseases, along with the cell damage caused directly by intracellular viral growth.

Differences are also observed with respect to the role of inducible host defenses. Circulating antibodies constitute an important defense against many bacterial pathogens, but are probably of little significance in other types of infections. Cell-mediated immunity, however, is of major importance in host resistance to fungi and viruses.

These conclusions have emerged from observations of persons exhibiting different states of *immunodeficiency,* either genetic or induced by immunosuppressive drugs. Persons deficient in the production of circulating antibodies are highly susceptible to respiratory infections by Gram-positive bacteria; persons deficient in T cell functions, however, tend to succumb to infections by fungi, viruses, and bacteria that grow predominantly intracellularly (e.g., tubercle bacilli and brucellae). There are not yet sufficient data concerning protozoan infections to evaluate the relative contributions of antibody-mediated and cell-mediated immunity.

In this chapter we shall describe some of the more important microbial diseases of man, and in so doing illustrate the variety of mechanisms that underlie host-pathogen relationships.

408

# bacterial diseases

Obligate bacterial parasites of man depend for their survival on transmission from one individual to another. All have evolved a characteristic *portal of exit, mode of transmission,* and *portal of entry.* In the following sections we shall group the bacterial diseases by their modes of transmission, because such a classification tends to link ecologically related pathogens.

### diseases transmitted by fecal contamination of food and drink

The intestinal tract is the natural habitat of many kinds of bacteria, most of which are harmless under ordinary conditions. However, a number of intestinal inhabitants are serious pathogens; these include the causative agents of typhoid and paratyphoid fevers, dysentery, cholera, and the *Salmonella* infections incorrectly referred to as "bacterial food poisoning." Some of these pathogens do their damage locally; others spread from the intestine to other parts of the body. All share two important attributes: they leave the host in excreted fecal matter, and they enter the next host via the mouth to reach the intestines once again.

The *enteric diseases,* as they are called, are thus acquired principally by swallowing food or drink contaminated with feces. Before the introduction of modern sanitation, water supplies were constantly subject to direct contamination by human feces. In the developed countries today, however, such contamination has become rare, and other modes of transmission have become relatively more important. For example, the common housefly is an effective agent of transmission because it visits both feces and food indiscriminately. Furthermore, many healthy individuals carry enteric pathogens (although frank clinical cases are rare), so anyone who handles food is a potential source of contamination. A large proportion of the world population, however, is still exposed at least intermittently to agents of enteric diseases via contaminated water.

Many animals, including cattle and fowl, may be naturally infected with *Salmonella,* which in man cause enteric infections. One can consequently become infected by eating contaminated meat or eggs.

Some important bacterial diseases transmitted by the fecal contamination of food and drink are described in Table 17.1.

### diseases transmitted by exhalation droplets

The transmission of disease by the respiratory route is called *droplet infection* because in such cases the pathogenic organisms are carried from person to person in microscopic droplets of saliva. In countries that practice modern methods of sanitation, droplet infection is by far the

MICROBIAL DISEASES OF HUMANS

**Table 17.1**
Some Human Diseases Transmitted by Fecal Contamination

| DISEASE | ETIOLOGIC AGENT | PATHOGENESIS |
|---|---|---|
| Typhoid fever | *Salmonella typhi,* a Gram-negative, peritrichously flagellated rod. Facultatively anaerobic; mixed-acid fermentation. | The organisms first multiply in the gastrointestinal tract. Invasion of the bloodstream leads to dissemination throughout the body. Pathogenicity is enhanced by the ability of the parasite to survive and even multiply within phagocytes. |
| Enteric fevers, gastroenteritis, septicemias ("bacterial food poisoning") | *Salmonella typhimurium, S. schottmülleri, S. choleraesuis:* Gram-negative, peritrichously flagellated rods. Facultatively anaerobic; mixed-acid fermentation. | Enteric fevers are diseases characterized by dissemination of the organism throughout the body. Enteric fever caused by salmonellae other than *S. typhi* are milder than typhoid fever and are called "paratyphoid fevers"; *S. schottmülleri* is the most common cause of enteric fevers in the United States. Gastroenteritis is a salmonellosis in which the organism remains localized in the gastrointestinal tract; *S. typhimurium* is the most common cause of salmonella gastroenteritis in the United States. *Salmonella* septicemias are most commonly caused by *S. choleraesuis.* |
| Cholera | *Vibrio cholerae,* a Gram-negative, polarly flagellated, curved rod. Facultatively anaerobic; mixed-acid fermentation. | The organism multiplies extensively in the small intestine. An exotoxin acts on the mucosal cells; loss of water and electrolyte leads to shock. In the absence of treatment, dehydration and shock can cause death in a few hours after onset of symptoms; intravenous replacement of fluid and electrolytes, however, almost invariably prevents death and allows recovery. Cholera is still a major health problem in areas in which war or poverty lead to inadequate sanitation, malnutrition, and inavailability of medical services. |
| Bacterial dysentery | *Shigella* species, a Gram-negative, immotile rod. Facultatively anaerobic; mixed-acid fermentation. | Lesions are formed in the large intestine and lower small intestine. Abdominal cramps, diarrhea, and fever are produced. Mortality rates are usually low, although some strains are particularly virulent; e.g., a South American strain of *S. dysenteriae* has caused epidemics in which the mortality rate in children has reached 20 percent. Chemotherapy usually speeds recovery, although an increasing number of cases are being caused by antibiotic-resistant shigellas. |

most important route by which disease is spread. Every time a person sneezes, coughs, or even speaks loudly, he exhales a cloud of tiny droplets of saliva. Each droplet contains a number of the microorganisms that inhabit the mouth and respiratory tract; the droplets quickly evaporate, leaving in the air great numbers of minute flakes that bear living bacteria. A person suffering from a respiratory infection will certainly contaminate every other person in whose presence he coughs, sneezes, or speaks.

Many highly important diseases involve infection of the respiratory tract and are spread by droplet exhalation. Some important bacterial respiratory diseases are described in Table 17.2.

### diseases transmitted by direct contact

There are a small number of pathogens for which the portal of entry is the skin or the mucous membranes, and they depend on direct contact for transmission. This group includes the causative agents of the venereal diseases syphilis and gonorrhea. The responsible organism in each case cannot survive for long outside the host and requires direct contact of mucous membranes for transmission. Sexual contact is the chief means of spreading these diseases, although syphilis can also be acquired before birth, and gonorrhea during birth, from an infected mother.

In the tropics there are several diseases caused by organisms closely related to the agent of syphilis that are normally not transmitted by sexual intercourse. These begin as skin infections and require direct contact for transmission. *Yaws* is an example of this group.

Three other nonvenereal diseases that are transmissible by direct contact are anthrax, tularemia, and brucellosis. All are diseases of animals that can be transmitted to man. *Brucellosis*, a disease of goats, cattle, and swine, constitutes a severe occupational hazard for animal handlers, including veterinarians, meat packers, and dairy workers. *Tularemia*, a disease of wild rodents, is often contracted by hunters or butchers who handle the carcasses of wild game. The principal bacterial diseases transmitted by direct contact are described in Table 17.3.

### diseases transmitted by animal vectors

Certain pathogens have become adapted to existence in two or more hosts. The plague bacillus, for example, can multiply in rats, fleas, and man; the flea is the vector* that carries it from rodent to rodent or from rodent to man; as a consequence, the plague bacillus never has to survive in environments unsuitable for growth.

The agents of plague and tularemia in addition to many rickettsial and some spirochetal diseases are transmitted by animal vector. Epidemics of typhus and plague spread in this fashion have radically altered the course of human history. The very property of having alternative hosts, which is such an advantage to parasites, has also led to the control of these diseases by man. By eliminating either the *vector* or the *reservoir of infection* (the species from which the vector derives the infection), man has been able to eradicate many such diseases from large areas.

The chief diseases transmitted by animal bite, together with their vectors and reservoirs, are listed in Table 17.4.

### diseases acquired by ingestion
### of bacterial toxins

Serious diseases are caused by the ingestion of food containing the toxins produced by *Clostridium botulinum, C. perfringens,* or *Staphylococcus aureus.* Although the disease is not further transmitted by the victim, an outbreak affecting many people can occur when a common food source becomes contaminated. In the early years of the canned food industry,

*A vector is an organism that carries a pathogen from one host to the next.

**Table 17.2**
Some Human Diseases Transmitted by Exhalation Droplets

| DISEASE | ETIOLOGIC AGENT | PATHOGENESIS |
| --- | --- | --- |
| Diphtheria | *Corynebacterium diphtheriae*, a Gram-positive, immotile rod; tends to be club-shaped. Postfission movements result in characteristic "palisade" arrangements. Facultatively anaerobic. | The organism establishes itself in the throat and remains localized in the upper respiratory tract. An exotoxin is produced that is disseminated by the bloodstream to all parts of the body. The exotoxin is encoded by a temperate phage; hence all virulent strains are lysogenic. Death may occur either from the systemic effects of the toxin (an inhibitor of protein synthesis) or by suffocation due to blockage of the air passage by local inflammation. Control has been largely accomplished by immunization with inactivated toxin; however, recent decreases in the proportion of the population receiving vaccinations has led to fears of possible epidemics. |
| Tuberculosis | *Mycobacterium tuberculosis*, a pleomorphic, immotile, acid-fast rod. Obligately aerobic. | The bacteria multiply both intracellularly and extracellularly within lesions of the lungs, called *tubercles*. An enlarging tubercle may discharge into a bronchus, promoting spread of the disease to other parts of the lung and (via exhalation droplets) to other individuals. Less frequently, organisms are disseminated via the bloodstream to set up secondary (metastatic) lesions in other organs. The toxins of *M. tuberculosis* have not been identified; at least some of the host damage results from hypersensitivity reactions of the delayed type. Tuberculosis has been one of the major causes of death in the West since the Industrial Revolution. Control has been accomplished only recently, with the advent of effective chemotherapy and improved living standards. |
| Plague | *Yersinia pestis*, a Gram-negative immotile, small rod. Facultatively anaerobic; mixed-acid fermentation. | A disease of rodents, plague is transmitted to man by flea bite. The disease is called *bubonic plague* when it is characterized by enlarged, infected lymph nodes ("buboes"). In severe cases the organism is disseminated to other organs; when the lungs become infected, the disease becomes transmissible from man to man by droplet infection. This form of the disease, called *pneumonic plague*, is highly contagious. Extremely high mortality rates in numerous epidemics and pandemics* throughout recorded history have made this one of the most feared diseases. Antibiotic therapy can reduce the mortality rate to a few percent if instituted early enough. Partial control may be accomplished by urban sanitation to keep rat populations small; however, complete elimination of the many rodents that can act as reservoir is impossible. |
| Meningococcal meningitis | *Neisseria meningitidis* (meningococcus), a Gram-negative, immotile coccus forming pairs of cells. | Meningococci are carried harmlessly in the nasopharynx by 25 percent or more of the population. For unknown reasons, they occasionally invade the bloodstream and localize in the meninges (the membranes surrounding the spinal cord). Without treatment, the mortality rate exceeds 80 percent; chemotherapy can reduce this to several percent. As with the pneumococci, virulence is directly dependent on the possession of a capsule, which protects the pathogen from phagocytosis. |
| Streptococcal infections | *Streptococcus pyogenes*, a Gram-positive coccus growing in chains. Lactic acid homofermentation. | The organism develops first in the pharynx ("strep throat"). Strains harboring a particular bacteriophage form a toxin; if the individual is sensitive to the toxin, a skin rash appears and the disease is called *scarlet fever*. Further spread of the organism may lead to mastoiditis, peritonitis, puerperal sepsis, cellulitis of the skin, or erysipelas. Rheumatic fever (characterized by inflammation of connective tissue, especially in the heart, joints, and nervous system) is a common sequel to recurrent streptococcal infections and is correlated with high titers of antibody to streptococcal antigens. No organisms can be isolated from the diseased heart tissue; damage results from immunological reaction. |

**Table 17.2 (cont.)**

| DISEASE | ETIOLOGIC AGENT | PATHOGENESIS |
|---|---|---|
| | | Most streptococcal infections yield readily to antibiotic therapy. Although immunity is frequently achieved if the infection is left untreated the possibility of immunological complications makes chemotherapy advisable. |
| Pneumococcal pneumonia | *Streptococcus pneumoniae* (pneumococcus), a Gram-positive coccus typically forming diplococci. Lactic acid homofermentation. | Between 40 and 70 percent of the adult population carry pneumococci in their throats. The organisms reach the lungs when normal barriers malfunction (e.g., during viral respiratory infections). Once a leading cause of death, especially as a complication of other diseases (e.g., influenza), control is now achieved by antibiotic therapy. |
| Other respiratory infections | *Hemophilus influenzae; Bordetella pertussis.* Small, Gram-negative, nonmotile rods. | *H. influenzae* causes respiratory infections in children; it is also the most common cause of bacterial meningitis in children. In adults, it may be a complicating factor in influenza epidemics but is rarely pathogenic alone. Antibiotic treatment of meningitis is effective. *B. pertussis* causes whooping cough, a disease principally of children. Although infrequently fatal, *B. pertussis* does not respond well to chemotherapy; early vaccination is thus advisable. |

*A pandemic is an unusually widespread epidemic, generally continental to worldwide in scope.

the technical problems of sterilization on an industrial scale were not fully appreciated, and canned foods were the source of a number of outbreaks of botulism. The development of adequate sterilization procedures, coupled with a vigorous inspection and control policy, has largely eliminated this health hazard. However, improperly sterilized home-canned and bottled foods remain a sporadic source of human deaths from botulism. Food handlers who have open staphylococcal skin lesions are still a common cause of outbreaks of staphylococcal food poisoning.

### wound infections

Whenever unsterilized foreign material penetrates a wound, microorganisms are introduced. If conditions are suitable for the growth of one or more contaminating microbes, an infection results that may eventually spread through the tissues or circulatory system.

Infection of wounds cannot be considered a "natural" route of transmission, because this method is too irregular and infrequent to ensure perpetuation of a parasitic species. Most often, infected wounds harbor ordinary soil-dwelling bacteria, such as the clostridia. The clostridia are obligate anaerobes that do not grow in healthy tissues. Deep wounds, however, form an ideal environment, because necrotic (dead) tissues are present, air is excluded, and tissue oxygenation is reduced as the result of impaired circulation. Clostridial spores are so ubiquitous in nature that any

**Table 17.3**
Some Human Diseases Transmitted by Direct Contact

| DISEASE | ETIOLOGIC AGENT | PATHOGENESIS |
|---|---|---|
| Anthrax | *Bacillus anthracis*, a Gram-positive, spore-forming, immotile rod. Obligately aerobic. | Anthrax is a disease of domestic and wild animals, including mammals, birds, and reptiles. Man acquires the disease primarily by direct contact with infected hides or carcasses, the organism gaining entry through a minor abrasion of the skin. A pustule forms at the site of infection and may be followed by a fatal septicemia. Chemotherapy is usually successful if applied before high concentrations of exotoxin accumulate. |
| Tularemia | *Francisella tularensis*, a short, immotile, Gram-negative rod. | Tularemia (discovered in Tulare County, California) is a disease of wild rodents, especially rabbits. Man usually acquires the disease by handling infected carcasses and skins, although it can also be transmitted by arthropod bite (ticks, deer flies), or by the bite of an animal (e.g., cat or dog) that has recently fed on an infected rabbit. The organism spreads throughout the body via the lymphatics and bloodstream; lesions develop in the lungs, liver, spleen, and brain. Growth is primarily intracellular. |
| Brucellosis | *Brucella melitensis, B. abortus, B. suis:* small, Gram-negative, immotile rods. | All three species are capable of infecting a wide range of mammals, although each has a preferred host. *B. melitensis,* goats; *B. abortus,* cattle; *B. suis,* swine. In their natural hosts, the brucellae localize in placental tissue of pregnant animals (and may cause abortion) and in mammary glands, from which they may be shed in milk. Infected milk and milk products were a principal source of human infection prior to widespread use of pasteurization and dairy herd inspections. In man, the organism is widely disseminated in the body and multiplies primarily within host phagocytic cells. The disease may undergo periodic remissions, in which case it is called *undulant fever*. Antibiotic therapy is not always successful in preventing relapse, presumably due to the intracellular location of the parasite. |
| Gonorrhea | *Neisseria gonorrhoeae* (gonococcus), a Gram-negative coccus, forming pairs of cells. | The organism is transmitted by sexual contact; newborn infants may acquire serious eye infections during passage through an infected birth canal. The routine use of bacteriocidal eyedrops immediately following birth has largely eliminated this problem. Following sexual contact, the organism penetrates the mucous membranes of the genitourinary tract; infection usually is restricted to the reproductive organs, although septicemia may occur. Penicillin treatment is usually successful, although penicillin-resistant strains are occurring with increasing frequency. Since the incubation period is short (a week or less) and there are no reservoirs other than man, eradication is theoretically possible. Men and (especially) women may be asymptomatic carriers. |
| Syphilis | *Treponema pallidum,* a spirochete. | The organism is transmitted by sexual contact; it may also be transmitted to the fetus during pregnancy. Following sexual contact, the organism penetrates the mucous membranes of the sexual organ, forming a local primary lesion, or "chancre." Several weeks later secondary lesions develop, including a skin rash and possible involvement of the eyes, bones, joints, or central nervous system. If untreated, the disease may progress several years later with the formation of tertiary lesions of the heart valves, central nervous system, eyes, bones, or skin. The toxins responsible for the disease are unknown; delayed hypersensitivity is believed to account for part or all of the damage occurring in the tertiary stage. Public health measures and treatment are, however, complicated by the fact that the primary lesions often go unnoticed, particularly in women. |

**Table 17.4**
Some Human Diseases Transmitted by Animal Bite

| ECOLOGY, WITH RESPECT TO MAN | DISEASE | TYPE OF MICRO-ORGANISM | VECTOR | RESERVOIR | REMARKS |
|---|---|---|---|---|---|
| Man serves as accidental host, not as a reservoir | Plague | Bacterium | Flea | Rat, other rodents | See Table 17.2. |
| | Tularemia | Bacterium | Tick | Wild rodents, tick[a] | See Table 17.3. |
| | Rabies | Virus | Dog, jackal, bat, etc. | Same as vector | See Table 17.10. |
| | Endemic typhus[b] | Rickettsia | Flea | Rat | See Table 17.6. |
| | Rocky Mountain spotted fever | Rickettsia | Tick | Wild rodents | See Table 17.6. |
| Man is one of two or more reservoirs | African sleeping sickness | Protozoon | Tsetse fly | Man, wild mammals | See Table 17.9. |
| | Yellow fever | Virus | Mosquito | Man, monkeys | See Table 17.10. |
| Man is sole reservoir | Malaria | Protozoon | Mosquito | Man | See Table 17.9. |
| | Epidemic typhus[b] | Rickettsia | Louse | Man | See Table 17.6. |
| | Dengue fever | Virus | Mosquito | Man | See Table 17.10. |

[a] Ticks act both as a vector and as a reservoir, because microorganisms multiply in the body of the tick and are transmitted through ovary and egg from one tick generation to the next.
[b] The two types of typhus fever are caused by closely related strains of rickettsiae.

deep wound into which clothing or soil is introduced has a high probability of being contaminated with one or another species of *Clostridium*. Many of these organisms produce potent exotoxins that kill the surrounding host tissues. One species, *C. tetani*, produces a toxin that affects the nerves and causes muscle spasms; infection, if not treated, is almost invariably fatal. This disease is called *tetanus*, or "lockjaw." Other clostridia cause severe local damage (*gangrene*) at the site of infection.

Although clostridia are the most dangerous wound pathogens, many other bacteria may become established in the damaged tissue. Common wound contaminants include staphylococci, streptococci, enterobacteria, and pseudomonads.

Leptospirosis, which often begins as a wound infection, is an occupational disease among workers who are exposed to frequent contact with polluted water. The leptospirae are parasites of pigs, dogs, and rodents and are excreted in the urine of infected animals. Man can be infected through minor wounds or breaks in the skin, or by ingestion of contaminated water, and the disease is thus most common among men who work in wet places, such as sewers, fish markets, wet fields, or canals. The common wound infections are described in Table 17.5.

**Table 17.5**
Some Common Wound Infections

| DISEASE | ETIOLOGIC AGENT | PATHOGENESIS |
|---|---|---|
| Tetanus | *Clostridium tetani*, a Gram-positive, spore-forming, peritrichously flagellated rod. Obligately anaerobic. | *C. tetani* is a common soil organism and is also common in the feces of animals (other than humans). It produces disease when accidentally introduced into wounds. Spore germination and growth require anaerobic conditions, which obtain when the wound is necrotic and vascular damage is severe. A neurotoxin is produced that is transported via the bloodstream and the peripheral nerves to the spinal cord. |
| Gas gangrene | *Clostridium perfringens, C. novyi, C. septicum:* Gram-positive, spore-forming, peritrichously flagellated rods. Obligately anaerobic. | Development of the organism in anaerobic wounds is accompanied by accumulation of hydrogen gas produced by fermentation. A variety of soluble toxins are produced. Tissue damage is extensive and progressive. Prevention may be accomplished by thorough cleaning of wounds (surgically if necessary) and prophylactic antibiotic treatment. Treatment usually involves surgical removal of the infected area. |
| Leptospirosis | *Leptospira icterohaemorrhagiae, L. canicola, L. pomona:* spirochetes. | The leptospirae cause mild, chronic infections of rodents and domestic animals, which shed the bacteria continuously in the urine. Man acquires the disease through contact with urine-contaminated water; the portal of entry is the skin. A disseminated infection results, during which organisms can be cultured from the blood. |

### rickettsial diseases

The rickettsias are extremely small, obligately parasitic bacteria; all are *intracellular* parasites (see Chapter 11).

One of the outstanding features of the rickettsias is their parasitic relationship to arthropods (lice, fleas, ticks, and mites); these are their natural hosts, in which they usually live without producing disease. The rickettsias have also become adapted to mammalian hosts, to which they are transmitted by arthropod bite. Thus, arthropod-mammal-arthropod chains of transmission are common. Several weeks after inoculation, clinical symptoms appear. These usually include chills, headache, fever (except for Q fever), and a rash. The disease may last several weeks. Mortality rates vary from nil (e.g., rickettsial pox) to very high (epidemic typhus). Chemotherapy is usually effective. Control measures usually focus on elimination of the vector with insecticides. In most cases, man is only an accidental host, not forming a part of a transmission chain; the one exception is louse-borne typhus. The principal rickettsial diseases of man are described in Table 17.6.

### chlamydial diseases

The chlamydias, like the rickettsias, are obligate intracellular parasites of birds and mammals (see Chapter 11). In their natural

**Table 17.6**
Some Representative Rickettsial Diseases of Man

| DISEASE | RESERVOIRS, VECTORS, AND TRANSMISSION TO MAN | REMARKS |
|---|---|---|
| Epidemic typhus | The rickettsia is carried from man to man by the body louse. | High mortality rates in man. |
| Endemic typhus | The rickettsia is normally a parasite of rats and fleas. It is maintained in nature by rat-flea-rat chains of transmission. Mild epidemic, involving man-louse-man chain of transmission, may follow bite by flea. | Milder disease than epidemic typhus. Epidemic and endemic typhus are caused by different, but closely related, rickettsias. |
| Scrub typhus (tsutsugamushi fever) | The true reservoir is the mite; the rickettsias are passed from one mite generation to the next via the eggs. They are also transmitted to rats by arthropod bite, and the rat is thus a secondary reservoir. Man is infected by the bite of a mite or flea. | Confined to the Far East. |
| Spotted fevers | The rickettsias are tick parasites and are passed from one tick generation to the next via the egg. They are also transmitted to a variety of mammalian hosts, including rodents and domestic animals, which thus form secondary reservoirs. Man is infected by tick bite. | There are several closely related diseases of this type, with differing geographical distributions (e.g., Rocky Mountain spotted fever, Mediterranean fever, South African tick-bite fever). |
| Rickettsial pox | The reservoirs are the house mouse and its mites; man is infected by mite bite. | The rickettsia is antigenically related to the spotted fever group. Rickettsia pox has only been observed in urban areas. |
| Q fever | The rickettsia is a parasite of numerous wild animals and ticks; the latter transmit it to goats, sheep, and cattle. Man acquires the disease by inhalation of infected dust or by direct contact with animals or animal products. | This is a unique rickettsial disease, in that it is transmissible to man by means other than arthropod bite. |

hosts they tend to produce prolonged, latent infections; overt disease is more characteristic of an infection acquired from a different host species.

Four human diseases are caused by chlamydias: psittacosis ("parrot fever" or ornithosis); lymphogranuloma venereum; and two diseases of the eye, trachoma and inclusion conjunctivitis. The last two are caused by closely related organisms, both of which are classified as *Chlamydia trachomatis*. The agent of inclusion conjunctivitis, however, normally inhabits the human genital tract, from which it occasionally is spread to the eye; the agent of trachoma normally inhabits the tissues of, and surrounding, the eye itself. All of the chlamydical diseases are amenable to antibiotic therapy. The principal chlamydial diseases of man are described in Table 17.7.

## fungal diseases

The fungal diseases of man are either *mycoses*, caused by true infection, or *toxicoses*, caused by the ingestion of toxic fungal metabolites.

**Table 17.7**
Chlamydial Diseases of Man

| DISEASE | AGENT | PRINCIPAL TRANSMISSION ROUTE | PATHOGENESIS |
|---------|-------|------------------------------|--------------|
| Psitticosis (ornithosis) | *Chlamydia psittaci* | Inhalation of feces from infected birds (many species) | Inflammation of the lungs; fever. Fatality rate may be as high as 20 percent in untreated cases. |
| Lymphogranuloma venereum | *C. trachomatis* (rarely, *C. psittaci*) | Sexual intercourse | Skin lesions and enlarged lymph nodes in genital regions of the body. |
| Trachoma | *C. trachomatis* | Mechanical spread by fingers or contaminated objects; possibly by flies. | Lesions form in tissues of and surrounding the eyes; may progress to blindness. The leading cause of blindness in the world. |
| Inclusion conjunctivitis | *C. trachomatis* | Newborn infants may be infected from the genital tract of the mother. Adults acquire the disease by sexual contact and subsequent finger-to-eye transfer. | Inflammation of the conjunctiva. |

## the mycoses

A small number of fungi are capable of causing human disease; for most of these, invasion of host tissue is accidental, their normal habitat being the soil. The exceptions are the *dermatophytes*, which inhabit the epidermis, hair, and nails; these are transmissible from person to person or from animal to person.

No toxins have yet been discovered to play a role in the pathogenesis of the mycoses, and it is believed that most of the host damage seen in fungal infections arises from hypersensitivity reactions, particularly of the delayed (cell-mediated) type. Localized lesions, resembling the classical delayed hypersensitivity responses of tuberculosis, are common, and represent areas where slow hyphal growth is taking place. Dissemination throughout the body usually involves the persistence of unicellular yeast forms within phagocytes.

The fungal diseases are generally grouped according to their depth of penetration. We will consider them as forming three such groups: the *dermatomycoses*; the *subcutaneous mycoses*; and the *deep*, or *systemic mycoses*.

## the dermatomycoses

The scaly, annular skin lesions caused by dermatophytes are called *tinea* (Latin, worm or insect larva), as they were originally thought to be caused by worms or by lice. They are generally classified according to the affected part of the body: *tinea pedis* (athlete's foot); *tinea capitis* (ringworm of the scalp); and *tinea corporis* (ringworm of the nonhairy skin of the body). Most tineas are caused by members of three genera of fungi: *Trichophyton*, *Microsporum*, and *Epidermophyton*. *Trichophyton* can grow in hair, skin, and nails; *Microsporum* can grow in hair and skin only; and *Epidermophyton* can grow in skin and occasionally in nails.

These organisms are transmitted by direct contact or by contact with infected hair clippings or epidermal scales. Animals form an additional reservoir; for example, over 30 percent of dogs and cats in the United States carry *M. canis*, an organism that can cause ringworm of the scalp in humans.

### the subcutaneous mycoses

Subcutaneous infections are initiated when certain soil-inhabiting fungi are introduced beneath the skin by thorns or splinters, or as contaminants of wounds. The diseases that develop are grouped into three categories: *sporotrichosis*, characterized by ulcerating skin lesions, is caused by a yeastlike fungus called *Sporotrichum schenkii*; *chromoblastomycosis*, characterized by skin lesions containing dark brown yeast cells, may be caused by any of several fungal species; and *maduramycosis*, characterized by a generalized destruction of the tissues of the foot or hand, may be caused by any of several fungi, as well as by several Actinomycetes.

### the systemic mycoses

A small number of fungal species produce deep lesions of the infected organ or widely disseminated lesions in the body. They include four soil inhabitants: *Blastomyces dermatitidis*, *Histoplasma capsulatum*, *Coccidioides immitis*, and *Cryptococcus neoformans*. They also include normally harmless inhabitants of the body such as *Candida albicans*, which become invasive only when the individual's normal antimicrobial defenses have been disturbed. For example, *Candida* will cause disease when the normal flora has been suppressed by antibiotic therapy, when immunosuppressive treatment is being given, or when the individual is severely debilitated by another disease. Most of these fungi proliferate within the body as yeasts. On agar media, however, *Geotrichum* and *Coccidiodes* are mycelial. *Blastomyces* and *Histoplasma* can be cultured as yeasts at 37°C or as mycelial fungi at 25°C. The major systemic mycoses are described in Table 17.8.

### the toxicoses

Many fungi produce poisonous substances, called *mycotoxins*, which cause serious—sometimes fatal—diseases if ingested. They also produce a variety of *hallucinogens*, such as lysergic acid. The mycotoxins of importance to man include the toxins produced by the poisonous mushrooms, the toxins of *Claviceps purpurea* (ergot, a pathogenic parasite of rye), and the *aflatoxins*.

**Table 17.8**
Some Important Systemic Mycoses

| ORGANISM | PATHOGENESIS | EPIDEMIOLOGY |
|---|---|---|
| *Candida albicans* | *Candida* is a harmless member of the normal flora of the mucous membranes in the respiratory, gastrointestinal, and female genital tracts. In debilitated patients it may produce systemic progressive disease or localized lesions of the skin, mouth (thrush), vagina, or lungs. | Most individuals harbor the organism, so transmission is not a factor in the disease. Prevention requires maintenance of the normal host defenses, including the normal flora. *Candida* infections often follow disturbances of the normal flora by antibiotic therapy. |
| *Cryptococcus neoformans* | Infection occurs via the respiratory tract. Most common clinical manifestation is chronic meningitis, which may be accompanied by lesions of the skin and lungs. Untreated cases are ultimately fatal. | Bird feces are the main source of infection; the disease is not transmissible from person to person. |
| *Blastomyces dermatitidis* | The infection may be limited to the lungs, or it may be widely disseminated to the skin, bones, viscera, and meninges. Lesions consist of small abscesses and tuberclelike granulomas (masses of macrophages). | *Blastomyces* appears to inhabit the soil and to be acquired by inhalation of dust. It is not transmissible from person to person. |
| *Histoplasma capsulatum* | The disease usually remains localized in the lungs, but in a minority of cases it becomes widely disseminated throughout the body. The tissue lesions resemble tubercles. | The organism occurs in the soil and grows abundantly in bird feces (chicken houses) and bat guano (caves). It is acquired by the inhalation of contaminated dust; it is not transmissible from person to person. |
| *Coccidioides immitis* | Respiratory infection, characterized by an influenzalike illness. Rarely, the disease progresses to a disseminated form closely resembling tuberculosis, with lesions in all organs and the central nervous system. | *Coccidiodes* is endemic in the soil and in rodents in the southwestern United States. It is acquired by inhaling contaminated dust; it is not transmissible from person to person. |
| *Geotrichum candidum* | Chronic bronchitis or lesions in the mouth. | The natural habitat is not known, but it may be the mouth and intestinal tract of man. |

The aflatoxins are formed by *Aspergillus flavus*, which grows in a variety of plant materials. Such crops as peanuts and grains, if not properly dried, may contain sufficient levels of aflatoxins to cause severe liver damage. In the United States rigid standards of food processing and storage, combined with the imposition of maximum permissive levels of toxin in foodstuffs, have effectively prevented diseases caused by aflatoxins. In India and many parts of Africa, however, aflatoxins are a serious problem.

Aflatoxins bind DNA and thereby inhibit RNA synthesis; they are also potent carcinogens, causing liver cancer in experimental animals. Indeed, a high correlation has been observed between human liver cancer and aflatoxin intoxication.

## protozoan diseases

Of the many thousands of protozoan species, only about 20 cause disease in man. Their impact on worldwide human health, however, is far out of proportion to their numbers; it has been estimated that at any given moment a quarter of mankind is afflicted with severely debilitating protozoan diseases. Malaria alone accounts for over 100 million cases a year, a million of which are fatal. In addition to the human diseases that they cause, protozoan parasites are indirectly responsible for widespread malnutrition in Africa, as a result of the diseases that they produce in domestic cattle.

Pathogens are found in each of the four main groups of protozoa; the principal ones are summarized in Table 17.9. In the following sections the protozoan diseases are grouped according to their modes of transmission, which are as varied as they are in the bacteria.

### diseases transmitted by
### the ingestion of cysts

Within each of the main groups of protozoa there are pathogens that are transmitted by the ingestion of cysts (thick-walled, resting cells). Four of these—*Entamoeba histolytica, Giardia lamblia, Balantidium coli,* and *Isospora* species—are parasites of the gastrointestinal tract; they have relatively simple life cycles, involving a proliferative state (*trophozoites*) and a cyst stage. The cysts are passed in the feces to the external environment, where they survive and contaminate food and water.

The cysts of *Toxoplasma gondii* may be transmitted in different ways. They are found in the skeletal muscle tissue of sheep and pigs, so that ingestion of incompletely cooked meat may be a source of infection. Toxoplasma cysts have also been seen in lung alveoli, so that inhalation of contaminated dust may be a second means of transmission. Domestic cats have recently been found to form another reservoir of this organism and to shed cysts in their feces. The ingestion or inhalation of cysts from this source is probably an important cause of infections in man.

Two novel mechanisms of host damage have been discovered in the intestinal protozoan parasites. *Giardia lamblia* has a disc on its ventral surface, by means of which it attaches itself to the intestinal epithelium of the host (Figure 17.1). In severe infections the inner surface of the upper small intestine may be totally covered by attached parasites, caus-

**Table 17.9**
Some Important Protozoan Diseases of Man

| GROUP | ORGANISM | DISEASE | PRINCIPAL TRANSMISSION ROUTE | PATHOGENESIS |
|---|---|---|---|---|
| Flagellates | *Giardia lamblia* | Flagellate diarrhea (Giardiasis) | Ingestion of cysts (fecal contamination) | Frequently asymptomatic; clinical disease is most common in children. Interference with intestinal function is the usual source of clinical symptoms, but extensive infection may lead to ulceration. Chronic cases of up to one-year duration are common. |
| | *Trichomonas vaginalis* | Genitourinary tract infections | Sexual intercourse | Causes an uncomfortable inflammatory vaginitis in females between puberty and menopause, especially during pregnancy. Usually asymptomatic in males, who thus act as a reservoir of the parasite. |
| | *Trypansoma gambiense* <br> *T. rhodesiense* | (African) sleeping sickness | Tsetse flies | The West African (*T. gambiense*) form is characterized by a long incubation period (up to four years) before onset of fever, chills, and chronic headache. Within a year, central nervous system involvement begins, leading progressively to apathy, weakness, personality changes (including psychoses), and ultimately coma and death after about two years. Mortality (frequently due to malnutrition or complicating infections) is high. The East African form (*T. rhodesiense*) follows roughly the same course as the West African form, but is much faster, the entire disease course often occurring within a year. Chemotherapy for both forms is frequently successful if initiated before nervous system involvement. |
| | *T. cruzi* | (American) Chagas' disease | Triatomid bugs (kissing bugs) | Found in the southern United States, Mexico, and Central and South America. Acute attacks involve chills, bone and muscle pain, and general debilitation. Chronic disease, frequently with only mild symptoms, may last decades. Mortality is high among children suffering acute attacks; otherwise low. |
| | *Leishmania donovani* | Kala-azar (infection of spleen, liver, lymph nodes, bone marrow) | Sandflies | Following an incubation period of several months, intermittent (often twice daily) fever, headache, and general malaise lasting for about two to four weeks. Visceral infection follows, and death usually occurs within two years, frequently from anemia, malnutrition, or complicating bacterial infection. |
| | *L. tropica* <br> *L. braziliensis* | Cutaneous lesions <br> Nasopharyngeal infections | Sandflies <br> Sandflies | Destruction of cutaneous or mucosal surfaces results in disfiguring lesions. The mortality rate is low, unless there is extensive involvement of mucosal surfaces (*L. braziliensis*) leading to complicating bacterial infection. |

422

**Table 17.9 (cont.)**

| GROUP | ORGANISM | DISEASE | PRINCIPAL TRANSMISSION ROUTE | PATHOGENESIS |
|---|---|---|---|---|
| Amebas | *Entamoeba histolytica* | Amebic dysentery; infections of liver, spleen, brain | Ingestion of cysts (fecal contamination) | Vegetative growth in the large intestine leads to ulceration of the colon and severe diarrhea. Penetration of the mucosa can lead to foci of infection elsewhere, especially the liver. A widespread debilitatory disease, especially common in areas of poor nutrition. Symptomless and subclinical cases are common. |
| Ciliates | *Balantidium coli* | Balantidial dysentery | Ingestion of cysts (fecal contamination) | Symptoms similar to amebic dysentery, but rarely spreads outside the intestine. Uncommon in man, common in hogs, which are probably the major source of infection. |
| Sporozoans | *Plasmodium falciparum* P. vivax P. malariae P. ovale | Malaria | Female *Anopheles* mosquito | Malaria is one of the most widespread human diseases, endemic in much of the world, particularly the tropics. It is characterized by repeated attacks of chills, high fever, headache, sweating, and general debility. *P. falciparum* malaria can be fatal; the others rarely are by themselves. Attacks are usually separated by latent periods of increasing duration after the first few months, until the infection dies out (usually after a few years, but up to 20 years in the case of *P. malariae*). Parasite resistance to chemotherapeutic agents and vector resistance to insecticides are increasingly serious public health problems. |
| | *Toxoplasma gondii* | Disseminated infections, particularly of the reticuloendothelial system | Transplacental infection of the fetus (congenital infection); ingestion of cysts; inhalation (?) | Frequently subclinical. Acute cases are rarely fatal, unless acquired transplacentally, in which case the fatality rate and incidence of developmental abnormalities in survivors are high. |
| | *Isospora belli* I. huminis | Intestinal infections | Ingestion of cysts | The parasite reproduces within epithelial cells of the small intestine, leading to dysentery, frequently accompanied by fever. Spontaneous recovery after about 3 to 4 weeks. |

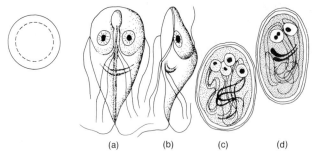

**Figure 17.1**
*Giardia intestinalis.* (a) "Face" and (b) "profile" of vegetative forms; (c) and
(d) cysts (binucleate and quadrinucleate stages). Circle represents red blood
cell for size comparison. From Jawetz, E., J. L. Melnick and E. A. Adelberg,
*Review of Medical Microbiology*, 12th edition. (Lange Medical Publications,
Los Altos, Calif., 1976.)

ing the mechanical blockage of fat absorption—a deficiency responsible for
some of the pathological effects of infection.

The second novel mechanism occurs in *Ent-
amoeba histolytica.* As its name implies, this organism causes the lysis of host
cells, particularly leucocytes. Leucocytolysis requires direct contact between
the leucocyte and the ameba; recent electron microscopic studies have shown
that *Entamoeba* possesses *surface lysosomes,* each with a vermiform appendage
that appears to serve as a trigger (Figure 17.2). Contact between the trigger
and a host cell may cause eversion of the lysosome, with destruction of the
leucocyte by the lysosomal enzymes.

### diseases transmitted by insect vectors

In three groups of pathogenic protozoa—the
trypanosomes, the leishmanias, and the malarial parasites of the genus *Plasmodium*—parts of the life cycle occur in an insect host. The insect serves as the
vector for transmission of the pathogen from human to human.

There are two groups of trypanosomes: the Afri-
can species, *T. gambiense* and *T. rhodesiense,* which are transmitted by the tsetse
fly (*Glossina* spp.) and cause sleeping sickness; and the American species,
*T. cruzi,* which is transmitted by triatomid bugs and causes Chagas' disease.

The leishmanias are transmitted by sandflies
(*Phlebotomus* spp.), and *Plasmodium* is transmitted by the *Anopheles* mosquito.
With the exception of the triatomid bugs, all the insect vectors pump the
parasites into the host's bloodstream in a stream of saliva while taking a blood
meal; only the female *Anopheles* mosquito transmits its parasite, for the male
does not suck blood. The triatomid bugs do not inject the trypanosome parasites, but instead deposit them in feces on the host skin while feeding. The
parasites are rubbed into the puncture wound or into the ocular conjunctiva

**Figure 17.2**
Surface lysosomes of *Entamoeba histolytica*. (a) Scanning electron micrograph of *E. histolytica* trophozo-ite, showing seven lysosomes in surface view, three of which have the trigger device in view, ×2,130. (b) Close-up view of a surface lysosome with a protruding trigger, ×6,800. (c) Electron micrograph of thin section of a surface lysosome, ×36,600. From R. D. P. Eaton, E. Meerovitch, and J. W. Costerton, "The Functional Morphology of Pathogenicity in *Entamoeba histolytica*," *Ann. Trop. Med. and Parasitol.* **64,** 299, (1970).

when the insect bite is scratched. The organisms then develop within macro-phages and ultimately within muscle cells.

The trypanosomes multiply in the bloodstream, the American forms multiplying in the reticuloendothelial cells as well. The damage they cause to the host appears to be mediated by toxins or by aller-gens, which affect principally the central nervous system and the heart.

*Leishmania donovani*, the agent of the disease called *Kala-azar*, multiplies in reticuloendothelial cells of spleen, liver, bone marrow, and lymph nodes. Other leishmanias tend to produce localized lesions of the skin and mucous membranes. The microbial products that damage the host have not been identified.

The malarial parasites, upon injection by the mosquito, develop first within liver cells and later within red blood cells. The pathological effects of infection are produced mainly during the latter phase. Anemia develops as a consequence of red cell destruction, but the anemia seen in malaria is not sufficient to account for the disease symptoms. Rather, it appears that factors are liberated during the red cell cycle that cause damage to bone marrow, spleen, kidney, and other organs. Whether these factors are products of the parasite or of the host is not yet known.

The diseases caused by insect transmitted parasites are restricted to regions of the world that represent the normal habitats of their insect vectors. Thus, sleeping sickness is found only in parts of Africa, and Chagas' disease is found in the southern United States, Mexico, and Central and South America. Malaria, however, may occur in almost any part of the world, following the widespread distribution of the *Anopheles* mosquito. Control or eradication of *Anopheles* has been accomplished in many regions of the temperate zones, however, so that today malaria is generally limited to the tropics and subtropics.

Within a given geographical region, the insect-transmitted diseases occur in patterns that reflect both the local habitats of the insects (e.g., the stagnant water breeding areas of *Anopheles*) and seasonal variations in their abundance.

### venereal disease

*Trichomonas vaginalis* infects the male urethra and prostate as well as the female vagina. The organisms do not form cysts and cannot survive for long outside the body; sexual contact is therefore the only means of transmission. The parasites, and the damage that they cause, are restricted to the genitourinary tracts.

### diseases caused by free-living
### soil amebas

Certain free-living soil amebas have been found to cause meningoencephalitis in man, apparently following their accidental introduction into the nasal mucosa. The organisms that have been isolated from the infected tissues have so far been classified in the genus *Naegleria;* members of a related genus, *Hartmannella,* have been found capable of producing a similar meningoencephalitis in experimental animals. All cases to date have been traced to infections acquired during swimming in freshwater rivers and lakes; extensive brain damage is produced, but the factors responsible are as yet unidentified.

# viral diseases

The principal viral agents of human disease are grouped according to their transmission routes in Table 17.10, which also indicates their major target organs (i.e., those in which the most damage is produced by viral replication). It should be emphasized that the subdivision of viruses in terms of their modes of transmission or their target organs is not correlated with the taxonomic subdivision of viruses in terms of their physio-chemical properties (Chapter 7, p. 150).

The ability of viruses to multiply in certain organs but not others reflects a high degree of *tissue specificity*. In a few cases this specificity has been traced to the presence or absence of *viral receptors* on the cell surface; for example, homogenates of susceptible organs can be shown to bind virions, but homogenates of nonsusceptible organs do not.

Viruses also show marked *host specificities*, being able to infect some animal species but not others; the presence or absence of viral receptors has been shown to be the determining factor in certain cases of host specificity also. For example, primate cells, but not rodent cells, have receptors for polio virions.

## host defenses against viral diseases

Different, although overlapping, sets of host defenses are involved in protection against viral diseases and bacterial diseases. In addition to some aspects of the inflammatory response (e.g., fever), defenses against viral infection include phagocytosis, the immune response and interferon production.

PHAGOCYTES. The phagocytic cells of the body constitute a primary line of defense against viruses as they do against cellular pathogens. Some viruses (like some bacterial pathogens) multiply within macrophages and are released in increased numbers to infect other types of cells.

INTERFERON. Practically all animal cells, but particularly bone marrow cells, spleen cells, and macrophages, are induced by viral infection to synthesize and secrete a protein that interferes with viral multiplication. This protein, called *interferon*, interferes with the multiplication of all viruses, not only with the type that induced its formation. Different animal species produce different interferons, which with few exceptions more effectively protect cells of the same animal species than cells of a different species.

Animal cells can be induced to synthesize interferon by *double-stranded RNA*, but not by any other form of nucleic acid. Since most RNA viruses produce double-stranded intermediates during their replication, their ability to induce interferon synthesis is easily explained. DNA

**Table 17.10**
Viruses of Human Disease, Grouped According to Their Transmission Routes

| PRINCIPAL TRANSMISSION ROUTE | VIRUS | MAJOR TARGET ORGANS | PATHOGENESIS |
|---|---|---|---|
| Respiratory | Influenza | Respiratory tract | Although usually confined to the upper respiratory tract, influenza may fatally involve the lungs; more common, however, as influenza-associated cause of death, are secondary bacterial pneumonias. Periodic pandemics of influenza occur. |
| | Parainfluenza | Respiratory tract | Causative agent of upper respiratory diseases, which may be quite severe in children. |
| | Respiratory syncytical | Respiratory tract | Causes mild upper respiratory disease ("colds") in adults and severe lower respiratory disease, sometimes fatal, in infants. |
| | Measles | Respiratory tract, skin | A common and distinctive disease, which is usually not serious. Rare involvement of the central nervous system can lead to death, epilepsy, or personality changes. Bacterial pneumonia is a frequent complication of measles. |
| | Mumps | Parotid glands, testes, meninges | Mumps usually involves salivary gland infection. Meningitis is not uncommon, however, although it is rarely serious. In males, testicular involvement is common, and testicular atrophy (although rarely total sterility) may occur in adults. |
| | Adenoviruses | Respiratory tract | Common cause of mild upper respiratory infections, although infection may sometimes be severe. In children, lower respiratory tract involvement may occur, with occasional fatal pneumonias. May also cause conjunctivitis, frequently contacted through swimming pool water. |
| | Rhinoviruses | Respiratory tract | One of the "common cold" viruses. Although normally confined to the upper respiratory tract, may cause bronchitis or even pneumonia, especially in children. |
| | Coxsackie viruses (some) | Respiratory tract, many others | Another agent of the common cold, and of pharyngeal disease. |
| | Coronaviruses | Respiratory tract | Severe coldlike symptoms characterize coronavirus infections, often with muscular aches, chills, and headache. |
| | Rubella | Skin, many others | A mild, measleslike disease ("German measles") is characteristic of rubella. The major concern is the high incidence of congenital defects in children born to mothers infected in their first trimester of pregnancy. |
| Enteric | Polio viruses | Intestinal mucosa, lymph nodes, central nervous system | Although usually causing pharyngeal and enteric infection, polio viruses may occasionally spread to the central nervous system where their multiplication may cause death or paralysis. Through vaccination, the disease has been largely eliminated in Europe and North America; it is still, however, endemic in the tropics. |
| | Coxsackie viruses (some) Echo viruses | Many Gastrointestinal tract; occasionally disseminated | A variety of syndromes are associated with different viral strains, including rare infections of heart tissue of infants, which may be fatal (Coxsackie viruses). Frequently involves short-lived general malaise. |

**Table 17.10 (cont.)**

| PRINCIPAL TRANSMISSION ROUTE | VIRUS | MAJOR TARGET ORGANS | PATHOGENESIS |
|---|---|---|---|
| | Hepatitis virus | Liver (also kidneys and spleen) | Hepatitis is characterized by weakness, occasional fever, nausea, diarrhea, and jaundice. The disease normally becomes chronic, and may be fatal. Two types of hepatitis, clinically very similar, are recognized: *infectious hepatitis* (caused by the enteric hepatitis virus) and *serum hepatitis* (caused by a virus transmitted directly with serum during blood transfusion or injection with nonsterile hypodermic needles). |
| Direct contact | Herpes simplex | Oral or genital mucous membranes | Primary infection of oral type is most often of young children, usually consisting of rapidly healing ulcers of the mouth and gums. The virus then becomes latent in sensory nerve ganglia, breaking out as "cold sores" or "fever blisters" upon stress (such as fever, sunburn, menstruation, anxiety, etc.). Genital herpes infects the genital system of both men and women, and has become the second most prevalent venereal disease (after gonorrhea) in the United States. |
| | Pox viruses | Skin, many others | Several skin diseases are caused by pox viruses, the most important of which is smallpox. Some strains of smallpox virus cause disease with up to 15 percent mortality rate; survivors may be terribly scarred. Formerly a worldwide scourge, vaccination has now confined the disease to a few pockets in eastern Africa, and total eradication is expected. |
| Animal bite | Arboviruses | Many | Arthropod-born viruses are transmitted by mosquito, fly, or tick bite. Most produce asymptomatic infections, or nonfatal infection characterized by headache, fever, chills, and severe muscle and joint pain. Hemorrhagic fevers are characterized by subcutaneous hemorrhage and bleeding from mucosal membranes, accompanied, in the case of yellow fever, by jaundice. These fevers can be fatal, especially yellow fever. Other fatal arbovirus diseases involve the central nervous system. |
| | Rabies | Central nervous system | Rabies involves muscle spasms, delirium, coma, and sometimes paralysis. Death almost inevitably results within a week after the onset of symptoms. Cure can usually be effected by the prompt application of antiserum at the wound site. Rabies virus infects all warmblooded animals and is lethal for most. |

viruses may also cause some double-stranded RNA to be produced in infected cells, although it is not known to be required for their replication or maturation. Such double-stranded RNA has been detected in cells infected with vaccinia virus (a DNA virus), and found to be effective in inducing the host to synthesize interferon.

Interferon does not act directly to inhibit viral multiplication; rather, it induces the formation of a second protein, called *antiviral protein,* which is the true inhibitor. Interferon induces antiviral protein synthesis not only in the cell in which it is itself synthesized, but also in the surrounding cells to which it diffuses. Its sphere of activity in host defense is thus greatly enlarged.

Antiviral protein appears to inhibit viral multiplication by blocking the translation of viral messenger RNA into viral proteins, which could account for its ability to protect against both RNA and DNA viruses.

IMMUNE RESPONSES. Most viruses possess polypeptides that are good antigens and can induce both neutralizing antibodies and cell-mediated immune mechanisms. Cellular immunity appears to play a more important role than humoral antibodies in protection against certain viruses, such as the pox viruses and the enveloped viruses. It also serves to eliminate host cells that have acquired new surface antigens as the result of harboring latent viruses or the genomes of tumor viruses.

The cell-mediated immune response brings into play not only an increased activity of macrophages, but also an increased level of interferon production, since antigen-activated lymphocytes release interferon along with the other factors described in Chapter 16.

### mechanisms of cell damage

Many viruses trigger hypersensitivity reactions of the delayed type, producing nonspecific tissue damage as described in Chapter 16. However, in most cases, the symptoms of viral disease reflect the pathological changes undergone by the infected cells themselves. Such changes, which often terminate in the death of the cell, are of two types: *morphological,* in which the membrane and internal structures of the cell are visibly disrupted; and *biochemical,* in which the synthetic processes and other physiological activities of the cell are impaired.

The virus-induced synthesis of new proteins appears to be necessary for most of these *cytopathic* effects; such proteins, which thus behave as intracellular toxins, may or may not be incorporated into the mature virion. The toxic effects of virus-induced proteins can be demonstrated by extracting them from infected cells and adding them to cultures of uninfected cells at high concentrations. For example, of two viral coat proteins that can be extracted from adenovirus-infected cells, one causes morphological damage to uninfected cells and the other causes the depression of macromolecular syntheses.

The precise nature of the biochemical lesions induced by viral products is, in most cases, not known. In phagocytic cells, much of the cell damage appears to result from the disruption of lysosomes, which discharge their lytic enzymes into the otherwise protected cytoplasm of the cell. In other types of cells, the virus-induced proteins appear to block host cell metabolism at a number of points, leading to the shut-off of normal macromolecular syntheses and the diversion of metabolic precursors into the pathways of viral synthesis. In some viral infections, large masses of virions or unassembled viral subunits accumulate in the nucleus or in the cytoplasm; these *inclusion bodies,* as they are called, may become large enough to cause mechanical damage to the cell.

### persistent and latent viral infections

Some viruses are able to *persist within host tissues* for long periods of time without producing overt disease. Such infections may eventually be converted into acute diseases by external factors that upset the equilibrium between host and parasite.

*Persistent infections* of two different types have been observed, depending on the particular virus involved. In some cases, a *carrier state* is established at the cellular level: at any given time only a few cells are infected which liberate virus; antibodies and other host defenses keep the extracellular virus population in check but do not eradicate it completely. In other cases, viral replication occurs at a moderate rate without harming the host cells; viral genomes are transmitted to daughter cells *intracellularly* by the process of cell division.

A number of formerly unexplained diseases (e.g., rare degenerative diseases of the brain) have now been found to follow prolonged periods (sometimes many years) of latent viral infection. These diseases are called *slow virus infections.*

Alternatively, viruses may persist in host cells in a nonreplicating form. Changes in the host physiology, frequently associated with physical or emotional stress, may trigger viral replication and an episode of acute disease. Such pathologies (termed *latent infections*) are characteristic of a few viruses, of which *herpes simplex* is the best known.

# INDUSTRIAL
# USES
# OF
# MICROORGANISMS

The role of microorganisms in the transformations of organic matter was not recognized until the middle of the nineteenth century. Nevertheless, microbial processes have been used by man since prehistoric times in the preparation of food, drink, and textiles; in many cases, these processes became controlled and perfected to an astonishing degree by purely empirical methods. Outstanding examples are the production of beer and wine; the pickling of certain plant materials; the leavening of bread; the making of vinegar, cheese, and butter; and the retting of flax. The rise of microbiology, which revealed the nature of these traditional processes, led not only to great improvements in many of them, but also to the development of entirely new industries based on the use of microorganisms.

## the use of yeast

Yeasts traditionally and still play an important role from a technical and industrial standpoint. Although many genera and species of yeast exist in nature and many are used industrially, the yeasts of greatest technical importance are strains of *Saccharomyces cerevisiae*. They are used in the manufacture of wine and beer and in the leavening of bread.

The manufacture of alcoholic beverages was already well established in early civilizations, many of which had myths about the origin of wine making that attributed the discovery to divine revelation. This fact suggests that the beginnings of the art were prehistoric. The use of yeast as a leavening agent for bread originated in Egypt about 6,000 years ago, and spread slowly from there to the rest of the Western world.

The discovery that alcohol can be distilled, and thus concentrated, originated either in China or the Arab world. Distilleries began to appear in Europe in the middle of the seventeenth century. At first the alcohol manufactured was used only for human consumption, but with the industrial revolution, the demand for alcohol as a solvent and chemical raw material developed and the distilling industry grew very rapidly.

### the making of wine

The making of wine involves the fermentation of the soluble sugars (glucose and fructose) of the juices of grapes into $CO_2$ and ethyl alcohol. After the grapes are harvested, they are crushed to yield a raw juice or *must*, a highly acidic liquid containing 10 to 25 percent sugar by weight. In many parts of the world the mixed yeast flora on the grapes serves as the inoculum. In such a natural fermentation, a complex succession of changes in the yeast population occurs; in the later stages the so-called true wine yeast, *Saccharomyces cerevisiae* var. *ellipsoideus*, predominates. In other areas, California, for example, the must is first treated with sulfur dioxide, which virtually eliminates the natural yeast flora; it is then inoculated with the desired strain of wine yeast. The fermentation proceeds vigorously, usually being completed in a few days. Must from both red and white wine grapes (*Vitis vinifera*) is white and results in a white wine. Since the color of red grapes is in the skin, skins are not separated from the must when a red wine is made. The alcohol developed during fermentation extracts the color into the wine.

Following fermentation the new wines must be clarified, stabilized, and aged to produce a satisfactory final product. These processes require months, and for high quality red wines, even years. During the first year, many wines (particularly red) undergo a second spontaneous fermentation, the *malo-lactic fermentation*, which can be caused by a variety of lactic acid bacteria (*Pediococcus*, *Leuconostoc*, or *Lactobacillus*). This fermentation converts malic acid, one of the two major organic acids of grapes, to lactic acid and $CO_2$, thus converting a dicarboxylic acid to a monocarboxylic acid and reducing the acidity of the wine. Although the malo-lactic fermentation proceeds spontaneously, slowly, and undramatically (sometimes even without the winemaker's knowledge), it is absolutely vital to produce red wines of good quality from grapes grown in cool districts, which yield wines with too high an initial acidity to be palatable.

Certain special types of wine must undergo additional microbial transformations. Sparkling wines (champagne types) undergo a second alcoholic fermentation, under pressure, at the expense of added sugar, either in the bottle or in bulk; the $CO_2$ thus produced carbonates the wine. This secondary fermentation is conducted with varieties of wine yeast that readily clump following fermentation and are consequently easily removed. Sherries (wines of the type produced in the Jerez district of Spain) are fortified with alcohol to about 15 percent, exposed to air, and allowed to develop a heavy surface growth of certain yeasts that impart the unique sherry flavor to the wine.

Some European sweet wines, notably those from the Sauternes district of France, undergo even more complex microbial transformations. Prior to picking, the grapes become spontaneously infected with a fungus, *Botrytis cinerea*. This infection causes water loss, thus increasing sugar content, and destruction of malic acid, thus decreasing the acidity of the grapes. Certain favorable changes of flavor and color occur. The resulting very sweet must from these infected grapes is fermented by so-called *glucophilic yeasts*, i.e., yeasts that rapidly ferment the glucose leaving residual fructose (the sweeter of the two sugars). The product is a sweet dessert-type wine.

Although the high alcohol content and low pH ($\sim$3.0) of wines make them unfavorable substrates for growth of most organisms, they are subject to microbial spoilage. The most serious spoilage problems are those that occur if wines are exposed to air. Film-forming yeasts and acetic acid bacteria grow at the expense of the alcohol, converting it to acetic acid, thus souring the wine. Serious problems can also be caused by fermentative organisms in the absence of air. Rod-shaped lactic acid bacteria can grow anaerobically at the expense of residual sugar and impart a "mousy" taste to the wine.

### the making of beer

Beers are manufactured from grains that, unlike grape or other fruit musts, contain no fermentable sugars. The starch of the grains must be hydrolyzed to fermentable sugars prior to fermentation by yeasts. The principal grains traditionally used for the production of beer were barley in Europe, and rice in the Orient. In each case, a different method of hydrolysis of starch was found. In the case of barley, starch-hydrolyzing enzymes (amylases) of the grain itself were used. Barley seeds contain little or no amylase, but upon germination, large amounts are formed. Hence, barley is moistened, allowed to germinate, and is then dried and stored for subsequent use. Such dried, germinated barley, called *malt*, is dark in color as a result of the exposure to increased temperatures during drying and has more flavor than untreated barley seeds. The starch of barley remains largely unaffected by the malting process. Hence, the first step in beer making is the grinding of malt and its suspension in water to allow hydrolysis of the starch. Malt itself is sometimes used as the total source of starch, or, if a lighter beer is desired, unmalted barley or some other cereal grain is added. Currently in the United States large quantities of rice are added to the malt in the manufacture of beer. After hydrolysis has reached the desired stage, the mixture of malt and rice is boiled to stop further enzymatic changes, and it is then filtered.

Hops (the female flower of the vine, *Humulus lupus*) are added to the filtrate (*wort*) and contribute a soluble resin, which imparts the characteristic bitter flavor of beer and also acts as a preservative against the growth of bacteria. The use of hops is a relatively recent modification of the art of beer making, having been introduced about the middle of the sixteenth century; even today, unhopped beer is made in some countries. After filtration, the hopped wort is ready for fermentation.

In contrast to wine fermentations, beer fermentations are always heavily inoculated with special strains of yeast derived from a previous fermentation. The fermentation proceeds at low temperatures for a period of 5 to 10 days. All yeasts used in making beer are *Saccharomyces cerevisiae*, but not all strains of *S. cerevisiae* can be used to make good beer. During the course of time, special strains with desirable properties have been selected; these are known as *brewer's yeasts*. Good brewer's yeasts were developed over a period of centuries; they cannot be found in nature. Like cultivated higher plants, they are a product of human art, and in recognition of this fact, the brewer refers to other yeasts that are unsatisfactory for making beer (including other strains of *S. cerevisiae*) as "wild yeasts."

Strains of brewer's yeast fall into two principal groups, known as *top* and *bottom* yeasts. Top yeasts are vigorous fermenters, acting best at relatively high temperatures (20°C), and are used for making heavy beers of high alcoholic content, such as English ales. Their name derives from the fact that during fermentation they are swept to the top of the vat by the rapid evolution of $CO_2$. In contrast, bottom yeasts are slow fermenters, act best at lower temperatures (12°C to 15°C), and produce lighter beers of low alcoholic content of the type commonly made in the United States. Their name derives from the fact that the slower rate of $CO_2$ evolution allows them to settle to the bottom of the vat during fermentation.

Although wines from grapes and beer from barley are the characteristic fermented beverages of the Western world, rice serves as the source of most fermented beverages (e.g., sake) in the Orient. The problem of hydrolyzing rice starch as a preliminary to fermentation has been solved by the use of amylases from molds, principally *Aspergillus oryzae*. In the manufacture of sake the first step is the preparation of a culture of the mold. Mold spores, saved from a previous batch, are sown on steamed rice and are allowed to grow until the mass of rice is thoroughly permeated with mycelium. This material (*koji*), which serves both as a source of amylase and as an inoculum, is added to a larger batch of steamed rice mixed with water. Hydrolysis of the starch proceeds, and when sufficient sugar has accumulated, a spontaneous alcoholic fermentation begins. Lactic acid bacteria as well as yeasts are present in koji, so lactic acid is produced in addition to alcohol and $CO_2$. The production of alcoholic beverages in the Orient thus differs from the Western process in two respects: hydrolysis of starch is effected by microorganisms, and hydrolysis proceeds simultaneously with fermentation.

### the making of bread

An alcoholic fermentation by yeasts is an essential step in the production of raised breads; this process is known as the *leavening of bread* (after the old word for yeast, "leaven"). The moistened flour is mixed with yeast and allowed to stand in a warm place for several hours. Flour itself contains almost no free sugar to serve as a substrate for fermentation, but there are some starch-splitting enzymes present that produce sufficient sugar to support leavening. In the highly refined flours, commonly used in the United States, these enzymes have been destroyed and sugar must be added to the dough. The sugar is rapidly fermented by yeast. The carbon dioxide produced becomes entrapped in the dough, causing it to rise; the alcohol produced is driven off during the baking process.

The yeasts used in bread making all belong to the species *Saccharomyces cerevisiae* and have been derived historically from strains of top yeasts used in brewing. Until the nineteenth century, yeasts for bread making were obtained directly from the nearest brewery. Today, yeast is produced commercially for use in bakeries and for individual baking needs. The commercial production of compressed yeast by industry was greatly stimulated by the application of mass production techniques to bread making. A large modern bakery may use many hundreds of pounds of yeast daily, for about 5 pounds of yeast are required to leaven 300 pounds of flour. Much of the bakers' yeast manufactured today is dried under controlled conditions that maintain viability of the yeast cells, a treatment that facilitates shipment and storage.

## microbes as sources of protein

Because of their rapid growth, high protein content, and ability to utilize organic substrates of low cost, microorganisms are potentially valuable sources of animal food. The growth of the science of animal nutrition has led to the development of a new industry, based on the cultivation of yeast for use as a supplement in animal feeds. Since the goal is to obtain cell material, the organisms are always grown under forced aeration to maximize the growth yield. However, there is the risk that when carbohydrates are used as substrates, part of the substrate will be diverted to alcohol, even if the oxygen supply is maintained at a high level. Consequently, strictly aerobic yeasts of the genus *Candida* are used in preference to fermentative yeasts.

### production of yeasts from petroleum

The cost of raw material is a factor of paramount importance in the production of microorganisms for use as food, and cheap sources of carbohydrate (e.g., whey, molasses, sulfite waste liquor) were initially used for growth of food yeasts. However, since aerobic growth conditions are always used to produce food yeast, any compound that can support respiratory metabolism may serve as a substrate for growth. This led to the development of processes that utilize petroleum as a substrate. Petroleum is still very cheap, compared with other possible substrates, and since hydrocarbons are the most highly reduced of organic compounds, growth yields on petroleum are extremely high.

The British Petroleum Corporation has built an industrial unit in France for the cultivation of *Candida lipolytica* in an aqueous emulsion of crude petroleum. This yeast can oxidize aliphatic, unbranched hydrocarbons of chain length $C_{12}$ to $C_{18}$, compounds that comprise part of the complex mixture of alkanes present in crude petroleum. Selective removal of certain classes of alkanes by the growth of *Candida lipolytica* produces a dewaxed petroleum that is much more easily refined. The economic feasibility of the British Petroleum process depends on its twofold function: simplification of refining and protein production.

Considerable efforts are also being made to develop processes for the production of microbial protein at the expense of other cheap substrates. One potential substrate is the gas methane, a major petrochemical product. However, many problems remain to be solved before an effective industrial process can be based on its use. The most important of these are the relatively low growth rates of methane-oxidizing bacteria and their tendency to produce large quantities of extracellular slime.

### production of specific amino acids

The great potential value of microorganisms as foods or feed supplements lies in their high protein content. This makes them the best agents for the rapid and efficient conversion of other more readily available organic compounds into protein, of which the world is becoming critically short. This point becomes evident when protein production by cattle and by yeast is compared. A bullock weighing 500 kg produces about 0.4 kg of protein in 24 hours. Under favorable growth conditions, 500 kg of yeast produce over 50,000 kg of protein in the same period.

Many plant foods contain sufficient protein to supply the quantitative needs of mammals, but they serve poorly as sole sources of dietary protein because their proteins are deficient in certain specific amino acids required by mammals. Wheat protein is low in lysine, rice protein in lysine and threonine, corn protein in tryptophan and lysine, bean and pea protein in methionine. The addition of the deficient amino acid(s) to diets that contain a single source of vegetable protein will render them adequate. The practicality of fortifying diets of vegetable protein with individual amino acids has been amply demonstrated in numerous experiments with both animals and humans. Thus, the world shortage of certain specific amino acids—notably, lysine, threonine, and methionine—is more critical than the shortage of total protein. The microbial production of *specific amino acids* has therefore been intensively studied.

Since the metabolism of microorganisms is precisely regulated (Chapter 5), microorganisms normally synthesize quantities of amino acids just sufficient to meet their growth requirements. However, naturally occurring and mutant strains of some microorganisms have defective mechanisms for the regulation of specific biosynthetic pathways and, as a consequence, excrete large amounts of certain amino acids into the medium. Methods for the microbial production of nutritionally important amino acids are now available and are constantly being improved.

## the use of acetic acid bacteria

When wine and beer are freely exposed to air, they frequently turn sour. Souring is caused by the oxidation of alcohol to acetic acid, mediated by the strictly aerobic acetic acid bacteria. The spontaneous souring of wine is the traditional method of manufacturing vinegar. The word *vinegar* is derived from the French "vinaigre" which literally means "sour wine."

The manufacture of vinegar still remains largely empirical. The principal modifications introduced during the past century concern the mechanical rather than the microbiological aspects of the process. In the traditional *Orleans Process,* which is still used in France, wooden vats are partially filled with wine, and the acetic acid bacteria develop as a gelatinous pellicle on the surface of the liquid. The conversion of ethanol to acetic acid takes several weeks—the rate of the process being limited by the slow diffusion of air into the liquid. The survival of this slow and inefficient method is attributable to the high quality of the product.

When the taste of the product is not of primary importance, vinegar is made by more rapid methods from cheaper raw materials (e.g., diluted distilled alcohol or cider). These methods are designed to accelerate oxidation by improved aeration and regulation of temperature, but they remain microbiologically uncontrolled. The oldest such method, developed early in the nineteenth century, utilizes a tank that is loosely filled with wooden shavings through which the liquid is circulated. The liquid is trickled into the tank, and air is blown countercurrent to the liquid flow. The acetic acid bacteria develop as a thin film on the wooden shavings, thus providing a large surface of cells, which are simultaneously exposed to the medium and to air. Once a bacterial population has become established on the shavings, successive batches of vinegar can be produced quickly; solutions initially containing 10 percent alcohol can be converted to acetic acid in 4 or 5 days. Much vinegar is still made by this method, but stirred, deep tank fermentors, similar to those used to produce antibiotics, are now being introduced.

The oxidation of ethanol to acetic acid is an example of the *incomplete oxidations* carried out by acetic acid bacteria. Certain other incomplete oxidative conversions by acetic acid bacteria are industrially important. Gluconic acid, which is used by the pharmaceutical industry, is made by oxidation of glucose by acetic acid bacteria. Many sugar alcohols are converted to sugars by acetic acid bacteria. One such reaction in commercial use is the production of sorbose from sorbitol. Sorbose is used as a suspending agent for certain pharmaceuticals, and it is an intermediate in the manufacture of L-ascorbic acid (vitamin C).

## the use of lactic acid bacteria

Lactic acid bacteria produce large amounts of lactic acid from sugar. The resulting decrease in pH renders the medium in which they have grown unsuitable for the growth of most other microorganisms. Growth of lactic acid bacteria, therefore, is a means of preserving food; in addition, they produce flavor components.

### milk products

The manufacture of such milk products as butter, cheese, and yogurt involves the use of microorganisms, among which the lactic acid bacteria are particularly important.

Many lactic acid bacteria occur normally in milk and are responsible for its spontaneous souring. Milk souring provides a means of preserving this otherwise highly unstable foodstuff, and the manufacture of cheese and other fermented milk products undoubtedly began largely as a means of preservation.

The manufacture of cheese involves two main steps: *curdling* the milk proteins to form a solid material from which the liquid is drained away; and the *ripening* of the solid curd by the action of various bacteria and fungi, although certain fresh cheeses are essentially unripened.

The curdling process may be exclusively microbiological, since acid produced by lactic acid bacteria is sufficient to coagulate milk proteins. However, an enzyme known as *rennin* (extracted from the stomachs of calves), which curdles milk, is also often used for this purpose.

The subsequent ripening of the curd is a very complex process, and is highly variable, depending on the kind of cheese being made. In the young cheese, all nitrogen is present in the form of insoluble protein, but as ripening proceeds, the protein is progressively cleaved to soluble peptides and ultimately to free amino acids. The amino acids can be further decomposed to ammonia, fatty acids, and amines. In certain cheeses, protein breakdown is restricted. For example, in Cheddar and Swiss cheese, only 25 to 35 percent of the protein is converted to soluble products. In soft cheeses, such as Camembert and Limburger, essentially all the protein is converted to soluble products. In addition to changes in the protein components, ripening involves considerable hydrolysis of the fats present in the young cheese. The enzymes present in the rennin preparation contribute somewhat to the ripening process, but microbial enzymes in the cheese play the major role. The types of microorganism involved are varied. Hard cheeses are ripened largely by lactic acid bacteria, which grow throughout the cheese, die, autolyze, and release hydrolytic enzymes. Soft cheeses are ripened by the enzymes from yeasts and other fungi that grow on the surface.

Some microorganisms play highly specific roles in the ripening of certain cheeses. The blue color and unique flavor of Roquefort cheese are a consequence of the growth of a blue-colored mold, *Penicillium roqueforti*, throughout the cheese.* The characteristic holes in Swiss cheese are formed by carbon dioxide, a product of the propionic acid fermentation of lactic acid by species of *Propionibacterium*.

Butter manufacture is also in part a microbiological process, since an initial souring of cream, caused by milk streptococci, is necessary for subsequent separation of butterfat in the churning process. These organisms produce small amounts of acetoin, which is spontaneously oxidized to *diacetyl*, the compound responsible for the flavor and aroma of butter. Since streptococci differ markedly in their ability to produce acetoin, it has become common practice to inoculate pasteurized cream with pure cultures of selected strains.

In many parts of the world, milk is allowed to undergo a mixed fermentation by lactic acid bacteria and yeasts, which produces a sour, mildly alcoholic beverage (e.g., kefir and kumiss).

### the lactic fermentation of
### plant materials

Certain lactic acid bacteria are found characteristically on plant materials. These organisms are responsible for the souring that occurs in the preparation of pickles, sauerkraut, and certain types of olives. In these lactic acid fermentations, sugars initially present in the plant materials serve as the fermentable substrates. The lactic acid produced imparts flavor to the product and protects it from further microbial attack.

The preservative value of a lactic acid fermentation is also exploited in the ensilaging of green cattle fodder. After plant materials have undergone fermentation in a silo, they may be kept indefinitely without risk of decomposition.

### dextran production

Some lactic acid bacteria belonging to the genus *Leuconostoc* produce large amounts of an extracellular polysaccharide known as *dextran*, when grown with sucrose. Dextran is a polyglucose of high but variable molecular weight (15,000 to 20,000,000); the average molecular

---

*In the United States a white mutant of *Penicillium roqueforti* is sometimes used to produce a mold-ripened cheese for those who find the flavor desirable but the color objectionable.

weight varies with the strain employed. These lactic acid bacteria first came to the attention of industrial microbiologists for their nuisance value; they occasionally develop in sugar refineries, and the large amounts of gummy polysaccharide produced may literally clog the works.

Dextran is now produced industrially, following the discovery that dextran derivatives that have been chemically cross-linked to make them insoluble in water can act as molecular sieves. Columns of such modified dextrans (marketed largely under the trade name of *Sephadex*) retard the passage of small molecules, and thus permit the physical fractionation of solutes that differ in molecular weight. Sephadex columns can be used for molecular weight determinations in the range of 700 daltons to 800,000 daltons, after calibration with compounds of known molecular weight.

Another class of microbial polysaccharides now being produced industrially are the chemically complex extracellular polysaccharides synthesized by aerobic pseudomonads of the Xanthomonas group. These substances have the physical property of forming thixotropic gels, and in addition, are stable at relatively high temperatures. As a result, they have a wide variety of industrial uses, notably as lubricants in the drilling of oil wells and as gelling agents in paints with a water base. They are also used to gel certain foods.

# the use of clostridia

### the retting process

*Retting* is a controlled microbial decomposition of plant materials, designed to liberate certain components of the plant tissue. The oldest retting process, which has been used for several thousand years, is the retting of flax and hemp to free the fibers used in the making of linen. These fibers, made up of cellulose, are held together in the plant stem by a cementing substance, pectin; their physical separation is difficult. The goal of retting is to bring about decomposition of the pectin, thus freeing the fibers without simultaneous decomposition of the fibers themselves. After the plant stems are immersed in water, aerobic microorganisms develop and use up the dissolved oxygen, making the environment suitable for the subsequent development of the clostridia. These organisms rapidly attack the plant pectin, freeing the fibers. If retting is unduly prolonged, cellulose-fermenting bacteria will also develop and destroy the fibers. An analogous retting process is used in the preparation of potato starch. Its purpose is to free the starch-containing cells in the potato tuber from the pectin in which they are embedded.

### the acetone-butanol fermentation

In the past 50 years, certain *Clostridium* spp. have been used on a very large scale for the production of the industrial solvents, acetone and butanol. Many clostridia carry out a fermentation of sugars with the formation of carbon dioxide, hydrogen, and butyric acid. Some carry out further reactions, converting the butyric acid to butanol and the acetic acid to ethanol and acetone. The commercial development of the so-called *acetone-butanol fermentation* mediated by *Clostridium acetobutylicum* began in England just before World War I and expanded rapidly during the war because acetone was needed as a solvent in the manufacture of explosives. After World War I, the demand for acetone diminished, but the process survived because another major product of the fermentation, butanol, found a use as a solvent for the rapid drying of nitrocellulose paints in the growing automobile industry. A byproduct of the fermentation, the vitamin riboflavin, also helped to maintain the commercial feasibility of the acetone-butanol fermentation.

Today, this industry has virtually disappeared as a result of the development of competing methods, only in part microbiological, for the synthesis of the major products. Both acetone and butanol are now produced in large amounts from petroleum; and a microbiological process based on the use of yeasts is the principal source of riboflavin.

The acetone-butanol fermentation made important technological contributions to industrial microbiology. It was the first large-scale process in which the exclusion of other kinds of microorganisms from the culture vessel was of major importance to the success of the operation. The medium used for the cultivation of *Clostridium acetobutylicum* is also favorable for the development of lactic acid bacteria; if these organisms begin to grow, they rapidly inhibit the further growth of the clostridia through lactic acid formation. An even more serious problem is infection with bacterial viruses, to which clostridia are highly susceptible. Thus, the acetone-butanol fermentation can be operated successfully only under conditions of careful microbiological control. The establishment of this industry led to the first successful use of *pure culture methods on a mass scale,* which were later improved and refined in connection with the industrial production of antibiotics.

# the microbial production of chemotherapeutic agents

The period since World War II has seen the establishment and extremely rapid growth of a major new industry, the use of microorganisms for the synthesis of chemotherapeutic agents, particularly antibiotics and hormones. The development of this industry has had a dramatic and far-reaching social impact. Nearly all bacterial infectious diseases that were, prior to the antibiotic era, major causes of human death have been brought under control by the use of these drugs. In the United States, bacterial infection is now a less frequent cause of death than suicide or traffic accidents.

## the rise of chemotherapy

The importance of acquired immunity as a means of protection against specific bacteriological diseases was recognized shortly after the discovery of the role of microorganisms as the etiological agents of infectious diseases (see Chapter 1). For several decades thereafter, control of infectious disease was based exclusively on the use of antisera and vaccines, and was largely preventative; usually, little could be done to cure infections after they had appeared.

A different kind of approach to the control of infectious disease was developed by the German chemist Paul Ehrlich, who initiated an empirical search for synthetic chemicals that possess *selective toxicity* for pathogenic microorganisms. He coined the word *chemotherapy* to describe this approach to the control of infectious disease. Ehrlich's efforts produced one limited success: in 1909 he discovered synthetic organic compounds containing arsenic, which were effective in the treatment of syphilis and other spirochetal infections, but had severe side effects.

The next significant advance in chemotherapy was also made empirically. Large numbers of aniline dyes were screened for antibacterial chemotherapeutic activity, and one substance of this class, *prontosil*, was found to be effective in 1935. In spite of its therapeutic value prontosil possessed no antibacterial action in vitro. Its antibacterial activity in vivo was then shown to be attributable to a colorless breakdown product, *sulfanilamide*, formed in the animal body. Sulfanilamide possesses antibacterial activity both in vitro and in vivo. D. D. Woods observed that the inhibition of bacterial growth by sulfanilamide can be reversed by a structural analogue, *p*-aminobenzoic acid (Figure 18.1).

Wood then made a brilliant series of deductions: that *p*-aminobenzoic acid is a normal constituent of the bacterial cell; that it has a coenzymatic function; and that this function is blocked by sulfanilamide as a result of its steric resemblance to *p*-aminobenzoic acid. In fact, *p*-aminobenzoic acid proved to be not a coenzyme, but a biosynthetic precursor of the coenzyme folic acid; sulfanilamide blocks its conversion to this end product. Sulfanilamide is selectively toxic because most bacteria must synthesize folic acid *de novo*, whereas mammals obtain it from dietary sources.

Wood's work appeared to offer a *rational approach to chemotherapy* through the synthesis of analogues of known essential metabolites. In succeeding years, thousands of structural analogues of amino acids, purines, pyrimidines, and vitamins were synthesized and tested; but very few useful chemotherapeutic agents were discovered.

The great modern advances in chemotherapy have come from the chance discovery that many microorganisms synthesize and excrete compounds that are selectively toxic to other microorganisms. These compounds, called *antibiotics*, have revolutionized modern medicine.

The first chemotherapeutically effective antibiotic was discovered by Alexander Fleming in 1929. He observed, as many

**Figure 18.1**
The structures of (a) sulfanilamide and (b) *p*-aminobenzoic acid.

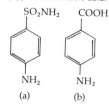

(a)   (b)

before him had done, that on a plate culture of bacteria that had become contaminated by a mold, bacterial growth in the vicinity of the mold colony was inhibited. He reasoned that the mold was excreting into the medium a chemical that prevented bacterial growth. Sensing the possible chemotherapeutic significance of his observation, he isolated the mold, which proved to be a species of *Penicillium,* and established that the culture filtrates contained an antibacterial substance, which he called *penicillin.*

Penicillin proved to be chemically unstable, and Fleming was unable to purify it. However, working with impure preparations, he demonstrated its remarkable effectiveness in inhibiting the growth of many Gram-positive bacteria, and he even used it with success for the local treatment of human eye infections. In the meantime, the chemotherapeutic effectiveness of sulfanilamides had been discovered, and Fleming, discouraged by the difficulties of penicillin purification, abandoned further work on development of penicillin as a chemotherapeutic agent.

Ten years later, a group of British scientists headed by H. W. Florey and E. Chain resumed the study of penicillin. Clinical trials with partly purified material were dramatically successful. By this time, however, England was at war; and the industrial development of penicillin was undertaken in the United States, where an intensive program of research and development was begun in many laboratories. Within 3 years, penicillin was being produced on an industrial scale, an astonishing achievement in view of the many difficulties that had to be overcome. Penicillin remains one of the most effective chemotherapeutic agents for treatment of many bacterial infections, in spite of the current availability of many other antibiotics.

Penicillin proved to be effective primarily against infections caused by Gram-positive bacteria. Its startling success in the treatment of such infections prompted intensive searches both at universities and in industry for new antibiotics. A second clinically important antibiotic, *streptomycin,* was discovered by A. Schatz and S. Waksman in 1945.

Streptomycin was the first example of an antibiotic possessing a *broad spectrum of activity,* effective against many Gram-positive and Gram-negative bacteria. Other antibiotics with even broader spectra of activity (for example, the tetracyclines) have been subsequently discovered. Antibiotics have proved to be less useful in the treatment of fungal infections: antifungal antibiotics such as nystatin and amphotericin B are considerably less effective therapeutically than their antibacterial counterparts, at least in part because their toxicity is far less selective. Good antiviral antibiotics are yet to be found.

Since 1945, thousands of different antibiotics produced by fungi, actinomycetes, or unicellular bacteria have been isolated and characterized. A small fraction of these are of therapeutic value; about 50 are currently produced on a large scale for medical and veterinary use. Their nomenclature is complicated, one antibiotic often being sold under several different names. There are two reasons for this proliferation of names. First, many antibiotics are members of a group of compounds, all of which possess similar structures; and a name is required for the *class of compounds,* as well as for each *individual representative.* Second, each manufacturer of an antibiotic assigns to it for marketing purposes a *trade name,* which, by law, only he can use. To protect a trade name for exclusive use, the law requires that another name, available for general use, be also assigned to the antibiotic in question; this is called the *generic name.* The generation of multiple names can be illustrated by the example of an antibiotic that in the United States is given the generic name, *rifampin.* The generic name of the same compound in Europe is *rifampicin.* Its class name is *rifamycin.* It is sold under the trade names *Rifactin* and *Rifadin,* among others.

The generic names, sources, uses, and mode of action of some antibiotics are shown in Table 18.1.

### mode of action of antibiotics

The search for new antibiotics remains an empirical enterprise, and their physiological significance for the microorganisms that produce them is obscure. However, the reasons for their selective toxicity

**Table 18.1**
Properties and Uses of Certain Antibiotics

| CHEMICAL CLASS | GENERIC NAME | BIOLOGICAL SOURCE | EFFECTIVE CHEMOTHERAPEUTICALLY AGAINST | MODE OF ACTION |
|---|---|---|---|---|
| β-lactams | Penicillins | *Penicillium* spp. | Gram-positive bacteria | Inhibit synthesis of bacterial cell wall (peptidogylcan) |
| | Cephalosporins | *Cephalosporium* spp. | Gram-positive and Gram-negative bacteria | |
| Macrolides | Erythromycin | *Streptomyces erythreus* | Gram-positive bacteria | Inhibit 50S ribosome function |
| | Carbomycin | *S. halstidii* | | |
| Aminoglycosides | Streptomycin | *S. griseus* | Gram-positive and Gram-negative bacteria | Inhibit 30S ribosome function |
| | Neomycin | *S. fradiae* | | |
| Tetracyclines | Tetracycline[a] | *Streptomyces aureofaciens* | Gram-positive and Gram-negative bacteria | Inhibits binding of aminoacyl-tRNAs to ribosomes |
| Polypeptides | Polymyxin G | *Bacillus polymyxa* | Gram-negative bacteria | Destroys cytoplasmic membrane |
| | Bacitracin | *B. subtilis* | Gram-positive bacteria | Inhibits synthesis of bacterial cell wall (peptidoglycan) |
| Polyenes | Amphotericin B | *S. nodosus* | Fungi | Inactivate membranes containing sterols |
| | Nystatin | *S. nouresii* | Fungi | |
| ——— | Chloramphenicol[b] | *S. venezuelae* | Gram-positive and Gram-negative bacteria | Inhibits translation step of ribosome function |

[a] Made microbiologically and by chemical dehydrochlorination of chlorotetracycline.
[b] Now made by chemical synthesis.

are in many cases now known. In general, antibiotics owe their selective toxicity to the fundamental biochemical differences between procaryotic and eucaryotic cells, their toxic effect being the consequence of their ability to inhibit one essential biochemical reaction specific either to the procaryotic or to the eucaryotic cell (see Chapter 3).

### the production of antibiotics

The antibiotics were the first industrially produced microbial metabolites that were not *major* metabolic end products. The yields, calculated in terms of conversion of the major carbon source into antibiotic, are low and are greatly influenced by the composition of the medium and by the other cultural conditions. These facts have encouraged intense research directed toward improving yields. For this purpose, genetic selection has proved remarkably successful. The wild type strain of *Penicillium chrysogenum* first used for penicillin production yielded approximately 0.1 gram of penicillin per liter. From this strain a mutant was selected that produced 8 grams per liter under the same growth condition, a 60-fold improvement in yield. Subsequent strain selection following chemical mutagenesis has led to the development of new strains with even greater capacity for antibiotic production. By such *sequential genetic selection,* improvements of antibiotic yield as great as a thousandfold have often been obtained. Most genetic improvement has been empirical; large numbers of mutagenized clones are evaluated for their abilities to produce larger quantities of the antibiotic. However, with increased knowledge of the pathways of biosynthesis of antibiotics, more rational approaches are being exploited. It is now possible to select strains in which control of the synthesis of known precursors of an antibiotic has been altered by mutation. Such strains produce larger amounts of the precursor, and sometimes also larger amounts of the antibiotic.

The synthesis of antibiotics begins only after growth of the organisms that produce them has virtually ceased (Figure 18.2). Thus, antibiotics belong to a class of microbial products called *secondary metabolites.* The control mechanisms that trigger the synthesis of secondary metabolites as growth ceases are a fascinating but almost completely unexplored aspect of biochemical regulation.

Although the microorganisms used to produce antibiotics are all aerobes and are grown under conditions of vigorous aeration, the production process is generally referred to in the technical literature as a "fermentation." Antibiotics are produced by so-called *submerged cultivation methods,* using deep stainless steel tanks that must be subjected to continuous forced aeration and rapid mechanical agitation. The provision of adequate aeration is of great importance to yield, and the energy expended for aeration contributes appreciably to the cost of production.

When a microorganism is grown aerobically in

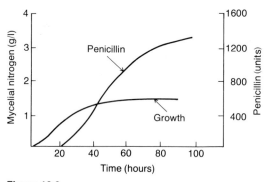

**Figure 18.2**
Temporal relationship between growth of *Penicillium chrysogenum* and its production of penicillin. After W. E. Brown and W. H. Peterson, "Factors Affecting Production of Penicillin in Semi-pilot Plant Equipment," *Ind. Eng. Chem.* **42,** 1769 (1950).

tanks with capacities of tens of thousands of gallons containing a rich, nonselective medium, the maintenance of a pure culture poses numerous special engineering problems. For the successful production of antibiotics, pure cultures are indispensable. This fact was first revealed during penicillin production. Many bacteria produce an enzyme, penicillinase, which catalyzes the hydrolytic cleavage of the four membered $\beta$-lactam ring of penicillin, with resulting loss of antibiotic activity. Consequently, contamination of a fermentor by penicillinase-producing bacteria can result in complete destruction of the accumulated penicillin.

In the manufacture of antibiotics the microbial product is sometimes subjected to subsequent chemical modification. One example is the chemical substitution of the acyl group of natural penicillins to produce a large variety of semisynthetic penicillins. Another example is the catalytic dehydrochlorination of chlorotetracycline to produce the more effective substance, tetracycline.

### microbial resistance to antibiotics

The antibiotic era of medicine began abruptly some 30 years ago. How long it will last has become an open question. Although the search for new antibiotics continues undiminished, the rate of their discovery has declined sharply; most of the really effective antibiotics have probably already been discovered. Furthermore, strains of pathogens resistant to antibiotics have begun to develop at an alarming rate. Although most strains of staphylococci were sensitive to penicillin when it was first introduced into medical practice, essentially all hospital-acquired staphylococcal infections are now resistant to this antibiotic. A problem of even greater concern is the appearance of bacterial strains that are simultaneously resistant to several antibiotics, the so-called *multiply-resistant strains*. Between 1954 and 1964 the frequency of multiply-resistant strains of *Shigella* in Japanese hospitals rose from 0.2 percent to 52 percent.

Bacterial resistance to an antibiotic is sometimes acquired by the mutation of a chromosomal gene, which modifies the structure of the cellular target. A good example is mutationally acquired streptomycin resistance. This antibiotic deranges bacterial protein synthesis by attachment to one of the proteins in the 30S subunit of the ribosome. Some mutations of the gene that encodes this ribosomal protein destroy the ability of the protein to bind streptomycin, but they do not substantially affect ribosomal function; the cell in which such a mutation occurs consequently becomes streptomycin-resistant.

Resistance can also be acquired as a result of infection of the bacterial cell by a plasmid belonging to the class of *resistance factors*. These plasmids often confer simultaneous resistance to several antibiotics. They carry genes that encode enzymes that catalyze chemical modifications of antibiotics, converting them to derivatives without antibiotic action. Multiply-resistant bacterial strains almost always owe their resistance to the presence of an R factor. Since these plasmids have wide host ranges, often being readily transferable between different bacterial genera, their increasing dissemination in natural bacterial populations is by far the most serious aspect of the problem of microbial resistance to antibiotics. Unless some solution to this problem can be found, the future therapeutic effectiveness of antibiotics is in jeopardy.

### microbial transformations of steroids

Cholesterol (Figure 18.3) and chemically related steroids are structural components of eucaryotic cellular membranes and therefore universal chemical constituents of eucaryotes. During the evolution of vertebrates, special pathways were evolved for the conversion of these universal cell constituents to new and functionally specialized steroids: the *steroid hormones,* which are potent regulators of animal development and metabolism. Steroid hormones are formed in specialized organs, through the secondary metabolism of cholesterol, a $C_{27}$ steroid. The adrenocortical hormones are synthesized in the adrenal gland and are all $C_{21}$ compounds, such as *cortisone* (Figure 18.3); the sex hormones are synthesized in the ovary or the testis and are $C_{18}$ or $C_{19}$ compounds (Figure 18.3). Accordingly, by relatively slight chemical modifications of the basic steroid structure, vertebrates have evolved two new subclasses of steroid molecules with highly specific physiological functions and of great potency.

The elucidation of the structures and general functions of mammalian steroid hormones was completed about 30 years ago, but it was only in 1950 that possible chemotherapeutic uses for them became

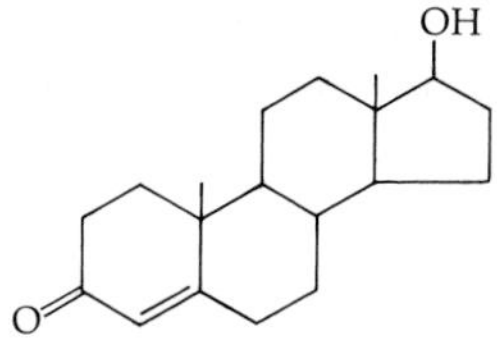

testosterone

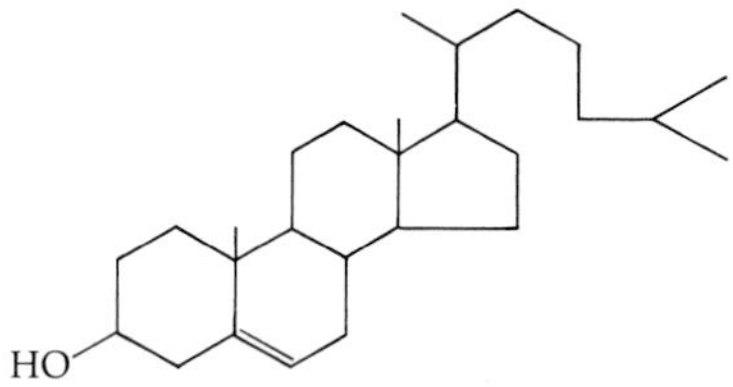

cholesterol

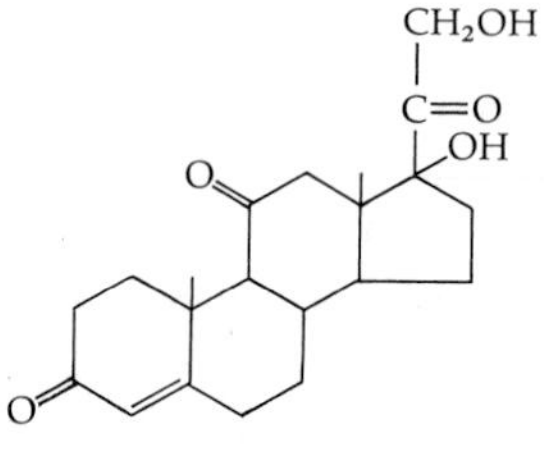

cortisone

**Figure 18.3**
The structure of cholesterol, a $C_{27}$ steroid, and of two mammalian steroid hormones for which cholesterol is a biosynthetic precursor—cortisone ($C_{21}$) and testosterone ($C_{19}$).

apparent, with the discovery that cortisone treatment can relieve dramatically the symptoms of rheumatoid arthritis. Today, cortisone and its derivatives are very widely used to treat a variety of inflammatory conditions, and additional uses for steroid hormones have emerged in the treatment of certain types of cancer and as oral contraceptives. The production of these compounds has now become a major industry.

Since the steroid hormones are produced by mammals in very small quantities, it was evident that their isolation from animal sources could not supply the clinical needs. Accordingly, the chemists turned their attention to the synthesis of those substances from plant sterols, which are abundant and can be cheaply prepared. One major chemical obstacle soon became apparent. All adrenocortical hormones are characterized by an organ-specific enzymatic hydroxylation of the biosynthetic precursor in the adrenal gland. Although it is easy to hydroxylate the steroid nucleus chemically, it is difficult to insert a hydroxyl group at a specific position.

The discovery was then made that many microorganisms—fungi, actinomycetes, and bacteria—are capable of performing limited oxidations of steroids, which cause small and highly specific structural changes. The positions and nature of these changes are often characteristic for a microbial species, so that by the selection of an appropriate microorganism as an agent, it is possible to bring about any one of a large number of different modifications of the steroid molecule.

# microbiological methods for the control of insects

In Chapter 12 the formation of crystalline inclusions in the sporulating cells of certain *Bacillus* species was described. These bacilli (*Bacillus thuringiensis* and related forms) are all pathogenic for the larvae (caterpillars) of insects belonging to the Lepidoptera (butterflies and related forms). Following the isolation of the crystalline inclusions from sporulating bacterial cells, it was shown that all the primary symptoms characteristic of the natural disease of insects could be reproduced by feeding larvae on leaves coated with the purified crystals. The crystals consist of a protein that can be dissolved in dilute alkali. The gut contents of larvae are, in general, alkaline, and when the ingested crystals reach the gut, they are dissolved and partially hydrolyzed. This modified protein attacks the cementing substances that keep the cells of the gut wall adherent, and as a consequence, the liquid in the gut can diffuse freely into the blood of the insect. The blood of the insect becomes highly alkaline, and this change in pH induces a general paralysis of the larva. Death, which ensues much later, appears to result from bacterial invasion of the body tissues.

The protein crystals possess a highly specific toxicity for the larvae of many Lepidoptera but are nontoxic for other animals (including all the vertebrates) and for plants. They thus provide an ideal agent for the control of many serious insect pests that damage plant crops. Recognition of this fact has led recently to the development of a new microbiological industry: the large-scale production of the toxic protein, for incorporation in dusting agents that can be used to protect commercial crops from the ravages of caterpillars. In industrial practice, the protein itself is not chemically isolated. Instead, the crystal-producing bacilli are grown on a large scale, harvested after the onset of sporulation with its accompanying crystal production, dried, and incorporated in a dusting powder.

# the production of other chemicals by microorganisms

The widespread use of microorganisms in the chemical and pharmaceutical industries has come about because of the recognition that it is often cheaper to use a microorganism for the synthesis of a complex organic compound (for example, an antibiotic) than to synthesize it chemically. Microbial syntheses also have distinct advantages in the preparation of optically active compounds, since chemical synthesis leads to racemic mixtures that must subsequently be resolved.

As previously discussed, the microbial production of acetone and butanol, once the major source of these chemicals, has now been largely superseded by chemical synthesis. Nevertheless, the microbial production of many relatively simple and cheap organic compounds remains competitive with chemical methods of synthesis. These compounds include gluconic acid, produced by *Aspergillis niger* and acetic acid bacteria, and citric acid, produced by *A. niger*.

The production of two vitamins, vitamin $B_{12}$ and riboflavin, provides an instructive lesson in the economics of industrial microbiology. Both are now produced commercially by microbial means. Vitamin $B_{12}$ is produced by certain *Pseudomonas* spp. Although the yields are very low, this process remains competitive because of the very high price of the product: the structural complexity of vitamin $B_{12}$ virtually precludes a commercially feasible chemical synthesis. Riboflavin (vitamin $B_2$) is a much simpler compound, which can be readily prepared by chemical synthesis. It is still produced microbiologically, as a result of the discovery that certain plant pathogenic fungi (*Ashbya gossypii* and *Eremothecium ashbyi*) overproduce this vitamin and excrete the excess into the medium. By further genetic selection and improvement of culture methods, strains have been developed that produce so much riboflavin that the vitamin crystallizes in the culture medium.

# the production of enzymes by microorganisms

The production of microbial enzymes, either pure or partly purified, is an important aspect of industrial microbiology. The uses of microbial enzymes in medicine and industry are remarkably diverse. For instance, proteases may be used as digestive aids, to promote the healing of wounds or burns, to digest the gelatin from photographic emulsions from which silver is to be recovered, and were, until widespread allergic reactions were discovered, used in the so-called "enzyme detergents."

# the use of microorganisms in bioassays

Microorganisms are extensively used for the performance of bioassays, to determine the concentration of certain compounds in complex chemical mixtures. Such assays can be used both for the

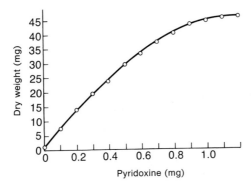

**Figure 18.4**
Microbial bioassay curve for the vitamin pyridoxine. The curve expresses the relationship between the amount of pyridoxine supplied and the amount of growth (dry weight) obtained. The organism used was a pyridoxine requiring mutant of the mold *Neurospora*. From J. L. Stokes, Alma Larsen, C. R. Woodward, Jr., and J. W. Foster, "A Neurospora Assay for Pyridoxine," *J. Biol. Chem.* **150,** 19 (1943).

quantitative determination of growth factors and for the quantitative determination of antibiotics and other specific growth inhibitors.

The principle of bioassays of growth factors is very simple. A medium is prepared that contains all nutrients required for growth of the test organism, except the substance to be assayed. If that substance is added to the medium in limiting amounts, the growth of the test organism will be proportional to the amount added. The first step in developing a bioassay is, accordingly, to measure the relationship between the amount of growth obtained and the amount of the limiting nutrient added (i.e., determine the yield coefficient, see Chapter 5). An example of the relationship obtained is shown in Figure 18.4.

If samples of a material suspected to contain the nutrient in question are added to the basal medium, the quantity present can be determined from the amount of growth that occurs. The lactic acid bacteria

**Figure 18.5**
A bioassay for the antibiotic penicillin. A tray of nutrient agar was seeded with the test bacterium, and differing amounts of penicillin were placed on the paper discs. The amount of penicillin increases from 10 units on the left-hand disc to 10,000 units on the right-hand disc. From the areas of the zones in which bacterial growth is inhibited, a curve relating penicillin concentration to the extent of inhibition may be derived.

are frequently used for bioassays because of their very extensive growth factor requirements. By modifying a single basal medium so that different nutrients are growth limiting, it is possible to assay a large number of different amino acids and vitamins with a single test organism. The vitamin content of foods is still determined by bioassay, but amino acids are now largely determined by chemical methods. Bioassay is also an indispensable tool in the detection and purification of new growth factors.

The quantitative determination of antibiotics and other antimicrobial compounds is conducted by the so-called *cup plate* assay originally developed for the assay of penicillin. An agar medium is densely seeded with the test bacterium, and a number of glass cylinders are placed on its surface. A known dilution of a solution of an antibiotic is added to each cup, and the plate is incubated until growth has occurred. The antibiotic diffuses into the surrounding agar during incubation and produces a zone of inhibition. The diameter of this zone is a function of the initial concentration of the antibiotic in the solution that was added to the cup. A standard curve, relating the diameter of the zone of inhibition to antibiotic concentration, permits the assay of solutions containing an unknown concentration of the antibiotic. A slight modification is the substitution of paper discs soaked in solutions of the antibiotic for the glass cylinders (Figure 18.5).

# INDEX

455

458

# 462